FOCUS ON SIGNAL TRANSDUCTION RESEARCH

FOCUS ON SIGNAL TRANSDUCTION RESEARCH

GAVIN MCALPINE
EDITOR

Nova Science Publishers, Inc.
New York

For permission to use material from this book please contact us:
Telephone 631-231-7269; Fax 631-231-8175
Web Site: http://www.novapublishers.com

Library of Congress Cataloging-in-Publication Data

Focus on signal transduction research / Gavin McAlpine (editor).
p. ; cm.
Includes bibliographical references and index.
ISBN 13 978-1-60021-376-2
ISBN 10 1-60021-376-6
1. Cellular signal transduction. 2. Cell interaction. 3. Cell division. I. McAlpine, Gavin.
[DNLM: 1. Signal Transduction--physiology. 2. Cell Communication. QU 375 F652 2006]
QP517.C45F63 2006
571.7'4--dc22 2006023221

Published by Nova Science Publishers, Inc. ✣New York

CONTENTS

PREFACE

Signal transduction is any process by which a cell converts one kind of signal or stimulus into another. Processes referred to as signal transduction often involve a sequence of biochemical reactions inside the cell, which are carried out by enzymes and linked through second messengers. In many transduction processes, an increasing number of enzymes and other molecules become engaged in the events that proceed from the initial stimulus. Responses of cells to environmental signals, toxins and stressors have profound implications for diverse aspects of human health and disease including development, cystic fibrosis, diabetes, asthma, heart, autoimmune diseases and cancer. The delineation of the signal transduction pathways affected in these and other complex human diseases are likely to present new avenues for therapeutic intervention and understanding of human disease mechanisms.

Zinc has long been regarded as a mere static component of proteins, either with catalytic or structural functions. Recent research indicates an additional role for zinc, in particular an intracellular pool of labile zinc, in signal transduction. Zinc homeostasis is tightly controlled, and a specialized system of transport proteins regulates entry and export through the plasma membrane, but also intracellular distribution. Within cells zinc can be stored in vesicular structures, so-called zincosomes, and a differential regulation of the levels of labile zinc in plasma and nucleus has been observed. Several reports in chapter 1, describe fluctuations of labile zinc after stimulation of cells, either of its total cellular concentration, or its intracellular distribution. An influence of zinc-fluctuations on several established signaling pathways, including protein kinase C, mitogen activated protein kinases, and calcium has been shown. What is even more important, it has been discovered that labile zinc is required for well known physiological signals, e.g. the tyrosine phosphorylation events that are triggered by insulin and insulin like growth factor-1. Investigations about the molecular mechanism by which zinc exerts these effects have identified that it directly interacts with several components of signaling pathways, including protein tyrosine phosphatases, interleukin receptor associated kinase 1, the transcription factor MTF-1 and cyclic nucleotide phosphodiesterases, and alters the signaling properties of all these proteins. These observations indicate that, in addition to calcium, zinc is a second bivalent metal ion which also has a function as a second messenger.

The ruminal epithelium is a stratified squamous epithelium that has evolved to display functions essential for the unique ability of cows and sheep to ferment dietary components like carbohydrates and protein, and to selectively absorb nutrients and minerals for the production of milk. A characteristic property of this tissue is its pronounced ability to transport magnesium against an electrochemical gradient. Absorption of magnesium is reduced by dietary elevation of ruminal potassium, leading to hypomagnesaemia that can reach clinical significance. Studies of the intact tissue and of isolated cells suggest that cellular magnesium uptake is decreased by apical depolarization of the ruminal membrane, resulting in both a lower cytosolic concentration and transepithelial transport of the element. Another characteristic feature of this unusual epithelium is the expression of a sodium-conducting channel with functional properties that are clearly distinct from the epithelial sodium channel (ENaC) found in most mammalian epithelia. Thus, it has not been possible to demonstrate direct regulation of ruminal sodium transport by aldosterone, and the effects of amiloride are clearly limited to an inhibition of the sodium proton exchanger (NHE3) expressed by this tissue. Studies at the level of the animal and the tissue suggest that sodium conductance is enhanced by depolarization of the apical membrane. Recent *in vitro* studies have demonstrated that ruminal epithelial cells express non-selective cation channels in the apical membrane that are regulated by changes in cytosolic magnesium. The authors in chapter 2, propose that the reduction in ruminal magnesium uptake observed after ingestion of high potassium fodder may be related to a role for magnesium in a signaling cascade that leads to an increase in the permeability of this non-selective cation channel for sodium, thus enhancing absorption of this ion from the rumen and restoring ruminal osmolarity, while contributing to the retention of potassium.

The protein Salvador (also named Shar-pei) has been shown to limit cell proliferation and stimulate apoptosis in *Drosophila melanogaster*. In vertebrates the function and expression pattern of the Salvador homolog WW45 under different physiological and pathological situations have not been characterized in detail. WW45 transcripts have been previously shown to be ubiquitously distributed in a variety of tissues of mouse and human origin and to be undetectable in several human cancer cell lines. In chapter 3 changes in the expression of WW45 upon T cell activation and the putative role of this protein in T cell mediated-bone resorption were investigated. WW45 transcript was significantly increased in human naïve $CD3^{+}$ T lymphocytes when activated with anti-CD3, anti-CD28 or the combination of both antibodies at 24 h, but not after 72 h of stimulation. WW45 mRNA expression was also high in mononuclear gingival infiltrates of patients with periodontal disease, whose loss of alveolar bone has been associated with the high content of activated/memory T cells in gingival tissues and the elevated expression of receptor activator of NFkB ligand (RANKL) that increases osteoclast differentiation and bone resorption. To explore the putative involvement of WW45 in T cell-induced inflammatory bone resorption, a rat model for experimental periodontal disease was used, in which local gingival stimulation of Th1-type memory T cells with antigen presenting cells (APC) and specific antigen increases alveolar bone resorption. When activation of these memory T cells with APC and their specific antigen was performed in the presence of antisense ($WW45^{ANT}$) or control scrambled oligonucleotides ($WW45^{SC}$) for WW45, the decreased WW45 expression in the former cells correlated with the decreased production of RANKL. Rats adoptively transferred with $WW45^{ANT}$-treated T cells exhibited and average of 90% lower experimental bone resorption and gingival RANKL expression than rats injected with $WW45^{SC}$-treated T cells. To summarize, WW45 expression appears to be transiently upregulated during T cell activation of human naïve T cells whereas exhibits a sustained upregulation upon antigen-specific stimulation of memory T cells. Furthermore, downregulation of WW45 in antigen-stimulated Th1-type memory T cells is associated with the reduction of RANKL expression and T cell-induced bone resorption *in vitro* and *in vivo*.

Therefore, careful manipulation of WW45 expression in diseases characterized by disrupted T cell homeostasis and excessive RANKL function may lead to a novel therapeutic approach to inhibit T cell-induced bone loss.

Glutamate is one of the most abundant neurotransmitters in the brain; it may be responsible for mediating synaptic neurotransmission at 40 percent or more of the brain's synapses. The glutamatergic signal is recognized by both metabotropic and ion channel receptors. The N-methyl-D-aspartic acid (NMDA) receptor is an example of a glutamate-gated ion channel receptor; the binding of glutamate to this receptor increases the likelihood that its channel will transiently assume an open-configuration, allowing the influx of calcium ions. Importantly, there are allosteric modulatory sites, including an obligatory co-agonist site that recognizes glycine, on the NMDA receptor complex that influence glutamate's ability to promote calcium ion conductance. Abnormalities of NMDA receptor-mediated neurotransmission are implicated in a variety of major neuropsychiatric disorders, especially schizophrenia. Specifically, NMDA receptor hypofunction (NRH) is implicated in the pathophysiology of schizophrenia because of the ability of phencyclidine (PCP) to precipitate a schizophreniform psychosis. PCP is a noncompetitive NMDA receptor antagonist that binds to a hydrophobic site within the channel; it is referred to as an open-channel blocker. Descriptively, the PCP-model of schizophrenia is the best pharmacological model of this disorder with symptoms manifest in all of the relevant domains of psychopathology, including positive (e.g., hallucinations), negative (e.g., affective flattening), cognitive (e.g., abnormalities of attention), mood (e.g., dysphoria) and motor symptoms (e.g., posturing). Because of PCP's pharmacological action as a noncompetitive NMDA receptor antagonist and the PCP-model of schizophrenia, their laboratory has been engaged in characterizing and quantifying behaviors in mice, including genetically-inbred strains, elicited by MK-801 (dizocilpine), a high-affinity analogue of PCP that binds to the same hydrophobic channel domain. Presumably, blockade of the NMDA receptor by MK-801 is a pharmacological strategy for creating NRH in the intact animal. The authors characterized two behavioral outcome measures: the dose-dependent ability of MK-801 to raise the threshold voltage required for the precipitation of tonic hindlimb extension and elicit irregular episodes of intense jumping behavior (referred to as "popping") in mice. Moreover, 24 hours after mice are forced to swim for up to 10 minutes in cold water, the ability of MK-801 to antagonize electrically precipitated seizures is reduced. Thus, stress affects the endogenous tone of NMDA receptor-mediated neurotransmission. Genetically inbred strains of mice differ in their behavioral sensitivity to MK-801 on these two measures; interestingly, the BALB/c inbred mouse strain is more sensitive than other inbred strains and the NIH Swiss outbred mouse strain. Stress and genetic strain differences interact to affect behavioral sensitivity to MK-801. Importantly, these animal models serve to test the ability of glycinergic interventions to modulate NMDA receptor-mediated neurotransmission in the intact animal. Thus, they are used as screening procedures for NMDA receptor agonist interventional strategies that may be developed as medications for the treatment of schizophrenia. A series of glycinergic interventions have been tested in these models. Also, these paradigms are useful in demonstrating that drugs, whose direct actions are on other neurotransmitter systems, affect NMDA receptor-mediated neurotransmission in the intact animal. Preliminary clinical trials of a few NMDA receptor agonist interventions have been conducted in patients with schizophrenia. In general, the medications are well-tolerated and the results suggest that the adjuvant therapeutic efficacy of specific glycinergic interventions may be influenced by state of illness and the specific antipsychotic medication chosen for maintenance pharmacotherapy (e.g., clozapine).Chapter 4 will concentrate on the results of our laboratory and clinical investigations, whose primary aim was modulation of NMDA receptor-mediated neurotransmission in mice and man.

The ubiquitous occurrence of bacteria relies on their ability to persist under various stressful conditions and changing environments. These bacteria initiate an adaptive response to the stresses they face by modulating their gene transcription programme. The rapidity of the physiological adaptation is critical for their survival in hostile ecological niches. Therefore bacteria have evolved early envelope-initiated signalling processes for responding to the imposition of environmental stresses and to directly monitor changes in the environment. The bacterial cell envelope comes into direct contact with the environment and consequently it is ideally suited for sensing environmental changes and triggering adaptive responses. Envelope stress responses have been described in a variety of Gram-negative bacteria and serve important roles in colonization and virulence. The two-component systems CpxAR and BaeSR, and the extracytoplasmic sigma factor σ^E are the best understood regulatory components involved in the maintenance of the bacterial envelope in response to envelope perturbations such as adhesion to abiotic surface or alteration in membrane protein folding and biogenesis (Raivio TL, Mol Microbiol. 2005; 56(5):1119-28). An emerging theme is the activation of a regulatory cascade as an outcome of changes in the cytoplasmic membrane fluidity. Many environmental stressors, such as xenobiotics, variations in temperature, pressure or pH, instantly alter the membrane fluidity by acting on the biochemical properties of the acyl components of the phospholipids. Certain regulatory systems are though to be triggered by a modification of membrane fluidity as exemplified by the multiple stress sensor kinase HiK33 of the cyanobacterium *Synechocystis* sp. PCC 6803 and the two-component system DesK/DesR of the Gram-positive bacterium *Bacillus subtilis* (Los DA and Murata N, Biochim Biophys Acta. 2004; 1666(1-2):142-57). New data suggest that membrane-based signal transduction systems are widely distributed in microorganisms. The authors have recently demonstrated that the Gram-negative bacterium *Pseudomonas aeruginosa* also recognizes alterations in its membrane fluidity as a trigger for adaptive cellular responses (Baysse C *et al.*, Microbiology. 2005; 151(Pt 8):2529-42). Chapter 5 will review the current knowledge of early stress detection mechanisms in bacteria and discuss novel fundamental issues in membrane-initiated signalling processes.

Many bacteria use small secreted signaling molecules called autoinducers to communicate and to coordinate expression of certain sets of genes in a process known as quorum sensing. Quorum sensing allows bacteria to launch a unified, coordinated response on a population-wide scale to accomplish tasks which would be impossible to achieve for a single bacterial cell. Work over the past years has provided evidence that bacterial quorum sensing signaling is exploited not only for intraspecies, but also for interspecies cell-to-cell communication with other bacteria occupying the same ecological niche and even for cross-talk with their eukaryotic hosts. As discussed in chapter 6, quorum sensing is a generic regulatory mechanism used by a broad spectrum of bacteria including human pathogens where it highly influences the ability to establish an infection. In the opportunistic human pathogen *Pseudomonas aeruginosa*, quorum sensing-dependent regulation of gene expression is essential for the synchronization of the production of a battery of virulence factors as well as for biofilm development. Given the essential role of this phenomenon for the pathogenic potential of bacteria, cell-to-cell communication represents an attractive target for the design of novel drugs.

Mitogen-activated protein kinases (MAPKs) function as serine/threonine kinases that regulate intracellular signaling cascades. The MAPK, p38, is stimulated by environmental stress, proinflammatory cytokines, oxidative stimuli, and UV irradiation. p38 MAPK is known to phosphorylate its downstream targets, mitogen and stress-activated protein kinase 1 (MSK1), MAP kinase-interacting kinase 1 (MNK1), and MAPK-activated protein kinase 2, 3 (MAPKAPK 2, 3). MAPKAPK 2 activates heat shock protein 27 (HSP27), cAMP response element-binding protein (CREB), and transcription factor-1 (ATF-1). Infection of cells by viruses has been shown to result in phosphorylation of p38 MAPK. In virus-infected cells, viral proteins have been shown to be key mediators of up- and down-regulation of p38 MAPK by experiments using specific inhibitors and over-expression of kinase-defective mutants. p38 MAPK is known to promote both cell death and survival, depending on the cell types and stimuli. The alterations in p38 MAPK activities by viral proteins influence apoptotic events and cytokine secretion in vivo and *in vitro*. Therefore, p38 MAPK may be a therapeutic target for suppressing viral replication and reducing host responses in virus-infected cells. Chapter 7 will highlight recent finding on the role of p38 MAPK signaling pathway including its up- and down-stream components by viral infection.

It has been known for many years that the innate immune response triggered by viral infections ultimately leads to the activation of a number of transcription factors involved in anti-viral mechanisms, such as NFκB and the interferon regulatory factor (IRF) family, via different signalling pathways. Recently, molecular details of some of the receptors responsible for viral recognition have been intensely studied, including the toll-like receptors (TLRs), and cytoplasmic helicase-domain containing proteins. The TLRs participate in the first step of host defence by triggering innate immunity in response to the recognition of pathogen-associated molecular patterns (PAMPs). TLRs signal via homotypic interactions between their intracellular Toll/Interleukin-1/Resistance (TIR) domains and those on TIR adaptor molecules such as MyD88, Mal/TIRAP and TRIF. Helicase-domain containing proteins, such as RIG-I and Mda5, lack TIR domains but signal via another protein-interacting domain, named caspase activation and recruitment domain (CARD). TIR- and CARD-mediated signalling pathways lead to the activation of NFκB and IRFs. Both viral proteins and nucleic acids can work as PAMPs. For instance, both TLRs and, more recently, RIG-I and Mda5 have been shown to sense dsRNA produced during viral infections. The specific signalling pathways triggered by each of these PAMPs have also been the focus of intense scrutiny. In chapter 8, the authors review current advances in viral recognition by cellular components such as TLRs and CARD-domain containing proteins and the increasingly complex signalling pathways triggered.

The discovery of the complex extrinsic regulation of mitochondrial functions by hormones and signal transduction systems is a new and important achievement of a mitochondriology. A number of hormones of all the chemical classes and with different mechanisms of activity stimulate a great plenty of processes; namely**,** Krebs cycle, respiratory chain, oxidative phosphorylation, energy dependent syntheses, and other mitochondrial enzymes. The realization of these effects are reproduced by signal transduction systems, including receptors, the second messengers (especially by cAMP and Ca^{2+}), protein and tyrosine kinases, protein and tyrosine phosphatases, anchor proteins, and transcription factors.

The receptors of lipophylic hormones and some kinase receptors of protein hormones (growth factors, neurotrophic_factors, growth hormone) are revealed in mitochondria. All the main cellular kinases are found in mitochondria. Protein kinases phosphorylate the protein 18 kDa from complex I, cytochrom *c*-oxidase, ATP synthase, cAMP/Ca^{2+} response element binding protein, voltage dependent anione channel, steroidogenic acute regulator, pro- and antiapoptotic proteins etc. Tyrosine kinases phosphorylate a number of mitochondrial proteins, including cytochrom *c*-oxidase and ATP synthase. As discussed in chapter 9, the pleiotropy of calcium regulation of mitochondrial functions is proved. Mitochondria as a whole realize the integrative function in cellular signaling. The main unsolved problems are noted. The present review is the first one, in which all the problem are considered as a whole and in a development.

In: Focus on Signal Transduction Research
Editor: Gavin McAlpine, pp.1-35

ISBN 1-60021-376-6

Chapter 1

THE PHYSIOLOGICAL ROLE OF ZINC IONS IN MAMMALIAN SIGNAL TRANSDUCTION

Hajo Haase and Lothar Rink*
Institute of Immunology, RWTH Aachen University Hospital, Aachen, Germany

ABSTRACT

Zinc has long been regarded as a mere static component of proteins, either with catalytic or structural functions. Recent research indicates an additional role for zinc, in particular an intracellular pool of labile zinc, in signal transduction. Zinc homeostasis is tightly controlled, and a specialized system of transport proteins regulates entry and export through the plasma membrane, but also intracellular distribution. Within cells zinc can be stored in vesicular structures, so-called zincosomes, and a differential regulation of the levels of labile zinc in plasma and nucleus has been observed. Several reports describe fluctuations of labile zinc after stimulation of cells, either of its total cellular concentration, or its intracellular distribution. An influence of zinc-fluctuations on several established signaling pathways, including protein kinase C, mitogen activated protein kinases, and calcium has been shown. What is even more important, it has been discovered that labile zinc is required for well known physiological signals, e.g. the tyrosine phosphorylation events that are triggered by insulin and insulin like growth factor-1. Investigations about the molecular mechanism by which zinc exerts these effects have identified that it directly interacts with several components of signaling pathways, including protein tyrosine phosphatases, interleukin receptor associated kinase 1, the transcription factor MTF-1 and cyclic nucleotide phosphodiesterases, and alters the signaling properties of all these proteins. These observations indicate that, in addition to calcium, zinc is a second bivalent metal ion which also has a function as a second messenger.

* Correspondence: Dr. Lothar Rink, Institute of Immunology and Transfusion Medicine, Ratzeburger Allee 160, D-23538 Lübeck, Germany. E-mail Rink@immu.mu-luebeck.de Email: for Hajo Haase: hhaase@ukaachen.de, Telephone +49 241 80 80205

INTRODUCTION

Zinc is an essential trace element. On average, a human body contains 2-3g, which, due to its low redox potential, is only present in one oxidation state in biological systems, Zn^{2+}. Zinc ions are present as a prosthetic group in more than 300 enzymes, where they have either structural roles, or participate in catalysis [Vallee and Falchuk, 1993].

An estimate of zinc binding sites, based on a search for structural motives in the human genome, reveals that zinc proteins are frequent. The classical Cys_2His_2 zinc finger domain from TFIIIA alone has been estimated to be present in up to 1% of all human gene products [Berg and Shi, 1996], and a more general search for the presence of zinc binding sites indicates that up to 10% of the human proteome could be zinc-binding proteins [Andreini et al. 2006]. This does not even include regulatory binding sites with lower affinity, since most of them have not been identified, so far.

Zinc has been shown to be of particular importance in tissues that have a high rate of proliferation and differentiation, as, for example, the immune system [Wellinghausen et al. 1997a]. Accordingly, zinc deficiency leads to increased susceptibility to infections, but it also causes mental and growth retardation, anemia, and impaired wound healing. For this reason, cells put considerable effort in selecting zinc and ensuring selectivity over a range of other metal ions for its specific binding sites [Williams, 1984].

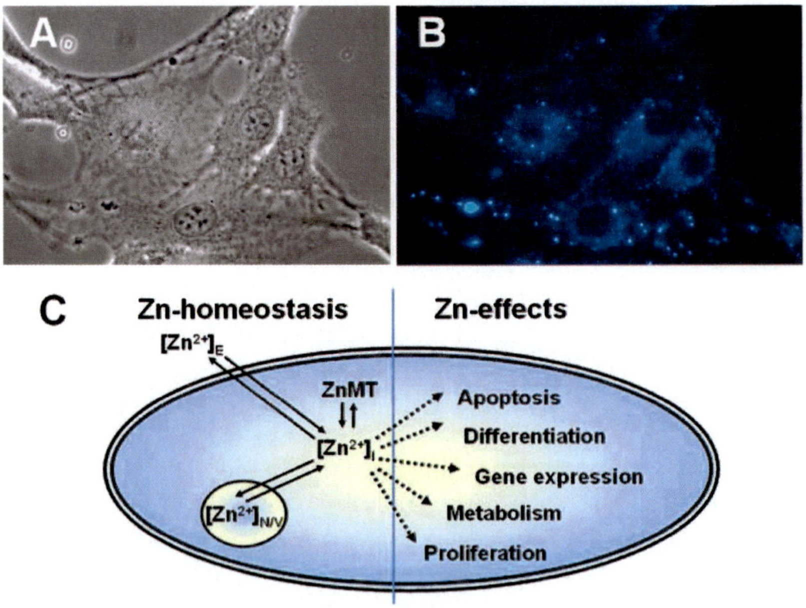

Figure 1: Intracellular labile zinc distribution and homeostasis. The distribution of zinc is tightly regulated. (A, B) Shown are microscopic pictures of a group of C6 rat glioma cells, either as phase contrast (A), or zinc-dependent fluorescence of the fluorophore Zinquin (B). The absence of nuclear fluorescence points towards a differential regulation of nuclear and cytosolic zinc, and the punctuate staining patterns within the cytosol indicate vesicular structures with high levels of labile zinc, so-called zincosomes. (C) The zinc homeostasis is mainly mediated by equilibria between the labile intracellular zinc [Zn2+]I, extracellular zinc [Zn2+]E, and zinc compartmentalized in the nucleus or vesicles (zincosomes) [Zn2+]N/V, or binding to metalloproteins, here represented by zinc-metallothionein [ZnMT]. This labile pool of zinc acts as a cellular signal, thereby regulating many cellular events.

In addition to its function as a static component of proteins, zinc has been considered to have an additional regulatory function: "*Inspection of zinc activities (...) leads one to consider the possibility that the internal homoeostasis of zinc (by metallothionein) is one of the dominant cytoplasmic growth hormones.*" [Frausto da Silva and Williams, 1991].

Since then, many interactions between zinc and different signaling pathways have been described [Beyersmann and Haase, 2001], and recent insights indicate that zinc indeed has a regulatory role in mammalian cells that can be best described as a second messenger.

The field of zinc biology, and in particular its role in signal transduction, has grown considerably in recent years, and this review will be primarily focused on giving a general overview on the effects of zinc on signaling events relevant to most cells. As discussed in more detail below, it has been found that the zinc concentration changes in response to different stimuli, and that this primarily affects a pool of loosely bound zinc, which fluctuates and is regulated by several mechanisms. This zinc binds to many proteins involved in signal transduction, regulating their activity, finally resulting in an effect of zinc on a multitude of biological functions (figure 1). Another role in which zinc acts as a regulator, neurotransmission, during which zinc is released and can reach concentrations up to 300 μM in the synaptic cleft [Assaf and Chung, 1984], will not be discussed here.

Effects of Zinc on the Cellular Level

Zinc is important for cellular growth and proliferation, which is not surprising for a component of many essential enzymes. However, several publications indicate that it might have a regulatory function, in which the level of intracellular zinc can affect growth, differentiation, and cell death (figure 2). Zinc deficiency reduces circulating levels of growth hormone and insulin like growth factor (IGF)-1, but these changes do not account for the entire growth retardation in zinc deficient animals, suggesting an additional role for zinc on the cellular level [MacDonald, 2000]. Chesters and coworkers have shown that a lack of zinc results in decreased thymidine kinase activity and mRNA expression, and suggested a zinc-regulated inhibitory protein that binds to the thymidine kinase promoter during zinc deficiency [Chesters et al. 1990; 1995]. Other potential mechanisms include a role for zinc in cyclin mRNA expression [Chesters and Petrie, 1999], or zinc acting as a "second messenger of mitogenic induction" that modulates DNA synthesis [Grummt et al. 1986]. Zinc has also been shown to be essential for differentiation of chick-embryo myoblasts to myotubes and zinc-deficiency lead to decreased mRNA expression and enzymatic activity of creatine kinase, an enzyme associated with myoblast differentiation [Petrie et al. 1991].

Not only proliferation and differentiation, but also programmed cell death is affected by zinc. While in general zinc depletion is regarded as an inducing factor for apoptosis, excessive zinc supplementation has also been shown to induce apoptosis, in some instances even in the same experimental system and in the absence of any additional apoptotic stimuli [Fraker and Telford, 1997; Haase et al. 2001; Wätjen et al. 2002]. One observation that could explain the antiapoptotic effect of zinc is its inhibition of caspase 3 at nanomolar concentrations [Perry et al. 1997; Maret et al. 1999]. Given the levels of intracellular labile zinc that are discussed below, zinc may be a physiological regulator of apoptosis via control of Caspase-3 activity. However, the effect of zinc on apoptosis has been intensely investigated, and many other mechanisms like the modulation of oxidative stress and several other caspases were suggested. An in-depth discussion of the role of zinc in apoptosis would exceed the scope of this chapter, and the reader is referred to the review of Truong-Tran et al. [2001a].

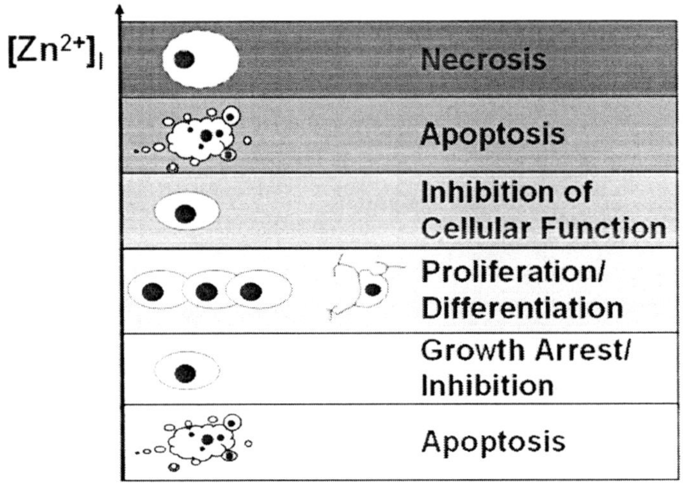

Figure 2. Effects of the intracellular labile zinc level. Several cellular functions seem to require different levels of labile zinc. While deregulation by both zinc deprivation and supplementation leads to apoptosis (excessive supplementation also results in uncontrolled, necrotic cell death), intermediate levels of labile zinc can regulate proliferation and differentiation.

In recent years, the use of microarray experiments to analyze the effects of zinc deprivation or supplementation on gene expression analysis has identified many zinc-dependent physiological processes, giving a first glimpse at the complexity of cellular events that depend on zinc. In cultured cells, namely the human monocytic cell line THP-1 [Cousins et al. 2003] or lymphoblastoid cells [Andree et al. 2004], a high number of zinc-regulated genes have been identified. Naturally, a clear effect was found on several genes involved in zinc homeostasis, like zinc transport proteins, but also metabolism, signal transduction and apoptosis, and, since immune cells were investigated, genes involved in the immune response and cytokine signaling were affected by zinc. When rats were subjected to nutritional zinc deficiency, metabolic pathways in rat liver, especially hepatic lipid metabolism were altered [tom Dieck et al. 2003; 2005]. Strikingly, changes in the zinc status can affect the expression of several hundred genes [Daniel and tom Dieck, 2004]. This leads to an important question: What is the molecular basis by which a single metal ion can regulate such a complexity of cellular events?

LABILE ZINC AND ITS FLUCTUATIONS

The total amount of cellular zinc is relatively high, and has been estimated to be several hundred micromolar [Arslan et al. 1985]. Most of this zinc is tightly bound to proteins and has a structural function, for example by stabilizing secondary structural elements like zinc fingers, or participates in reactions at the catalytic sites of enzymes. The availability of cellular zinc may affect loading of the apoproteins, but, once incorporated, this zinc is tightly bound and therefore does not participate in signal transduction. In addition, there is a smaller pool of cellular zinc, so-called "labile zinc", which is only loosely bound to small molecules or associated with low affinity binding sites in proteins. Fluctuations in the concentration of this pool, which can interact with regulatory binding sites in proteins, constitute an intracellular zinc signal. When the size of this labile zinc pool was investigated, contradicting results have been obtained and the concentration of labile cellular zinc was found to be between femtomolar and micromolar, spanning 9 orders of magnitude. Femtomolar concentrations in a mammalian cell would come down to only a fraction of an ion of labile zinc per cell, which would surely not be sufficient to have an impact on signaling. It has been the subject of debate, if labile zinc does exist at all and can directly bind proteins, or if zinc has to be bound to chaperone proteins, as in the case of copper [Tottey et al. 2005], and zinc can only be transferred from one of those chaperones to zinc-binding apo-proteins [Finney and O'Halloran, 2003]. This would of course prohibit a role for labile zinc in signal transduction.

The uncertainty results from the several different approaches that have been employed to estimate the concentration of labile intracellular zinc. The group of O'Halloran used the *in vitro* binding constants of zinc-regulated proteins as a measure to estimate labile zinc, and concluded that the concentration could only be femtomolar, according to binding affinities of the zinc sensor proteins Zur and ZnTR [Outten and O'Halloran, 2001; Finney and O'Halloran, 2003]. However, it has to be noted that the experiments that found femtomolar zinc were performed with *E. coli*, and the absence of free zinc in the cytosol of eukaryotic cells can not be concluded from these data [Outten and O'Halloran, 2001].

A great deal of knowledge was gained about labile zinc in mammalian cells by using fluorescent probes, many of which have been introduced in recent years and have already contributed significantly to our understanding of labile zinc [Thompson et al. 2005]. Fluorescence microscopy with several of these probes, as in the example with Zinquin in figure 1, shows noteworthy zinc-dependent fluorescence, which is a clear indicator for the existence of labile zinc in mammalian cells. When the fluorescence was quantified, e.g. by flow cytometry [Haase et al. 2006], and the values transformed into concentrations using the formula introduced in the classic paper by Grynkiewicz et al. (1985) for the quantification of calcium, it was found that the cellular concentration of labile zinc in mammalian cells is significantly higher than in prokaryotes, most values ranging from high picomolar to low nanomolar concentrations. Even higher, micromolar values have been measured with the fluorescent probe Zinquin [Brand and Kleineke, 1996]. However, Zinquin forms both a 1:1 complex and a 1:2 complex, and in this paper only the dissociation constant for the (weaker) 1:1 complex was used. In addition, it has been shown that Zinquin does not only detect labile zinc, but also zinc that is bound to the protein metallothionein, and probably also to other proteins, which would lead to an overestimation of the labile zinc concentration when this probe is used for quantification [Coyle et al. 1994].

Different experimental approaches that do not depend on fluorescent probes also find higher than femtomolar concentrations of labile zinc in eukaryotic cells. Using phosphoglucomutase activity, the zinc in rabbit muscle cells was estimated to be 100 pM [Peck and Ray, 1971], and with ^{65}Zn 24 pM were found in red blood cells [Simons, 1991]. Another point is that the binding constants for intracellular zinc proteins are between 10 and

12 (pK_D) [Maret, 2004], and the average dissociation constant of the zinc finger protein TFIIIA was determined to be even higher [Huang et al. 2004]. Thermodynamically, femtomolar availability of labile zinc would mean that all these enzymes and zinc finger containing transcription factors would be in the apo-form, and hence inactive, which contradicts numerous *in vivo* observations.

Table 1. Labile zinc concentrations in mammalian cells

Labile Zinc [nM]	Detection Method	Cell Type	Reference
0.1	phosphoglucomutase activity	rabbit skeletal muscle	Peck and Ray, 1971
0.024	^{65}Zn [1]	human erythrocytes	Simons, 1991
1	Fura-2	rat cardiac muscle	Atar et al. 1995
1,260	Zinquin	rat hepatocytes	Brand and Kleineke, 1996
2	MagFura-2	murine cortical neurons	Sensi et al. 1997
0.5	5-F-BAPTA, NMR	E367 rat neuroblastoma cell line	Benters et al. 1997
0.005	Modified carbonic anhydrase and dapoxyl sulfonamide	PC12 rat pheochromocytome cell line	Thompson, 2005
0.52	Fura-2	rat cardiomyocytes	Ayaz and Turan, 2006
0.17 (monocytes) 0.35 (lymphocytes)	FluoZin-3, flow cytometry	human peripheral blood mononuclear cells	Haase et al. 2006

[1]null-point measurement of labile zinc exchange through the plasma membrane in the presence of an ionophore

Today, most estimates point to high picomolar to low nanomolar labile intracellular zinc (table 1), but the precise determination of this concentration is still an intricate task. So is it difficult to apply binding constants of the fluorophores, which have been determined in buffers *in vitro*, to the conditions within cells. Even if the binding constants do not differ significantly, a high intracellular concentration of probe may exceed and concurrently deplete the labile zinc pool, conditions under which the fraction of probe that is bound to zinc is no longer determined by the affinity of the probe, and hence, this affinity is no longer useable for quantification [Dineley et al. 2002]. Also, all metal-sensing fluorescent probes do so by chelating metal ions, and may therefore interfere with the events that they are being used to investigate. Another issue is that fluorescent probes are never specific for a certain metal ion, but in the best case selective to the point that under the conditions they are used, other metal ions do not interfere with the measurement. Early reports already indicated that the calcium indicator quin-2 detects not only calcium, but is also quenched by labile cellular zinc, leading to an underestimation of calcium levels [Arslan et al. 1985]. The calcium indicator Fura-2 even binds zinc with a higher affinity than calcium, leading to a similar fluorescent signal. Some examples where Fura-2 has been used to quantify zinc can be found in tables 1 and 2. It is intriguing to think that some fluctuations that have been ascribed to calcium may in fact have been caused by zinc. Likewise, it also requires adequate controls to avoid that calcium fluctuations interfere with zinc measurements.

In order for a metal ion to transduce signals, its intracellular concentration has to fluctuate in response to stimulation. Many examples of zinc fluctuations can be found in the literature, some of which are listed in table 2. In addition to the stimulations mentioned in that table, in many reports cultured cells were exposed to zinc supplementation to the culture medium, resulting in influx and a rise in labile zinc. This is a convenient way to increase labile intracellular zinc to investigate its effect on signal transduction, but probably not a source of zinc fluctuations *in vivo*. The physiological level of extracellular labile zinc is only between 12 to 16 μM, and plasma contains high amounts of zinc-buffering proteins like albumin, α2-macroglobulin, and transferrin. For albumin alone, the concentration in human serum has been estimated to be 0.6 mM [Peters, 1995] and its dissociation constant for zinc is 2.95×10^{-8} [Masuoka et al. 1993]. Therefore, a maximum of nanomolar concentrations of extracellular labile zinc can be expected, and it is not clear if changes in the overall extracellular zinc concentration that could induce significant intracellular fluctuations do occur *in vivo*, or if these would be buffered.

To investigate the impact that zinc fluctuations have on cellular signaling, several tools have been developed that can be used to modulate labile zinc. On the one hand, ionophors are used to increase labile zinc, including pyrithione, PDTC (pyrrolidine dithiocarbamate), and A23187. The latter is widely used as a calcium ionophore, but is also very effective for other divalent cations; in fact it binds zinc with the highest affinity of several metal ions that have been investigated, including calcium [Pfeiffer and Lardy, 1976]. On the other hand, a reduction of the concentration of cellular labile zinc can be achieved by chelators. This can be done either by depleting the culture medium of zinc, or by direct intracellular chelation of labile zinc. Depletion can be achieved by chelex treatment, or addition of DTPA (Diethylenetriaminepentaacetic acid), EGTA (Ethylene glycol-bis(2-aminoethylether)-N,N,N′,N′-tetraacetic acid), EDTA (Ethylenediaminetetraacetic acid), or BAPTA (1,2-bis(2-aminophenoxy)ethane-N,N,N',N'-tetraacetic acid). In any case these chelators would also bind other metal ions besides zinc, which has to be taken into account for the experimental design. BAPTA has also been used as an acetoxymethyl ester, which is membrane permeable and can be used to load cell with this chelator. For the intracellular chelation of zinc, membrane permeable chelators like oPA (1,10-orthophenanthroline) and the frequently used TPEN (N,N,N′,N′-Tetrakis(2-pyridylmethyl)ethylenediamine) can be used. TPEN is highly selective, having a dissociation constant of 0.26 fM for zinc, while calcium and magnesium are bound with far lower affinities of 4 μM and 20 mM, respectively [Arslan et al. 1985]. A recent report suggesting that TPEN might not only act by chelation of zinc but could also affect other metal ions [Thompson, 2005] deserves more attention in the future. TPEN binds iron with 2.4 fM, so an interference with this metal has to be considered. By using all of these substances it should be kept in mind that modulating the labile zinc strongly affects cellular viability, and zinc supplementation as well as chelation are potent inducers of apoptosis and due caution has to be used to ensure that an effect that is blocked by zinc chelation is really due to a requirement for zinc in this specific process, and not by a detrimental effect on cellular viability.

Table 2. Fluctuations of labile zinc in mammalian cells

Stimulus	Fluctuation	Detection Method	Cell Type	Reference
electrical stimulation	<0.1 nM resting level to 1 nM after stimulation	Fura-2	rat cardiomyocytes	Atar et al. 1995
glucagon, cAMP	decrease in intracellular zinc	Zinquin	rat hepatocytes	Brand and Kleineke, 1996
nitric monoxide	increase in nuclear and cytosolic fluorescence	Zinquin, fluorescence microscopy, flow cytometry	L929 cell line, mouse splenocytes, rat endothelial cells	Berendji et al. 1997
hypochlorus acid, selenite	from 0,2 nM resting level to 7 nM after stimulation	Fura-2	rabbit ventricular myocytes	Turan et al. 1997
nitric monoxide	vesicular fluorescence, total zinc-dependent fluorescence doubled	TSQ, microscopy, fluorescence spectrometry	C6 rat glioma cell line	Haase and Beyersmann, 1999
IGF-1, serum withdrawal	increase in nuclear fluorescence	Zinquin	myoblast cell lines	Apostolova et al. 2000
2,2`-dithiodipyridine	intracellular zinc release	Newport Green, fluorescence imaging	rat neurons	Aizenman et al. 2000
PMA, H_2O_2	increase in cytoplasmic fluorescence	TSQ, confocal microscopy	NIH 3T3	Korichneva et al. 2002
VP-16 [1]	increase in intracellular zinc	Zinquin	human lymphoma cell lines	Smith et al. 2002
cytokines	increase in nuclear fluorescence	Zinquin	murine aortic endothelial cells	Spahl et al. 2003
NMDA, mitochondrial depolarization	1-3 nM increase in labile zinc	FluoZin-3	rat cortical neurons	Sensi et al. 2003
2,2`-dithiodipyridine	intracellular zinc release from MT	Mag-Fura-2	rat astrocytes	Malaiyandi et al. 2004
PDGF, EGF, IGF-1	increase due to higher retention	^{65}Zn	3T3	Simpson and Xu, 2006
PMA	intracellular labile zinc doubled	FluoZin-3, flow cytometry	primary human monocytes	Haase et al. 2006

[1] VP-16 : topoisomerase II inhibitor, induces DNA damage

REGULATION OF ZINC HOMEOSTASIS

Zinc is required in many different tissues and in different parts of cells, but the human body does not have a specialized zinc storage system. This necessitates a constant nutritional supply of zinc, and at the same time a complex mechanism for transport through biological membranes and the distribution of zinc into cellular compartments. Within the last ten years, many proteins have been discovered that mediate this transport. In eukaryotic cells, two main groups have been found, the ZnT proteins, which are members of the cation diffusion facilitator family, and the Zrt,Irt-like proteins (ZIP). While ZIP proteins mediate the transport of zinc into the cytoplasm, either from the exterior or from cellular organelles, the ZnT proteins have the opposite task, transporting zinc from the cytoplasm into the extracellular space, or into intracellular compartments [for a review see Eide, 2006]. Labile zinc is not evenly distributed in the cells. Fluorescence microscopy with the probe Zinquin showed a punctuate appearance in human chronic lymphatic leukemia cells and rat thymocytes and splenocytes [Zalewski et al. 1993], an observation that has been confirmed in many other cells types. A typical picture is shown in figure 1B, showing cytosolic, but nearly no nuclear fluorescence, which indicates differences in cytosolic and nuclear labile zinc, since the presence of the fluorescent probe in both compartments has been confirmed [Haase and Beyersmann, 2002]. No ion transport proteins for the regulation of nuclear zinc were found so far. On the other hand, transporter like ZnT-2 have been identified as the proteins that pump zinc into vesicles [Palmiter et al. 1996], so-called zincosomes, which can accumulate large amounts of labile zinc and correspond to the bright spots of fluorescence in the cytosol in figure 1B. Sequestration into zincosomes is reversible, because it has been shown that during serum deprivation labile zinc accumulates in vesicular structures, which were no longer detectable 24 h after serum had been replaced [Haase and Maret, 2005a]. Hitherto, it is not known to what type of vesicles these zincosomes belong.

Recent observations indicate another potential source for fluctuations of intracellular labile zinc. In cultured rat cortical neurons, stimulation with N-methyl-D-aspartate or mitochondrial depolarization with carbonyl cyanide p-(trifluorormethoxy)phenylhydrazone resulted in an increase of labile zinc concentration by 1-3 nM, due to a release of zinc from the mitochondria [Sensi et al. 2003].

Another important aspect of zinc homeostasis is its binding to proteins. Metallothioneins (MT) are small, cysteine rich proteins that can bind up to seven atoms of zinc, tetraedrically coordinated by a total of 20 cysteine residues. Several functions have been discussed for MT, including protection from heavy metal intoxication and against oxidative stress, but also a function in the homeostasis of zinc. An essential role for MT could not be confirmed in knockout experiments, since mice with a double knockout of the main forms MT-1 and -2 are viable. They show normal development, and are phenotypically unremarkable and fertile [Michalska and Choo, 1993; Masters et al. 1994]. However, a later study did report the occurrence of obesity and elevated plasma leptin levels with increasing age [Beattie et al. 1998]. So far, not all forms of MT have been knocked out. In addition, there may be a backup system, proteins with a similar function that can compensate for the loss of MT, at least when the mice are investigated under laboratory conditions. Experiments with zinc-deficient nursing MT-1 and -2 null mice showed impaired kidney development in the offspring, indicating that MT may be essential for zinc supply under conditions of limited availability

[Kelly et al. 1996]. Transgenic mice overexpressing MT-1 or MT-3 show no indication of phenotypical alterations [Palmiter et al. 1993; Erickson et al. 1995]. In contrast, a recent study using adenoviral overexpression of MT-2 in primary rat astrocytes showed two functions for this protein, acting as a zinc buffer when cells are confronted with an acute increase in cellular zinc, and a pool from which zinc can be released by oxidation [Malaiyandi et al. 2004]. Another function for MT may be the controlled transfer of zinc to and from proteins. The apoprotein thionein has been reported to reactivate a variety of zinc-inhibited enzymes [Maret et al. 1999]. In contrast, several apo-enzymes, including *E. coli* alkaline phosphatase, thermolysin, yeast aldolase, and bovine erythrocyte carbonic anhydrase did receive zinc from MT, indicating that MT could have a function in supplying zinc to newly synthesized zinc-requiring apo-enzymes [Udom and Brady, 1980].

Another role for MT could be as a reservoir for oxidant releasable zinc. The major zinc-binding amino acids in proteins are glutamate, aspartate, histidine, and cysteine. Zinc is bound to cysteine-sulfur as a zinc-thiolate. This renders zinc-binding sensitive to the redox state of its ligand. Oxidation to disulfides, sulfenes, sulfines, or, if the oxidizing agent is nitric monoxide (NO), S-nitroso-thiols leads to forms that do not bind zinc, hence the oxidation results in a release of zinc. Hereby, the per-se redox inert zinc ion can be influenced indirectly by the cellular redox state via its ligands [Maret, 2000; 2006]. This enables a redox-control of zinc binding to certain proteins.

Without the oxidative regulation of the number of high affinity zinc binding sites, MT, which has a high affinity for zinc (the overall binding constant for mammalian MT and zinc at pH 7 is 2×10^{12} M^{-1} [Vasak and Kägi, 1994]), would likely chelate the entire labile zinc, given that the cellular amount of MT would be high enough. Using a fluorescent labeling method that can discriminate between free and zinc bound thiols in metallothionein, it was found that under normal physiological conditions, a notable fraction of the cellular MT is present in its metal free form, thionein [Yang et al. 2001]. At first it seems unlikely to have high levels of thionein and still see labile zinc. Nonetheless, this method does not discriminate between thionein and the oxidized form of the protein, thionin, which can not bind zinc. This indicates an equilibrium between MT and thionein, but also thionin (the oxidized form of MT) in the cells, by which the cellular redox state controls the number of thiol zinc binding sites. Moreover, the flavoenzyme sulfhydryl oxidase has been found to oxidize thionein *in vitro* [Haase and Maret, 2004]. Although an oxidation of MT by sulfhydryl oxidase remains to be shown *in vivo*, it indicates that not only oxidative stress may release zinc, but that enzymatic mechanisms might exist that control the oxidation state of zinc binding sites, and thereby levels of labile zinc. Then again, zinc turned out to be an inhibitor of sulfhydryl oxidase, opening the possibility for a negative feedback by which labile zinc affects the oxidation of zinc binding sites [Brohawn et al. 2003].

A mechanism by which zinc release from proteins into the labile pool can be induced is the NO-mediated oxidative release of zinc from MT and zinc finger containing transcription factors, leading to an intracellular increase in labile zinc [Kröncke et al. 1994, Berendji et al. 1997]. Using a protein that consists of MT, sandwiched between two mutated green fluorescent proteins, it was possible to measure the conformational changes induced by NO-mediated metal release through fluorescence resonance energy transfer [Pearce et al. 2000]. Also, a NO-generated increase in labile zinc was not observed in cells derived from MT knockout mice, but was reconstituted by adenoviral MT gene transfer [St Croix et al. 2002].

Taken together, these experiments confirm that MT is a source for increased labile zinc after exposure of cells to NO.

A translocation of MT, and in some cases also labile zinc, from the cytoplasm to the nucleus occurs during proliferation and differentiation of adipocytes, myoblasts, and fibroblasts [Schmidt and Beyersmann, 1999; Apostolova and Cherian, 2000; Apostolova et al. 2000]. It was shown that stimulation of endothelial cells with a combination of the pro-inflammatory cytokines interleukin-1β, tumor necrosis factor-α, and interferon-γ induces nuclear translocation of MT and an increased production of NO, synthesized by the inducible nitric oxide synthase, leading to an increase in labile nuclear zinc [Spahl et al. 2003]. This increase was not observed in cells lacking MT, again pointing to MT as the source of zinc released by NO. This may be a mechanism to supply zinc to nuclear proteins during certain phases of the cell cycle. So far, relatively little is known about the mechanisms involved in nuclear translocation of MT. It has no classical nuclear translocation signal in its peptide sequence [Ogra and Suzuki, 2000]. It is known that the translocation of MT depends on several signaling pathways, and inhibitors for phosphatidylinositol 3-kinase (PI3-K), mitogen activated protein kinase (MAPK), and p70S6 kinase blocked IGF-1 induced nuclear translocation of MT. The MAPK inhibitor PD 98059 only blocked nuclear translocation of MT and proliferation, while differentiation was increased. Both other inhibitors blocked proliferation as well as differentiation of myoblasts [Apostolova et al. 2000]. This indicates that, at least in this experimental system, MT may be more important for nuclear events during proliferation than differentiation.

Effects of Zinc on Signal Transduction

Many reports describe an effect of zinc supplementation or deprivation on signal transduction pathways. In many cases it is unclear if zinc directly activates the proteins investigated, or if the activation is caused by an effect of zinc upstream in the signaling cascade.

Protein Kinases

A common feature in the transduction of signals along signaling pathways is reversible protein phosphorylation, especially on tyrosine, serine, and threonine residues. Their phosphorylation is governed by a balance of the activities of kinases, enzymes that transfer the terminal phosphate group of ATP to these amino acids, and phosphatases, which cleave it. When looking at the effect of zinc on these signals, it can be distinguished if a change in phosphorylation is due to a direct regulation of kinase or phosphatase activity by zinc, or if the effect originated upstream, leading to changed phosphorylation patterns of the entire pathway. Not all of the proteins mentioned below are directly affected by zinc, but rather can the regulation be attributed to such upstream events in many instances.

Most upstream phosphorylations, like the ones of the insulin receptor, the epidermal growth factor (EGF) receptor, and the Src-family kinases are tyrosine phosphorylations. While a lot of zinc effects may be based on phosphatase inhibition, which is discussed below, a direct effect of zinc on kinases has been shown *in vitro*. Increased phosphorylation of the Src-family kinase Lck has been described. This cannot be attributed to phosphatase inhibition alone, since the effect was also observed after purification of the kinase by immunoprecipitation. Also was it observed in the presence of magnesium, indicating that the role of zinc was not simply as a substitute for magnesium in complex formation with ATP. These data indicate a direct effect of zinc on the autophosphorylation and activation of Lck [Pernelle et al. 1991].

Zinc imitates the effects of insulin, the so-called insulinomimetic effect, activating the same signaling pathways, including tyrosine phosphorylation of the insulin receptor (IR) β subunit, phosphoinositol-3-kinase and Akt [Tang and Shay, 2001]. Hereby, zinc can induce tyrosine phosphorylation of the IRs activating tyrosine residues 1158, 1162, and 1163 and the adaptor protein IRS-1, and induces downstream activation of p21-activated kinase 1 and 2 phosphorylation. Zinc also has a dual effect on extracellular regulated kinase (ERK)-1/2 phosphorylation, activating at low concentrations, but inhibiting at higher ones. The IR, IRS-1 and both kinases were only affected when zinc was added together with pyrithione, indicating an intracellular site of action for zinc [Haase and Maret, 2003; 2005a]. After stimulation with insulin, membrane permeable zinc chelators were able to abrogate insulin induced signals, an indication that cellular zinc is involved in these signaling pathways. So did addition of TPEN inhibit the activating phosphorylation of the IR [Haase and Maret, 2003] and zinc was also shown to be required for insulin stimulated phosphorylation of p70S6 kinase, since this was abrogated in the presence of oPA. The authors ascribed this effect to a direct activation of mTOR by zinc [Lynch et al. 2001], although a very high concentration of zinc had to be used, especially when compared to the concentrations that did regulate other signaling proteins (table 3).

Many other reports describe an activation of kinases after zinc incubation, frequently involving the activation of MAP kinases, but also numerous other kinases. So does an early report indicate that an incubation with zinc leads to tyrosine phosphorylation and ERK activation [Hansson et al. 1996]. In mouse cortical cells, zinc activated ERK and hereby the transcription factor Egr-1, an effect which was inhibited by an inhibitor of the mitogen-activated/extracellular signal-regulated kinase kinases (MEK), which act upstream of ERK [Park and Koh, 1999]. Extracts of zinc-stimulated cells showed that zinc had induced kinase activity of PKC, and the MAPKs ERK, p38, and c-Jun N-terminal kinase (JNK) [LaRochelle et al. 2001]. When oligodendrocytes were treated with peroxynitrite, activation of ERK-1/2, 12-lipoxygenase, and reactive oxygen species were a consequence of zinc release, since all these effects were blocked by TPEN [Zhang et al. 2006]. In neuroblastoma cells, zinc treatment also induced phosphorylation of p70S6 kinase, GSK-3β, and the MAP kinases ERK-1/2, JNK, and p38 via mTOR [An et al. 2005; Bjorkdahl et al. 2005]. Treatment of cultured cells with zinc resulted in phosphorylation and activation of p70S6 kinase, which was mediated by PI3kinase, but was independent of calcium or PKC [Kim et al. 2000].

When peripheral blood mononuclear cells are incubated with zinc, this stimulates the secretion of cytokines like interleukin (IL)-1β and tumor necrosis factor-α. Studies with low molecular weight inhibitors showed that the zinc-induced activation involved protein tyrosine kinases and cyclic nucleotide dependent kinases [Wellinghausen et al. 1996]. However, the effect of zinc on kinases is not only an activating one, zinc ions have also been shown to specifically inhibit IL-1 receptor associated kinase 1 (IRAK-1), abrogating subsequent signaling events, a mechanism by which zinc can inhibit the stimulation of T-cells by IL-1 [Wellinghausen et al. 1997b].

Table 3. Signaling proteins directly affected by zinc *in vitro*

Protein	Zinc Concentration[1]	Effect	Reference
SHPTP-1	98 nM (IC_{50})	inhibition	Haase and Maret, 2003
PTP 1B	17 nM (IC_{50})	inhibition	Haase and Maret, 2003
	200 μM	inhibition	Samet et al. 1999
TC PTP	200 nM (IC_{50})	inhibition	Maret et al. 1999
SHPTP-2	1-2 μM (IC_{50})	inhibition	Haase and Maret, 2005a
LAR	20 μM (IC_{50})	inhibition	Haase and Maret, 2005a
Ca^{2+}-dependent endonuclease	1 μM (IC_{50})[2]	inhibition	Lohmann and Beyersmann, 1993
PDE 5	0.05-0.4 μM	activation	Francis et al. 1994
	> 1μM	inhibition	
PDE 4A	0.5 μM (EC_{50})	activation	Percival et al. 1997
	> 5 μM	inhibition	
PDE 6	0.2 μM (EC_{50})	activation	He et al. 2000
	> 1μM	inhibition	
Caspase 3	<10 nM (IC_{50})	inhibition	Maret et al. 1999
	100 nM (IC_{50})	inhibition	Perry et al. 1997
calcineurin	250 nM (IC_{50})[3]	inhibition	Takahashi et al. 2003
mTOR	10-100 μM	activation	Lynch et a. 2001
IKK	8.7 μM (IC_{50})	inhibition	Jeon et al. 2000

[1] Values are either given as IC_{50} or EC_{50} (concentration of zinc that causes 50% inhibition or activation, respectively), or, if no such value was given, as the lowest concentrations at which a significant effect was observed.

[2] Values varied, depending on the concentration of calcium. Value given was measured in the presence of 10 μM Ca^{2+}.

[3] IC_{50} for Ni^{2+}-activated calcineurin, in the presence of the physiological activator Ca/CaM an IC_{50} of 7 μM was found.

A wealth of data on the effects of zinc exists in airway epithelial cells, which have been extensively investigated by the group of Samet to elucidate in which way zinc as an airborne pollutant may affect lung tissue. Incubation with zinc induces Src-mediated, dimerization independent phosphorylation of the epidermal growth factor (EGF) receptor at several tyrosines [Samet et al. 2003], resulting in the activation of downstream signaling pathways. By this mechanism, zinc activates the protein kinase Akt by at least two actions, an activation of phosphatidyinositol 3-kinase (PI3-K), and the ubiquitinylation and subsequent proteolytic degradation of PTEN, a negative regulator of Akt [Wu et al. 2003; 2005]. Also does zinc-induced Src activation and subsequent tyrosine phosphorylation of residue 845 of the EGF receptor cause Ras activation, which in turn leads to the activation of the MEK/ERK pathway [Wu et al. 2002a; b]. Mechanistically, zinc has been shown to inhibit phosphatases that dephosphorylate Src and the EGF receptor [Samet et al. 1999; Tal et al., 2006]. In these cells, zinc treatment was shown to induce IL-8 expression by a mechanism involving ERK, JNK, and c-fos and c-Jun (the latter two form the transcription factor AP-1) phosphorylation and activation, with the ERK and JNK phosphorylation resulting from an inhibition of MAPK phosphatases, suggesting dual specificity MAPK phosphatase (MKP)-1 to be one of them [Kim et al. 2006].

Treatment of several lung cell lines with zinc chelators reduced the TNF-α-induced mRNA production of Eotaxin, Rantes, and MCP-1, indicating that labile zinc is involved in the expression of C-C chemokines. Conversely, zinc chelation had no effect on the CXC chemokine IL-8 [Richter et al. 2003].

Protein Phosphatases

Zinc has been reported to be an inhibitor of protein tyrosine phosphatases (PTP) [Brautigan et al. 1981]. Later publications investigated the inhibition of specific phosphatases, like PTP 1B [Samet et al. 1999; Tal et al. 2006] and indicate that zinc is a very potent inhibitor of PTP, with nanomolar inhibition constants for T cell PTP and PTP 1B [Maret et al. 1999; Haase and Maret, 2003; Haase and Maret, 2005a; b]. These constants indicate that the labile intracellular zinc constitutively inhibits a part of PTP activity. This is supported by the finding that chelation of labile zinc interrupts insulin and IGF-1-induced tyrosine phosphorylation of the receptor and downstream signals [Haase and Maret, 2003; 2005a] indicating that cellular zinc really does inhibit a part of PTP under normal conditions, and that this inhibition is essential for signal transduction.

The expression of PTP is also regulated by zinc. Culture of osteoblastic cells in the presence of zinc increased PTP activity, an effect that was sensitive to protein synthesis inhibition, and zinc enhanced an IGF-1-induced rise in PTP activity [Yamaguchi and Fukagawa, 2005].

The mechanism by which zinc inhibits PTP enzymatic activity is still unresolved. Since the truncated form of ΔSH2-SHP-1, which consists only of the catalytic domain, was still inhibited by nanomolar concentrations of zinc, the inhibitory binding site is somewhere on this domain [Haase and Maret, 2003]. The catalytic domain is highly conserved between the PTPs, indicating that zinc may affect a considerable number of these enzymes [Haase and Maret, 2005a]. It is noteworthy that there are several potential ligands for zinc-binding within the catalytic domain, and zinc may interact with the catalytically active cysteine residue, leading to a direct competition for binding to the active site between zinc and the substrate [Haase and Maret, 2003]. MKP use a comparable catalytic mechanism, involving a highly conserved HCX_5R motif [Farooq and Zhou, 2004]. This suggests that these phosphatases could be regulated by zinc in a similar fashion than PTP, and a recent publication indicates that this seems to be the case at least for MKP-1 [Kim et al. 2006]. On the other hand, serine-threonine phosphatases employ a different catalytic mechanism, in which the nucleophilic attack on the phosphate involves a metal bound water molecule [Barford, 1996], so the presumed mechanism for PTP-inhibition by zinc can not occur in serine/threonine phosphatases. This does not mean that these phosphatases can not be affected by zinc via a different mechanism. So has it been reported that calcineurin, a phosphatase that acts on the transcription factor NFAT, is inhibited by zinc [Tanaka et al. 2005]. It is conceivable that this effect could be an interaction of zinc with the activation of NFAT by calcium and calmodulin.

Protein Kinase C

Addition of zinc stimulates protein kinase C (PKC). One central event during the activation of PKC is its translocation to the plasma membrane. *In vitro* experiments showed that micromolar concentrations of zinc can induce binding of PKC to the cytoskeleton by increasing its interaction with actin filaments [Forbes et al. 1990a; Zalewski et al. 1991]. In intact cells, zinc (in the presence of pyrithione) induces the translocation of PKC to the cytoskeleton [Zalewski et al. 1990]. Also, zinc increases the binding of phorbol esters to PKC, which is abrogated by chelation of zinc [Csermely et al. 1988b; Forbes et al. 1990b].

The chelator TPEN inhibits N-methyl-D-aspartate (NMDA)- and A23187-induced membrane translocation of PKC, and this effect is reversed by addition of zinc [Baba et al. 1991]. Zinc chelators also abrogated activation of PKC in lymphocytes that had been treated with phorbol ester or antigens [Csermely et al. 1988a; Forbes et al. 1991]. PKC-α activity is downregulated in zinc-deficient cells, but on the other hand, zinc-deficiency induced apoptosis depends on PKC-δ in the same experimental system [Chou et al. 2004].

Analysis of the primary structure of PKC identified two identical zinc-binding motifs at the N-terminus, which is the regulatory region of this enzyme [Parker et al. 1986], and experiments employing NMR spectroscopy of PKC-α demonstrated that two non-consecutive sets of zinc-binding residues form two separate metal-binding sites [Hommel et al. 1994]. Experiments involving several deletion or point mutations of these zinc binding sites indicated that at least one of these sequences is required for the binding of phorbol esters and the resulting activation of PKC [Ono et al. 1989; Szallasi et al. 1996]. Later, PKCβI was shown to bind four zinc ions per molecule and EXAFS spectra indicated a complexation of each of the four zinc atoms by one nitrogen (histidine) and three sulfur (cysteine) ligands [Hubbard et al. 1991]. The binding of four atoms of zinc per molecule has since been confirmed for several other isoforms of PKC, including α, β II, and γ [Quest et al. 1992]. From those data it could be deduced that these four zinc binding sites mediate the regulatory effects of zinc on PKC. However, the zinc in these zinc finger-like structures is tightly bound and was not removed by high affinity heavy metal ion chelators, making it unlikely that intracellular labile zinc could regulate PKC via these sites [Hubbard et al. 1991]. Hence, the possibility has to be considered that the activation of PKC by a chelatable pool of zinc could also be mediated by binding sites that are not identical with the zinc finger-like structures.

Not only does labile cellular zinc influence PKC activation, when PKC is stimulated this also leads to fluctuations in intracellular labile zinc. The stimulation of lymphocytes with phorbol esters leads to a translocation of zinc from nuclear and mitochondrial compartments into cytosol and microsomes, and it has been postulated that this is part of lymphocyte signal transduction [Csermely et al. 1987; Csermely and Somogyi, 1989]. Activation of PKC by phorbol esters leads to an increase in labile zinc [Korichneva et al. 2002; Haase et al. 2006]. It has been shown that oxidative release of protein bound zinc is involved and that oxidation of the zinc-binding cysteine residues in PKC can even lead to zinc release from PKC itself. Oxidation of the zinc finger structures leads to a release of zinc and to an increase of the autonomous PKC-activity, i. e. the activity in the absence of activating cofactors, but to a loss of sensitivity to these cofactors [Knapp and Klann, 2000]. It seems that zinc, the intracellular redox state, and PKC form a network which can affect downstream signaling pathways, and interacts in several ways in what seems to be a feedback loop.

Calcium Signaling

Extracellular zinc affects calcium signaling, which leads to a crosstalk between these two metal ions in cellular signaling [Maret, 2001]. In several mammalian cell lines, a so-called "Orphan-Receptor", whose physiological stimulating agent is still unknown, was found. It is triggered by cadmium ions, which leads to the formation of inositol(1,4,5)trisphosphate via G protein mediated phospholipase C activation that results in an intracellular calcium release. This effect is inhibited by physiological concentrations of zinc [Smith et al. 1994]. Zinc has also been shown to inhibit UTP-induced intracellular calcium fluctuations, but only in the presence of extracellular calcium [Koizumi et al. 1995], and inhibits the rise in intracellular calcium triggered by complement fragment C5a in a monocytic cell line [Monk and Partridge, 1993]. In contrast to the inhibition of intracellular calcium fluctuations, there are also several reports describing that zinc may trigger a rise in the intracellular concentration of this ion. Treatment with extracellular zinc induced an increase in resting intracellular calcium in Madin Darby canine kidney cells, caused by Ca^{2+} influx into the cells, but not efflux from stores in the endoplasmatic reticulum [Jan et al. 1999]. The authors suggested a zinc-activated receptor. In the same system zinc inhibited ATP-induced capacitative Ca^{2+} entry, an effect that had been observed in several different cell types [Jan et al. 1999, and references therein]. Incubation of primary hepatocyte cultures with zinc caused an increase of intracellular calcium through stimulation of hormone sensitive intracellular calcium stores [McNulty and Taylor, 1999]. The authors of this report postulated that zinc may have a function as a local hormone that is secreted with insulin from pancreatic β-cells, and participates in the regulation of liver metabolism by interacting with a heavy metal ion receptor that is specifically expressed on hepatocytes. On several different other cell types, Hershfinkel et al. have described a plasma membrane receptor that is specifically triggered by extracellular zinc, but not by copper, manganese, nickel, or iron, and has therefore been named "zinc-sensing receptor". It causes a phospholipase C-dependent release of calcium from thapsigargin-sensitive intracellular calcium stores [Hershfinkel et al. 2001]. Stimulation of the zinc-sensing receptor leads to ERK1/2 phosphorylation and subsequent activation of a Na^+/H^+ exchanger [Azriel-Tamir et al. 2004]. The zinc-induced calcium fluctuations were not sensitive to pertussis toxin, suggesting that PLC was activated by a pertussis-independent class of G proteins, which was recently confirmed by showing that an inhibitor of Gαq could block zinc-triggered rise in intracellular calcium [Hershfinkel et al. 2001; Sharir and Hershfinkel, 2005].

The zinc content of the serum changes during infections, in response to nutritional supply, and with aging, which may affect calcium signaling. Also, zinc release together with insulin from pancreatic beta-cells has been demonstrated with confocal fluorescence microscopy using the zinc specific fluorophore Zinquin [Qian et al. 2000]. This suggests that extracellular zinc is a regulator for calcium signaling. Still, the receptors have not been cloned so far, so they could also be receptors for other, yet unidentified ligands, whose signals are modulated by zinc.

In addition to the action of extracellular zinc, intracellular zinc also interferes with calcium homeostasis. It inhibits the uptake and increases the release of nuclear calcium in isolated nuclei from rat hepatocytes, a mechanism which might influence nuclear calcium signaling [Yamaguchi, 1993].

Another aspect by which zinc affects calcium signaling is by its interaction with calmodulin (CaM), a calcium binding protein that confers calcium sensitivity to many different target proteins. Binding of calcium to CaM leads to a change in the proteins conformation that can be monitored by an increase in tyrosine fluorescence. Such an effect was also observed after addition of zinc, indicating that binding of both ions induces a similar conformational change. Also did zinc activate CaM-dependent phosphodiesterase activity. Since zinc was only effective in the presence of CaM in these experiments and a CaM inhibitor blocked the zinc-mediated activation, it was shown that this activation was indeed based on an interaction of zinc with CaM [Chao et al. 1984]. However, zinc does not seem to occupy the same binding sites as calcium. It was found that, in addition to the four calcium binding sites, CaM also has six additional binding sites to which zinc and other bivalent metal ions can bind [Milos et al. 1989]. Also, there is only a weak competition between zinc and calcium for CaM binding, and binding of zinc can cause a potentiating conformational change of CaM which occurs only when calcium is bound [Mills and Johnson, 1985].

Zinc has a dual effect on calcium/CaM-dependent protein kinase II (CaMPK-II), where low concentrations can inhibit the calcium/CaM stimulated activity, while concentrations above 100 μM caused an activation that was independent of CaM [Weinberger and Rostas, 1991]. In a later report, it was indicated that zinc might bind to CaMPK-II, hereby altering its conformation, leading to phosphorylation of several autophosphorylation sites that affect kinase activity; an additional activation of the enzyme by zinc/CaM was also reported [Lengyel et al. 2000].

Transcription Factors

The transcription factor nuclear factor kappa B (NF-κB), which is a dimer of proteins from the Rel/NF-κB family, predominantly a p50/p65 heterodimer, is associated with its inhibitory subunit inhibitor of kappa B (IκB), which keeps NFκ-B in an inactive state in the cytosol. Upon stimulation, the IκB kinases (IKK) phosphorylate serine residues of IκB, targeting the protein for ubiquitinylation and subsequent degradation. As a result the released NF-κB is translocated into the nucleus, where it can bind to its consensus sequences. Zinc has been found to either activate or inhibit NF-κB, depending on the experimental system investigated.

It has been reported that DNA binding of NF-κB is zinc dependent. The chelator oPA inhibited DNA binding of the transcription factor *in vitro*, which was reversed by the addition of zinc [Zabel et al. 1991]. However, most assays for NF-κB activity are routinely performed in the presence of millimolar concentrations of EDTA, which is a chelator with high affinity for zinc. Another publication finds the same effect of oPA, but other zinc chelators like EDTA, EGTA, or D-penicillamine were ineffective [Yang et al. 1995], suggesting that the zinc free form of oPA might disturb NF-κB DNA binding, rather than remove any zinc that is essential for DNA-binding. Besides, NF-κB is not a zinc finger protein, and the crystal structures of the p50 homodimer and the p50/p65 heterodimer bound to DNA do not contain zinc [Müller et al. 1995; Chen et al. 1998]. This is also supported by a study using four newly synthesized zinc chelators. While all of them did inhibit DNA-binding of HIV-EP1 (a transcription factor that contains two Cys_2His_2 zinc fingers), and this effect was reversed by addition of zinc, only two of these substances did inhibit NF-κB DNA binding. This effect was not reversible by addition of zinc, indicating a zinc-independent mechanism [Otsuka et al. 1995]. Taken together, these results make it seem more likely that the requirement for zinc originates upstream, affecting the signals that regulate NF-κB activation, but not NF-κB directly.

Conversely, there are also reports that zinc antagonizes NF-κB activation. This is supported by *in vivo* studies that showed that after lipopolysaccharide stimulation of Kupffer cells in mice zinc inhibits the NF-κB signaling pathway, including IκB phosphorylation [Zhou et al. 2004], and by the finding from our group that zinc abrogates LPS-induced NF-κB activation by a mechanism depending on cyclic nucleotide-mediated inhibition of Raf kinase (unpublished observations). Another report describes that zinc attenuates TNF-induced activation of NF-κB and AP-1 in porcine endothelial cells [Connell et al. 1997]. This effect may be related to an inhibition of IKK by zinc which has been reported to have a half maximal inhibition at 8.7 μM [Jeon et al. 2000]. A direct inhibition of NF-κB DNA binding by thiol-binding metals, including zinc has also been described, but in these experiments several hundred micromolar of zinc were necessary, an unphysiological concentration that does not occur *in vivo* [Yang et al. 1995; Shumilla et al. 1998]. The effect of zinc on NF-κB activity is most likely dependent on upstream signal transduction and many of the signaling pathways affected by zinc are known to stimulate NF-κB. MT seems also to be involved in NF-κB signaling, since the specific activity of this transcription factor was much higher in MT knockout mice [Butcher et al. 2004].

While NF-κB is an example for a transcription factor that does not contain any zinc, many others do, mainly for structural stabilization of zinc fingers and related domains [Berg and Shi, 1996]. Most of these motives bind zinc very tightly, so that under normal conditions they will be in the zinc-bound state, and not subjected to regulation by zinc fluctuations. Nevertheless, zinc can be removed from these binding sites by oxidation, changing their DNA-binding capability, but this is an effect of redox signaling, and not of regulation by zinc.

Even though zinc does not seem to have a direct effect on NF-κB, labile cellular zinc can directly regulate transcription factor activity. The promoters for several genes contain so-called metal response elements (MRE) which render their expression sensitive to metal ion concentrations. The core consensus sequence (TGCRCNC) binds the metal-response element-binding transcription factor-1 (MTF-1) [Radtke et al. 1993]. MTF-1 is involved in the basal expression of MT, but upon increased zinc availability it comes to augmented nuclear translocation and binding to DNA. By this mechanism MTF-1 acts as a cellular metal ion sensor that directly interacts with zinc [for reviews see Andrews, 2001; Lichtlen and Schaffner, 2001]. MTF-1 contains six highly conserved Cys_2His_2 zinc finger domains which are involved in DNA binding. It has been shown that zinc-regulated interaction with DNA is mediated by reversible binding of zinc to some of these structural motives, hereby stabilizing the zinc-finger secondary structures consisting of two beta sheets and one alpha helix. There exist contradictory reports as to which zinc finger(s) are constitutively zinc-bound, and which ones are sensitive to changes in labile zinc [Chen et al. 1999; Koizumi et al. 2000; Jiang et al. 2003; Potter et al. 2005].

MTF-1 controls the expression of proteins that are essential for zinc homeostasis, like MT [Heuchel et al. 1994], in particular the isoforms I and II, and the transporter ZnT-1 [Langmade et al. 2000], but also of proteins that are not directly involved with zinc transport or storage, like the heavy chain subunit of the γ-glutamyl-cysteine synthetase [Günes et al.

1998]. A thorough search in MTF-1 knockout mice, employing a combination of different methods, has pointed out 40 additional potential targets for MTF-1 [Lichtlen et al. 2001].

Zinc does not only control MTF-1 DNA binding activity by direct interaction, it also regulates MTF-1-controlled transcription via effects on calcium signaling and phosphorylation pathways including PKC, JNK, PI3-K, and tyrosine kinases [LaRochelle, et al. 2001]. Phosphorylation of MTF-1 at serine and tyrosine residues increases in response to metal exposure, and PKC, casein kinase II, tyrosine kinases, and Ca/CaM seem to be involved in the regulation of MTF-1 [Saydam et al. 2002; Adams et al. 2002]. The phosphorylation state of MTF-1 is not important for its interaction with DNA, but changes in the phosphorylation pattern seem to control nuclear translocation and transcriptional activation of MTF-1 [LaRochelle et al. 2001; Saydam et al. 2002; Adams et al. 2002], leading to a model in which zinc binding to zinc fingers is responsible for DNA binding, and phosphorylation regulates transcriptional activity. However, there are alternative models regarding the control of MTF-1 mediated transcription that involve additional proteins. One study showed no significant changes in the phosphorylation of MTF-1 after metal treatment *in vivo*, although the transcription factor was confirmed to be a target for several kinases *in vitro* [Jiang et al. 2004]. Since inhibition of these kinases did inhibit metal induced transcriptional activation, the authors suggested that other proteins (cofactors) that interact with MTF-1 may be the actual targets for phosphorylation. In another investigation it has been postulated that zinc binding releases MTF-1 from an inhibitor protein, while MTF-1 itself is constitutively active [Palmiter, 1994]. So far, none of these proteins suspected to interact with MTF-1 could be identified.

Metals other than zinc, in particular cadmium, can increase MT transcription in a MTF-1 dependent fashion *in vivo*, but do not increase nuclear translocation or DNA binding of the transcription factor *in vitro* [Bittel et al. 1998; Smirnova et al. 2000]. The activation of MTF-1 by other heavy metal ions and reactive oxygen species could either be due to an interaction with the signaling pathways that modulate the activity of the transcription factor, or to release of zinc by cysteine oxidation or displacement by other ions, which then binds to and activates MTF-1. Experimental evidence for zinc displacement is provided by the observation that the activation of MTF-1 DNA binding in a cell-free transcription system by cadmium, copper, and hydrogen peroxide was due to zinc-release from MT [Zhang et al. 2003]. Also, NO donors can activate MTF-1 in endothelial cells, but this effect is not observed in MT knock out mice, indicating that nitrosation of MT and the resulting zinc-release can activate MTF-1 [Stitt et al. 2006].

Another transcription factor whose DNA binding has been suggested to be directly modulated by labile zinc is p53 [Meplan et al. 2000a]. Chelation of intracellular zinc with a membrane permeable chelator reduced DNA binding of p53 [Verhaegh et al. 1998]. This effect was reversed after removal of the chelator, but only if sufficient extracellular zinc was present [Meplan et al. 2000b]. The binding of zinc to p53 is highly temperature dependent, and it easily dissociates from the p53 DNA binding domain at 37°C, indicating that a considerable amount of cellular p53 may exist in the zinc-free state. Interestingly, in a number of p53 mutations identified in human tumors, this dissociation is even faster. The zinc free form shows reduced DNA binding and, in addition, leads to the aggregation of zinc-free as well as active p53, and may hereby inactivate remaining zinc-bound p53 [Butler and Loh, 2003]. Moreover, p53 activity is affected by MT. While high expression of MT deactivated p53, presumably by chelating zinc, moderate overexpression of MT resulted in an activation of p53 DNA-binding capacity, indicating that MT may also be involved in supplying zinc to this transcription factor [Meplan et al. 2000b].

Cyclic Nucleotide Signaling

The cellular levels of the second messengers cyclic adenosine monophosphate (cAMP) and cyclic guanosine monophosphate (cGMP) are regulated by their synthesis by adenylate (AC) or guanylate cyclases (GC) and their degradation by cyclic nucleotide phosphodiesterases (PDE). Both groups of enzymes have been shown to be directly affected by low micromolar concentrations of zinc *in vitro*, and incubation of cultured cells with zinc can change the intracellular levels of cyclic nucleotides. Initial observations described an inhibition of cAMP and cGMP hydrolysis by zinc in PDE from bovine heart [Donnelly, 1978] or cardiac membranes [Dubois et al. 1985]. Since then there have been several reports both of activating and inhibiting effects of zinc on PDE and the enzymatic activity of different isoforms.

The cGMP-binding cGMP-specific PDE 5 is optimally activated by low zinc concentrations between 0.05-0.4 μM with a K_D value for zinc of approximately 0.5 μM. This observation, together with the identification of two tandem histidine containing amino acid sequences ($HX_3HX_{24\text{-}26}E$) in the catalytic domain of mammalian PDE, indicating the complexation of catalytic zinc ions, led to the assumption that PDEs may be zinc-hydrolases [Francis et al. 1994; Omburo et al. 1998]. Similar results were also found for recombinant human PDE 4A [Percival et al. 1997] and retinal PDE 6 [He et al. 2000]. Interestingly, all three studies reported inhibitory effects on the same PDEs at slightly higher zinc concentrations. Site directed histidine to serine mutations in one of these motifs of human monocyte PDE 4A resulted in a loss of catalytic activity and the resulting enzymes were no longer activated by divalent metals [Omburo et al. 1998]. Activation by zinc has not been found in all types of PDE. cGMP-inhibited PDE did not show an activation upon addition of zinc, but an inhibition at doses above 1 μM [Omburo et al. 1995]. Also, purified cAMP specific PDE 4D was activated by Mg^{2+}, Co^{2+}, and Mn^{2+}, but not Zn^{2+}, indicating that this PDE may not be a zinc hydrolase [Kovala et al. 1997]. However, given the high number of observations that support a requirement for zinc, it can not be excluded that zinc had remained bound to the enzymes during the isolation process, which could also explain the lack of zinc-activation in the latter two studies.

Zinc does not only inhibit purified PDE, but does also modulate the levels of cyclic nucleotides in intact cells. In PC 12 rat pheochromocytoma cells that had been incubated with zinc, an increase of the cellular cGMP-concentration occurred. In the homogenate of these cells a zinc-mediated inhibition of cGMP hydrolysis was identified as the cause for this observation [Wätjen et al. 2001]. In human monocytes, degradation of cyclic nucleotides is mediated by PDEs 1, 3 and 4, and an inhibition of all PDE subtypes has been observed in the lysate of these cells. Although zinc inhibited the cleavage of cAMP and cGMP alike, only an increase in the cellular cGMP was found after zinc-treatment of intact cells, because zinc did also interfere with the synthesis of cAMP [von Bülow et al. 2005]. In these experiments zinc acted as a negative regulator of lipopolysaccharide-induced secretion of inflammatory cytokines via its effect on cellular cGMP [von Bülow et al. 2005].

Zinc has a dual effect on the activity of PDEs, acting as a component of the enzyme that is involved in catalytic activity, while labile zinc acts as a negative regulator. This does not limit the effect of zinc on cyclic nucleotide signaling, the mRNA expression of several PDEs is also inhibited by increased labile zinc [von Bülow et al. 2005]. In addition to the effects on PDEs, several other components of the cyclic nucleotide signaling system are also modulated by zinc. So is the AC inhibited by zinc, too [Klein et al. 2002], whereas at least the soluble GC is not affected [von Bülow et al. 2005]. Two mechanisms have been suggested for AC inhibition, on the one hand an influence of zinc on the conformation of AC [Klein et al. 2004], and on the other an inhibition of the GTPase activity of the alpha subunit of the stimulatory GTP-binding protein of AC [Gao et al. 2005]. What makes the system even more complex is that soluble GC is activated by NO, and the NO synthases are dimers whose dimerization requires zinc as a crosslink with two zinc binding cysteines contributed from each monomer [Munro et al. 2000]. This zinc tetrathiolate center constitutes a redox switch by which peroxinitrite can oxidize the cysteines, subsequently releasing zinc and uncoupling nitric oxide synthesis [Maret, 2006]. As discussed above, NO also releases protein bound zinc, e.g. from MT, which in turn may bind to and inhibit PDEs. By all these mechanisms zinc and NO interact to regulate the synthesis and degradation of cGMP.

ZINC AND DISEASE

With respect to the impact that zinc has on signal transduction, it can be expected that disturbances of the zinc homeostasis are associated with several kinds of diseases. A well characterized example is *Acrodermatitis enteropathica*, which results from a mutation in the zinc transporter hZIP-4 [Kury et al. 2002], leading to severe zinc deficiency due to reduced uptake of zinc from the food. As previously mentioned, zinc deficiency affects in particular the immune system, and one consequence of this disease is a severe impairment of immune function, leading to infections and death [Wellinghausen et al. 1997a]. Interestingly, several diseases that are associated with autoimmune or inflammatory events, like diabetes [Chausmer, 1998], rheumatoid arthritis [Zoli et al. 1998], asthma [Truong-Tran et al. 2001b], and chronic obstructive pulmonary disease [Karadag et al. 2004] are accompanied by significantly reduced levels of serum zinc. The reduced zinc levels could be regarded as a consequence of a disease, as in the case of increased urinary excretion that has been observed during diabetes. However, the development of spontaneous or chemically induced diabetes in an animal model of type one diabetes, the non obese diabetic mouse, is prevented by zinc supplementation [Schott-Ohly et al. 2004]. Given the interaction of zinc with insulin receptor signaling, it is conceivable that local zinc deficiencies could contribute to development of insulin resistance and diabetes [Haase and Maret, 2005b].

It is frequently observed that elderly people have at least marginal zinc deficiency. Many of the immunological alterations of the aging immune system are similar to what is observed during zinc deficiency. Since a number of these alterations, like T cell function, response to vaccination, and the release of several cytokines was counteracted by nutritional zinc supplementation, a causal relationship between zinc status and immunological alterations seems probable [Ibs et al. 2003].

CONCLUSION

When zinc is compared to calcium, which has been recognized as a second messenger for more than 20 years, several parallels emerge. Both are bivalent cations, the labile concentrations of which are tightly controlled, kept on a low intracellular level by intricate homeostatic mechanisms, and a deregulation of their homeostasis frequently leads to apoptosis. Differences are that, in contrast to calmodulin for calcium, no such adaptor protein has been described for zinc; it rather seems to directly interact with several proteins from different signaling cascades. Also, the intracellular concentration of labile zinc may in some instances be a permissive signal rather than a second messenger, meaning that labile zinc has to reach a certain level to sufficiently inhibit phosphatases or phosphodiesterases for stimulus-induced signals to prevail. Another difference is the kind of ligands the two metals bind to. This is of particular relevance since zinc is tightly bound by cysteine thiols, enabling a cross-talk between zinc and redox signals, which does not occur with calcium.

From all metal ions that could have been made use of, why zinc? Firstly, it can not be excluded that other metal ions do also contribute to signaling. Yet, zinc shares an additional advantage with calcium. In contrast to several other heavy metal ions that could have been utilized, like copper or iron, zinc is redox inert in biological systems, and does not contribute to oxidative stress. This is important for an ion that is released in the cell during fluctuations, especially nuclear fluctuations that bring it into the immediate vicinity of DNA.

Given all these similarities between zinc and calcium, it seems peculiar that the role of zinc in signaling is being recognized so much later. One reason may be that zinc has two important biological functions that seem to be largely independent. On the one hand it is a component of proteins and on the other hand it is a signaling ion. Its role in proteins was recognized starting in the 1940s, when it was identified as a component of carbonic anhydrase [Keilin and Mann, 1940], and from there, zinc has been found in many more proteins by methods like atomic absorption spectrometry. This alone could account for the essentiality of zinc, so it was not necessary to search for an additional function of zinc in signal transduction. In contrast, labile zinc was difficult to analyze, especially due to the lack of appropriate fluorescent probes prior to the introduction of Zinquin [Zalewski et al. 1993], which is why fluctuations were, and sometimes still are, hard to observe. In addition, the changes of the very small pool of labile zinc in the presence of the high concentration of structural zinc in proteins are not detectable with conventional methods that monitor total cellular zinc.

Prokaryotes seem to have significantly lower labile zinc (if any), which raises the question if zinc is only a component of mammalian signal transduction. Its homeostasis is tightly regulated in prokaryotes, too, and it surely is an essential component of proteins, but probably not in signaling. The main roles identified for zinc in mammalian signaling so far include neurotransmission, the immune system, and signaling in response to growth factors. All these are features of complex organisms. Potentially, when cells started to organize into larger structures and form organisms, this was accompanied by a need for intracellular transmission of intercellular communication, leading to signaling functions for a metal ion that had previously only been used as a structural component of proteins. In this context it should be noted that even for calcium a signaling function in bacteria is discussed, but is still not conclusively shown [Dominguez, 2004].

Intracellular labile zinc fluctuates, and several targets have been found by which these fluctuations affect different levels of signal transduction from receptors down to directly regulating transcription factor activity (figure 3). Although some information is now available and a picture starts to emerge, the research about the signaling function of labile zinc is still in its beginnings and many questions remain to be solved. It is highly unlikely that zinc only interacts with the proteins identified so far, especially its role in stabilizing protein-protein interactions may yield additional targets. Also, much remains to be found out about the mechanisms that regulate the labile zinc level: transporter activity and expression, intracellular sequestration, redox signals, and the coordination of the interaction of all these aspects. Finally, when zinc signaling will be sufficiently understood and the biological processes that depend on zinc are identified, the administration of zinc or drugs which modulate zinc signaling, e.g. during the onset of diabetes, could be added to the list of therapeutically useful modulators of signal transduction.

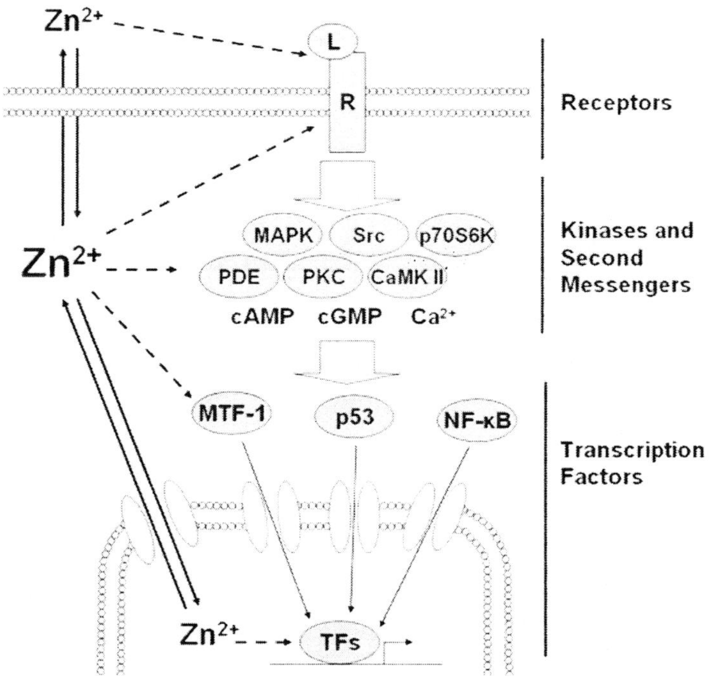

Figure 3. Multiple targets for zinc in cellular signal transduction. Levels of labile extracellular, intracellular, and nuclear zinc affect cellular signaling by acting (either directly or indirectly) on receptors, kinases, phosphatases, second messengers, and transcription factors, which are discussed in the text. Hereby zinc exerts control over cellular signaling pathways at all levels.

References

Adams TK, Saydam N, Steiner F, Schaffner W, Freedman JH. (2002) Activation of gene expression by metal-responsive signal transduction pathways. *Environ. Health Perspect.* 110: 813-7.

Aizenman E, Stout AK, Hartnett KA, Dineley KE, McLaughlin B, Reynolds IJ. (2000) Induction of neuronal apoptosis by thiol oxidation: putative role of intracellular zinc release. *J. Neurochem.* 75(5): 1878-88.

An WL, Bjorkdahl C, Liu R, Cowburn RF, Winblad B, Pei JJ. (2005) Mechanism of zinc-induced phosphorylation of p70 S6 kinase and glycogen synthase kinase 3beta in SH-SY5Y neuroblastoma cells. *J. Neurochem.* 92(5): 1104-15.

Andree KB, Kim J, Kirschke CP, Gregg JP, Paik H, Joung H, Woodhouse L, King JC, Huang L. (2004) Investigation of lymphocyte gene expression for use as biomarkers for zinc status in humans. *J. Nutr.* 134(7): 1716-23.

Andreini C, Banci L, Bertini I, Rosato A. (2006) Counting the Zinc-Proteins Encoded in the Human Genome. *J. Prot. Res.* 5(1): 196-201.

Andrews GK. (2001) Cellular zinc sensors: MTF-1 regulation of gene expression. *BioMetals* 14(3-4): 223-37.

Apostolova MD, Cherian MG. (2000) Delay of M-phase onset by aphidicolin can retain the nuclear localization of zinc and metallothionein in 3T3-L1 fibroblasts. *J. Cell. Physiol.* 183(2): 247-53.

Apostolova MD, Ivanova IA, Cherian MG. (2000) Signal transduction pathways, and nuclear translocation of zinc and metallothionein during differentiation of myoblasts. *Biochem. Cell Biol.* 78(1): 27-37.

Arslan P, Di Virgilio F, Beltrame M, Tsien RY, Pozzan T. (1985) Cytosolic Ca2+ homeostasis in Ehrlich and Yoshida carcinomas. A new, membrane-permeant chelator of heavy metals reveals that these ascites tumor cell lines have normal cytosolic free Ca2+. *J. Biol. Chem.* 260(5): 2719-27.

Assaf SY, Chung SH. (1984) Release of endogenous Zn2+ from brain tissue during activity. *Nature* 308(5961): 734-6.

Atar D, Backx PH, Appel MM, Gao WD, Marban E. (1995) Excitation-transcription coupling mediated by zinc influx through voltage-dependent calcium channels. *J. Biol. Chem.* 270(6): 2473-7.

Ayaz M, Turan B. (2006) Selenium prevents diabetes-induced alterations in [Zn2+]i and metallothionein level of rat heart via restoration of cell redox cycle. *Am. J. Physiol. Heart Circ. Physiol.* 290(3): H1071-80.

Azriel-Tamir H, Sharir H, Schwartz B, Hershfinkel M. (2004) Extracellular zinc triggers ERK-dependent activation of Na+/H+ exchange in colonocytes mediated by the zinc-sensing receptor. *J. Biol. Chem.* 279(50): 51804-16.

Baba A, Etoh S, Iwata H. (1991) Inhibition of NMDA-induced protein kinase C translocation by a Zn2+ chelator: implication of intracellular Zn2+. *Brain Res.* 557(1-2): 103-8.

Barford D. (1996) Molecular mechanisms of the protein serine/threonine phosphatases. *Trends Biochem. Sci.* 21(11): 407-12.

Beattie JH, Wood AM, Newman AM, Bremner I, Choo KH, Michalska AE, Duncan JS, Trayhurn P. (1998) Obesity and hyperleptinemia in metallothionein (-I and -II) null mice. *Proc. Natl. Acad. Sci. USA* 95(1): 358-63.

Benters J, Flögel U, Schäfer T, Leibfritz D, Hechtenberg S, Beyersmann D. (1997) Study of the interactions of cadmium and zinc ions with cellular calcium homoeostasis using 19F-NMR spectroscopy. *Biochem. J.* 322: 793-9.

Berendji D, Kolb-Bachofen V, Meyer KL, Grapenthin O, Weber H, Wahn V, Kroncke KD. (1997) Nitric oxide mediates intracytoplasmic and intranuclear zinc release. *FEBS Lett.* 405(1): 37-41.

Berg JM, Shi Y. (1996) The galvanization of biology: a growing appreciation for the roles of zinc. *Science* 271(5252): 1081-5.

Beyersmann D, Haase H. (2001) Functions of zinc in signaling, proliferation and differentiation of mammalian cells. *BioMetals* 14(3-4): 331-341.

Bittel D, Dalton T, Samson SL, Gedamu L, Andrews GK. (1998) The DNA binding activity of metal response element-binding transcription factor-1 is activated in vivo and in vitro by zinc, but not by other transition metals. *J. Biol. Chem.* 273(12): 7127-33.

Bjorkdahl C, Sjogren MJ, Winblad B, Pei JJ. (2005) Zinc induces neurofilament phosphorylation independent of p70 S6 kinase in N2a cells. *Neuroreport* 16(6): 591-5.

Brand IA, Kleineke J. (1996) Intracellular zinc movement and its effect on the carbohydrate metabolism of isolated rat hepatocytes. *J. Biol. Chem.* 271(4): 1941-9.

Brautigan DL, Bornstein P, Gallis B. (1981) Phosphotyrosyl-protein phosphatase. Specific inhibition by Zn. *J. Biol. Chem.* 256(13): 6519-22.

Brohawn SG, Miksa IR, Thorpe C. (2003) Avian sulfhydryl oxidase is not a metalloenzyme: adventitious binding of divalent metal ions to the enzyme. *Biochemistry* 42(37): 11074-82.

Butcher HL, Kennette WA, Collins O, Zalups RK, Koropatnick J. (2004) Metallothionein mediates the level and activity of nuclear factor kappa B in murine fibroblasts. *J. Pharmacol. Exp. Ther.* 310(2): 589-98.

Butler JS, Loh SN. (2003) Structure, function, and aggregation of the zinc-free form of the p53 DNA binding domain. *Biochemistry* 42(8): 2396-403.

Chao SH, Suzuki Y, Zysk JR, Cheung WY. (1984) Activation of calmodulin by various metal cations as a function of ionic radius. *Mol. Pharmacol.* 26(1): 75-82.

Chausmer AB. (1998) Zinc, insulin and diabetes. *J. Am. Coll. Nutr.* 17(2): 109-15.

Chen X, Chu M, Giedroc DP. (1999) MRE-Binding transcription factor-1: weak zinc-binding finger domains 5 and 6 modulate the structure, affinity, and specificity of the metal-response element complex. *Biochemistry* 38(39): 12915-25.

Chen FE, Huang D-B, Chen Y-Q, Gosh G (1998) Crystal structure of p50/p65 heterodimer of transcription factor NF-kB bound to DNA. *Nature* 391: 410-413.

Chesters JK, Boyne R, Petrie L, Lipson KE. (1995) Role of the promoter in the sensitivity of human thymidine kinase to lack of Zn2+. *Biochem. J.* 308(Pt 2): 659-64.

Chesters JK, Petrie L. (1999) A possible role for cyclins in the zinc requirements during G1 and G2 phases of the cell cycle. *J. Nutr. Biochem.* 10(5): 279-90.

Chesters JK, Petrie L, Travis AJ. (1990) A requirement for Zn2+ for the induction of thymidine kinase but not ornithine decarboxylase in 3T3 cells stimulated from quiescence. *Biochem. J.* 272(2): 525-7.

Chou SS, Clegg MS, Momma TY, Niles BJ, Duffy JY, Daston GP, Keen CL. (2004) Alterations in protein kinase C activity and processing during zinc-deficiency-induced cell death. *Biochem. J.* 383: 63-71.

Connell P, Young VM, Toborek M, Cohen DA, Barve S, McClain CJ, Hennig B. (1997) Zinc attenuates tumor necrosis factor-mediated activation of transcription factors in endothelial cells. *J. Am. Coll. Nutr.* 16(5): 411-7.

Cousins RJ, Blanchard RK, Popp MP, Liu L, Cao J, Moore JB, Green CL. (2003) A global view of the selectivity of zinc deprivation and excess on genes expressed in human THP-1 mononuclear cells. *Proc. Natl. Acad. Sci. USA* 100(12): 6952-7.

Coyle P, Zalewski PD, Philcox JC, Forbes IJ, Ward AD, Lincoln SF, Mahadevan I, Rofe AM. (1994) Measurement of zinc in hepatocytes by using a fluorescent probe, zinquin: relationship to metallothionein and intracellular zinc. *Biochem. J.* 303: 781-6.

Csermely P, Gueth S, Somogyi J. (1987) The tumor promoter tetradecanoyl-phorbol-acetate (TPA) elicits the redistribution of zinc in subcellular fractions of rabbit thymocytes measured by X-ray fluorescence. *Biochem. Biophys. Res. Commun.* 144(2): 863-8.

Csermely P, Somogyi J. (1989) Zinc as a possible mediator of signal transduction in T lymphocytes. *Acta Physiol. Hung.* 74(2): 195-9.

Csermely P, Szamel M, Resch K, Somogyi J. (1988a) Zinc can increase the activity of protein kinase C and contributes to its binding to plasma membranes in T lymphocytes. *J. Biol. Chem.* 263(14): 6487-90.

Csermely P, Szamel M, Resch K, Somogyi J. (1988b) Zinc increases the affinity of phorbol ester receptor in T lymphocytes. *Biochem. Biophys. Res. Commun.* 154(2): 578-83.

Daniel H, tom Dieck H. (2004) Nutrient-gene interactions: a single nutrient and hundreds of target genes. *Biol. Chem.* 385(7): 571-83.

Dineley KE, Malaiyandi LM, Reynolds IJ. (2002) A reevaluation of neuronal zinc measurements: artifacts associated with high intracellular dye concentration. *Mol. Pharmacol.* 62(3): 618-27.

Dominguez DC. (2004) Calcium signalling in bacteria. *Mol. Microbiol.* 54(2): 291-7.

Donnelly TE. (1978) Effects of zinc chloride on the hydrolysis of cyclic GMP and cyclic AMP by the activator-dependent cyclic nucleotide phosphodiesterase from bovine heart. *Biochim. Biophys. Acta.* 522(1): 151-60.

Dubois M, Prigent AF, Nemoz G, Fougier S, Pacheco H. (1985) Modulation of the cardiac membrane-bound cyclic nucleotide phosphodiesterase: inhibition due to a contamination of TLC purified S-adenosyl-L-[methyl-3H] methionine by Zn++ ions. *Life Sci.* 37(4): 315-24.

Eide DJ. (2006) Zinc transporters and the cellular trafficking of zinc. *Biochim. Biophys. Acta.* 1763(7): 711-22.

Erickson JC, Masters BA, Kelly EJ, Brinster RL, Palmiter RD. (1995) Expression of human metallothionein-III in transgenic mice. *Neurochem. Int.* 27(1): 35-41.

Farooq A, Zhou MM. (2004) Structure and regulation of MAPK phosphatases. *Cell. Signal.* 16(7): 769-79.

Finney LA, O'Halloran TV. (2003) Transition metal speciation in the cell: insights from the chemistry of metal ion receptors. *Science* 300(5621): 931-936.

Forbes IJ, Zalewski PD, Giannakis C. (1991) Role for zinc in a cellular response mediated by protein kinase C in human B lymphocytes. *Exp. Cell Res.* 195(1): 224-9.

Forbes IJ, Zalewski PD, Giannakis C, Betts WH. (1990a) Zinc induces specific association of PKC with membrane cytoskeleton. *Biochem. Int.* 22(4): 741-8.

Forbes IJ, Zalewski PD, Giannakis C, Petkoff HS, Cowled PA. (1990b) Interaction between protein kinase C and regulatory ligand is enhanced by a chelatable pool of cellular zinc. *Biochim. Biophys. Acta.* 1053(2-3): 113-7.

Fraker PJ, Telford WG. (1997) A reappraisal of the role of zinc in life and death decisions of cells. *Proc. Soc. Exp. Biol. Med.* 215(3): 229-36.

Francis SH, Colbran JL, McAllister-Lucas LM, Corbin JD. (1994) Zinc interactions and conserved motifs of the cGMP-binding cGMP-specific phosphodiesterase suggest that it is a zinc hydrolase. *J. Biol. Chem.* 269(36): 22477-80.

Frausto da Silva JJR, Williams, RJP. (1991) The biological chemistry of the elements. Clarendon press, Oxford

Gao X, Du Z, Patel TB. (2005) Copper and zinc inhibit Galphas function: a nucleotide-free state of Galphas induced by Cu2+ and Zn2+. *J. Biol. Chem.* 280(4): 2579-86.

Grummt F, Weinmann-Dorsch C, Schneider-Schaulies J, Lux A. (1986) Zinc as a second messenger of mitogenic induction. Effects on diadenosine tetraphosphate (Ap4A) and DNA synthesis. *Exp. Cell Res.* 163(1): 191-200.

Grynkiewicz G, Poenie M, Tsien RY. (1985) A new generation of Ca2+ indicators with greatly improved fluorescence properties. *J. Biol. Chem.* 260(6): 3440-50.

Günes C, Heuchel R, Georgiev O, Müller KH, Lichtlen P, Blüthmann H, Marino S, Aguzzi A, Schaffner W. (1998) Embryonic lethality and liver degeneration in mice lacking the metal-responsive transcriptional activator MTF-1. *EMBO J.* 17(10): 2846-54.

Haase H, Beyersmann D. (1999) Uptake and intracellular distribution of labile and total Zn(II) in C6 rat glioma cells investigated with fluorescent probes and atomic absorption. *BioMetals* 12(3): 247-54.

Haase H, Beyersmann D. (2002) Intracellular zinc distribution and transport in C6 rat glioma cells. *Biochem. Biophys. Res. Comm.* 296: 923-8.

Haase H, Hebel S, Engelhardt G, Rink L. (2006) Flow cytometric measurement of labile zinc in peripheral blood mononuclear cells. *Anal. Biochem.* 352(2): 222-30.

Haase H, Maret W. (2003) Intracellular zinc fluctuations modulate protein tyrosine phosphatase activity in insulin/insulin-like growth factor-1 signaling. *Exp. Cell Res.* 291, 289-98.

Haase H, Maret W. (2004) An HPLC assay for the redox- and metal binding states of metallothionein. *Anal. Biochem.* 333(1): 19-26.

Haase H, Maret W. (2005a) Fluctuations of cellular, available zinc modulate insulin signaling via inhibition of protein tyrosine phosphatases. *J. Trace Elem. Med. Biol.* 19, 37-42.

Haase H, Maret W. (2005b) Protein tyrosine phosphatases as targets of the combined insulinomimetic effects of zinc and oxidants. *BioMetals* 18, 333-8.

Haase H, Wätjen W, Beyersmann D. (2001) Zinc induces apoptosis that can be suppressed by lanthanum in C6 rat glioma cells. *Biol. Chem.* 382, 1227-34.

Hansson A. (1996) Extracellular zinc ions induces mitogen-activated protein kinase activity and protein tyrosine phosphorylation in bombesin-sensitive Swiss 3T3 fibroblasts. *Arch. Biochem. Biophys.* 328(2): 233-8.

He F, Seryshev AB, Cowan CW, Wensel TG. (2000) Multiple zinc binding sites in retinal rod cGMP phosphodiesterase, PDE6alpha beta. *J. Biol. Chem.* 275(27): 20572-7.

Hershfinkel M, Moran A, Grossman N, Sekler I. (2001) A zinc-sensing receptor triggers the release of intracellular Ca2+ and regulates ion transport. *Proc. Natl. Acad. Sci. USA* 98(20): 11749-54.

Heuchel R, Radtke F, Georgiev O, Stark G, Aguet M, Schaffner W. (1994) The transcription factor MTF-1 is essential for basal and heavy metal-induced metallothionein gene expression. *EMBO J.* 13(12): 2870-5.

Hommel U, Zurini M, Luyten M. (1994) Solution structure of a cysteine rich domain of rat protein kinase C. *Nat. Struct. Biol.* 1(6): 383-7.

Huang M, Krepkiy D, Hu W, Petering DH. (2004) Zn-, Cd-, and Pb-transcription factor IIIA: properties, DNA binding, and comparison with TFIIIA-finger 3 metal complexes. *J. Inorg. Biochem.* 98(5): 775-85.

Hubbard SR, Bishop WR, Kirschmeier P, George SJ, Cramer SP, Hendrickson WA. (1991) Identification and characterization of zinc binding sites in protein kinase C. *Science* 254(5039): 1776-9.

Ibs KH, Gabriel P, Rink L. (2003) Zinc and the immune system of elderly. *Adv. Cell Aging Gerontol.* 13: 243-59.

Jan CR, Wu SN, Tseng CJ. (1999) Zn2+ increases resting cytosolic Ca2+ levels and abolishes capacitative Ca2+ entry induced by ATP in MDCK cells. *Naunyn Schmiedebergs Arch. Pharmacol.* 360(3): 249-55.

Jeon KI, Jeong JY, Jue DM. (2000) Thiol-reactive metal compounds inhibit NF-kappa B activation by blocking I kappa B kinase. *J. Immunol.* 164(11): 5981-9.

Jiang H, Daniels PJ, Andrews GK. (2003) Putative zinc-sensing zinc fingers of metal-response element-binding transcription factor-1 stabilize a metal-dependent chromatin complex on the endogenous metallothionein-I promoter. *J. Biol. Chem.* 278(32): 30394-402.

Jiang H, Fu K, Andrews GK. (2004) Gene- and cell-type-specific effects of signal transduction cascades on metal-regulated gene transcription appear to be independent of changes in the phosphorylation of metal-response-element-binding transcription factor-1. *Biochem. J.* 382: 33-41.

Karadag F, Cildag O, Altinisik M, Kozaci LD, Kiter G, Altun C. (2004) Trace elements as a component of oxidative stress in COPD. *Respirology* 9(1): 33-7.

Keilin D, Mann T. (1940) Carbonic anhydrase. Purification and nature of the enzyme. *Biochem. J.* 34(8-9): 1163-76.

Kelly EJ, Quaife CJ, Froelick GJ, Palmiter RD. (1996) Metallothionein I and II protect against zinc deficiency and zinc toxicity in mice. *J. Nutr.* 126(7): 1782-90.

Kim S, Jung Y, Kim D, Koh H, Chung J. (2000) Extracellular zinc activates p70 S6 kinase through the phosphatidylinositol 3-kinase signaling pathway. *J. Biol. Chem.* 275(34): 25979-84.

Kim YM, Reed W, Wu W, Bromberg PA, Graves LM, Samet JM. (2006) Zn2+-induced IL-8 expression involves AP-1, JNK, and ERK activities in human airway epithelial cells. *Am. J. Physiol. Lung Cell. Mol. Physiol.* 290(5): L1028-35.

Klein C, Heyduk T, Sunahara RK. (2004) Zinc inhibition of adenylyl cyclase correlates with conformational changes in the enzyme. *Cell Signal.* 16(10): 1177-85.

Klein C, Sunahara RK, Hudson TY, Heyduk T, Howlett AC. (2002) Zinc inhibition of cAMP signaling. *J. Biol. Chem.* 277(14): 11859-65.

Knapp LT, Klann E. (2000) Superoxide-induced stimulation of protein kinase C via thiol modification and modulation of zinc content. *J. Biol. Chem.* 275(31): 24136-45.

Koizumi S, Nakazawa K, Inoue K. (1995) Inhibition by Zn2+ of uridine 5'-triphosphate-induced Ca(2+)-influx but not Ca(2+)-mobilization in rat phaeochromocytoma cells. *Br. J. Pharmacol.* 115(8): 1502-8.

Koizumi S, Suzuki K, Ogra Y, Gong P, Otuska F. (2000) Roles of zinc fingers and other regions of the transcription factor human MTF-1 in zinc-regulated DNA binding. *J. Cell. Physiol.* 185(3): 464-72.

Korichneva I, Hoyos B, Chua R, Levi E, Hammerling U. (2002) Zinc release from protein kinase C as the common event during activation by lipid second messenger or reactive oxygen. *J. Biol. Chem.* 277(46): 44327-31.

Kovala T, Sanwal BD, Ball EH. (1997) Recombinant expression of a type IV, cAMP-specific phosphodiesterase: characterization and structure-function studies of deletion mutants. *Biochemistry* 36(10): 2968-76.

Kröncke KD, Fehsel K, Schmidt T, Zenke FT, Dasting I, Wesener JR, Bettermann H, Breunig KD, Kolb-Bachofen V. (1994) Nitric oxide destroys zinc-sulfur clusters inducing zinc release from metallothionein and inhibition of the zinc finger-type yeast transcription activator LAC9. *Biochem. Biophys. Res. Commun.* 200(2): 1105-10.

Kury S, Dreno B, Bezieau S, Giraudet S, Kharfi M, Kamoun R, Moisan JP. (2002) Identification of SLC39A4, a gene involved in acrodermatitis enteropathica. *Nat. Genet.* 31(3): 239-40.

Langmade SJ, Ravindra R, Daniels PJ, Andrews GK. (2000) The transcription factor MTF-1 mediates metal regulation of the mouse ZnT1 gene. *J. Biol. Chem.* 275(44): 34803-9.

LaRochelle O, Gagne V, Charron J, Soh JW, Seguin C. (2001) Phosphorylation is involved in the activation of metal-regulatory transcription factor 1 in response to metal ions. *J. Biol. Chem.* 276(45): 41879-88.

Lengyel I, Fieuw-Makaroff S, Hall AL, Sim AT, Rostas JA, Dunkley PR. (2000) Modulation of the phosphorylation and activity of calcium/calmodulin-dependent protein kinase II by zinc. *J. Neurochem.* 75(2): 594-605.

Lichtlen P, Schaffner W. (2001) The "metal transcription factor" MTF-1: biological facts and medical implications. *Swiss Med. Wkly.* 131(45-46): 647-52.

Lichtlen P, Wang Y, Belser T, Georgiev O, Certa U, Sack R, Schaffner W. (2001) Target gene search for the metal-responsive transcription factor MTF-1. *Nucleic Acids Res.* 29(7): 1514-23.

Lohmann RD, Beyersmann D. (1993) Cadmium and zinc mediated changes of the Ca(2+)-dependent endonuclease in apoptosis. *Biochem. Biophys. Res. Commun.* 190(3): 1097-103.

Lynch CJ, Patson BJ, Goodman SA, Trapolsi D, Kimball SR. (2001) Zinc stimulates the activity of the insulin- and nutrient-regulated protein kinase mTOR. *Am. J. Physiol. Endocrinol. Metab.* 281(1): E25-34.

MacDonald RS. (2000) The role of zinc in growth and cell proliferation. *J. Nutr.* 130(5S): 1500S-8S.

Malaiyandi LM, Dineley KE, Reynolds IJ. (2004) Divergent consequences arise from metallothionein overexpression in astrocytes: zinc buffering and oxidant-induced zinc release. *Glia* 45(4): 346-53.

Maret W. (2000) The function of zinc metallothionein: a link between cellular zinc and redox state. *J. Nutr.* 130(5S): 1455S-8S.

Maret W. (2001) Crosstalk of the group IIa and IIb metals calcium and zinc in cellular signaling. *Proc. Natl. Acad. Sci USA* 98(22): 12325-7.

Maret W. (2004) Zinc and sulfur: a critical biological partnership. *Biochemistry* 43(12): 3301-9.

Maret W. (2006) Zinc coordination environments in proteins as redox sensors and signal transducers. *Antioxid. Redox Signal.* 8(9-10): 1419-41.

Maret W, Jacob C, Vallee BL, Fischer EH. (1999) Inhibitory sites in enzymes: zinc removal and reactivation by thionein. *Proc. Natl. Acad. Sci. USA* 96(5): 1936-40.

Masters BA, Kelly EJ, Quaife CJ, Brinster RL, Palmiter RD. (1994) Targeted disruption of metallothionein I and II genes increases sensitivity to cadmium. *Proc. Natl. Acad. Sci. USA* 91(2): 584-8.

Masuoka J, Hegenauer J, Van Dyke BR, Saltman P. (1993) Intrinsic stochiometric equilibrium constants for the binding of zinc(II) and copper(II) to the high affinity site of serum albumin. *J. Biol. Chem.* 268(29): 21553-7.

McNulty TJ, Taylor CW. (1999) Extracellular heavy-metal ions stimulate Ca^{2+} mobilization in hepatocytes. *Biochem. J.* 339(3): 555-561.

Meplan C, Richard MJ, Hainaut P. (2000a) Redox signalling and transition metals in the control of the p53 pathway. *Biochem. Pharmacol.* 59(1): 25-33.

Meplan C, Richard MJ, Hainaut P. (2000b) Metalloregulation of the tumor suppressor protein p53: zinc mediates the renaturation of p53 after exposure to metal chelators in vitro and in intact cells. *Oncogene.* 19(46): 5227-36.

Michalska AE, Choo KHA. (1993) Targeting an germ-line transmission of a null mutation at the metallothionein I and II loci in mouse. *Proc. Natl. Acad. Sci. USA* 90: 8088-92.

Mills JS, Johnson JD. (1985) Metal ions as allosteric regulators of calmodulin. *J. Biol. Chem.* 260(28): 15100-5.

Milos M, Comte M, Schaer JJ, Cox JA. (1989) Evidence for four capital and six auxiliary cation-binding sites on calmodulin: divalent cation interactions monitored by direct binding and microcalorimetry. *J. Inorg. Biochem.* 36(1): 11-25.

Monk PN, Partridge LJ. (1993) Characterization of a complement-fragment-C5a-stimulated calcium-influx mechanism in U937 monocytic cells. *Biochem. J.* 295: 679-84.

Müller CW, Rey FA, Sodeoka M, Verdine GL, Harrison SC (1995) Structure of the NF-kB p50 homodimer bound to DNA. *Nature* 373: 311-317.

Munro AW, Taylor P, Walkinshaw MD. (2000) Structures of redox enzymes. *Curr. Opin. Biotechnol.* 11(4): 369-76.

Ogra Y, Suzuki KT. (2000) Nuclear trafficking of metallothionein: possible mechanisms and current knowledge. *Cell. Mol. Biol. (Noisy-le-grand)* 46(2): 357-65.

Omburo GA, Brickus T, Ghazaleh FA, Colman RW. (1995) Divalent metal cation requirement and possible classification of cGMP-inhibited phosphodiesterase as a metallohydrolase. *Arch. Biochem. Biophys.* 323: 1-5.

Omburo GA, Jacobitz S, Torphy TJ, Colman RW. (1998) Critical role of conserved histidine pairs HNXXH and HDXXH in recombinant human phosphodiesterase 4A. *Cell Signal.* 10(7): 491-7.

Ono Y, Fujii T, Igarashi K, Kuno T, Tanaka C, Kikkawa U, Nishizuka Y. (1989) Phorbol ester binding to protein kinase C requires a cysteine-rich zinc-finger-like sequence. *Proc. Natl. Acad. Sci. USA* 86(13): 4868-71.

Otsuka M, Fujita M, Aoki T, Ishii S, Sugiura Y, Yamamoto T, Inoue J. (1995) Novel zinc chelators with dual activity in the inhibition of the kappa B site-binding proteins HIV-EP1 and NF-kappa B. *J. Med. Chem.* 38(17): 3264-70.

Outten CE, O'Halloran TV. (2001) Femtomolar sensitivity of metalloregulatory proteins controlling zinc homeostasis. *Science* 292(5526): 2488-92.

Palmiter RD. (1994) Regulation of metallothionein genes by heavy metals appears to be mediated by a zinc-sensitive inhibitor that interacts with a constitutively active transcription factor, MTF-1. *Proc. Natl. Acad. Sci. USA* 91(4): 1219-23.

Palmiter RD, Cole TB, Findley SD. (1996) ZnT-2, a mammalian protein that confers resistance to zinc by facilitating vesicular sequestration. *EMBO J.* 15(8): 1784-91.

Palmiter RD, Sandgren EP, Koeller DM, Brinster RL. (1993) Distal regulatory elements from the mouse metallothionein locus stimulate gene expression in transgenic mice. *Mol. Cell. Biol.* 13(9): 5266-75.

Park JA, Koh JY. (1999) Induction of an immediate early gene egr-1 by zinc through extracellular signal-regulated kinase activation in cortical culture: its role in zinc-induced neuronal death. *J. Neurochem.* 73(2): 450-6.

Parker PJ, Coussens L, Totty N, Rhee L, Young S, Chen E, Stabel S, Waterfield MD, Ullrich A. (1986) The complete primary structure of protein kinase C--the major phorbol ester receptor. *Science* 233(4766): 853-9.

Pearce LL, Wasserloos K, St Croix CM, Gandley R, Levitan ES, Pitt BR. (2000) Metallothionein, nitric oxide and zinc homeostasis in vascular endothelial cells. *J. Nutr.* 130(5S): 1467S-70S.

Peck EJ, Ray WJ. (1971) Metal complexes of phosphoglucomutase in vivo. Alterations induced by insulin. *J. Biol. Chem.* 246(4): 1160-7.

Percival MD, Yeh B, Falgueyret JP. (1997) Zinc dependent activation of cAMP-specific phosphodiesterase (PDE4A). *Biochem. Biophys. Res. Commun.* 241(1): 175-80.

Pernelle JJ, Creuzet C, Loeb J, Gacon G. (1991) Phosphorylation of the lymphoid cell kinase $p56^{lck}$ is stimulated by micromolar concentrations of Zn^{2+}. *FEBS Lett.* 281(1,2): 278-82.

Perry DK, Smyth MJ, Stennicke HR, Salvesen GS, Duriez P, Poirier GG, Hannun YA. (1997) Zinc is a potent inhibitor of the apoptotic protease, caspase-3. A novel target for zinc in the inhibition of apoptosis. *J. Biol. Chem.* 272(30): 18530-3.

Peters T. (1995) All about albumin: Biochemistry, Genetics, and medical applications. (Academic Press, New York).

Petrie L, Chesters JK, Franklin M. (1991) Inhibition of myoblast differentiation by lack of zinc. *Biochem. J.* 276: 109-11.

Pfeiffer DR, Lardy HA. (1976) Ionophore A23187: The effect of H^+ concentration on complex formation with divalent and monovalent cations and the demonstration of K^+ transport in mitochondria mediated by A23187. *Biochemistry* 15(5): 935-43.

Potter BM, Feng LS, Parasuram P, Matskevich VA, Wilson JA, Andrews GK, Laity JH. (2005) The six zinc fingers of metal-responsive element binding transcription factor-1 form stable and quasi-ordered structures with relatively small differences in zinc affinities. *J. Biol. Chem.* 280(31): 28529-40.

Qian WJ, Aspinwall CA, Battiste MA, Kennedy RT. (2000) Detection of secretion from single pancreatic beta-cells using extracellular fluorogenic reactions and confocal fluorescence microscopy. *Anal. Chem.* 72(4): 711-7.

Quest AF, Bloomenthal J, Bardes ES, Bell RM. (1992) The regulatory domain of protein kinase C coordinates four atoms of zinc. *J. Biol. Chem.* 267(14): 10193-7.

Radtke F, Heuchel R, Georgiev O, Hergersberg M, Gariglio M, Dembic Z, Schaffner W. (1993) Cloned transcription factor MTF-1 activates the mouse metallothionein I promoter. *EMBO J.* 12(4): 1355-62.

Richter M, Cantin AM, Beaulieu C, Cloutier A, Larivee P. (2003) Zinc chelators inhibit eotaxin, RANTES, and MCP-1 production in stimulated human airway epithelium and fibroblasts. *Am. J. Physiol. Lung Cell. Mol. Physiol.* 285(3): L719-29.

Samet JM, Dewar BJ, Wu W, Graves LM. (2003) Mechanisms of Zn(2+)-induced signal initiation through the epidermal growth factor receptor. *Toxicol. Appl. Pharmacol.* 191(1): 86-93.

Samet JM, Silbajoris R, Wu W, Graves LM. (1999) Tyrosine phosphatases as targets in metal-induced signaling in human airway epithelial cells. *Am. J. Respir. Cell. Mol. Biol.* 21(3): 357-64.

Saydam N, Adams TK, Steiner F, Schaffner W, Freedman JH. (2002) Regulation of metallothionein transcription by the metal-responsive transcription factor MTF-1: identification of signal transduction cascades that control metal-inducible transcription. *J. Biol. Chem.* 277(23): 20438-45.

Schmidt C, Beyersmann D. (1999) Transient peaks in zinc and metallothionein levels during differentiation of 3T3L1 cells. *Arch. Biochem. Biophys.* 364(1): 91-8.

Schott-Ohly P, Lgssiar A, Partke HJ, Hassan M, Friesen N, Gleichmann H. (2004) Prevention of spontaneous and experimentally induced diabetes in mice with zinc sulfate-enriched drinking water is associated with activation and reduction of NF-kappa B and AP-1 in islets, respectively. *Exp. Biol. Med. (Maywood)* 229(11): 1177-85.

Sensi SL, Canzoniero LM, Yu SP, Ying HS, Koh JY, Kerchner GA, Choi DW. (1997) Measurement of intracellular free zinc in living cortical neurons: routes of entry. *J. Neurosci.* 17(24): 9554-64.

Sensi SL, Ton-That D, Sullivan PG, Jonas EA, Gee, KR, Kaczmarek LK, Weiss JH. (2003) Modulation of mitochondrial function by endogenous Zn^{2+} pools. *Proc. Natl. Acad. Sci. USA* 100(10): 6157-62.

Sharir H, Hershfinkel M. (2005) The extracellular zinc-sensing receptor mediates intercellular communication by inducing ATP release. *Biochem. Biophys. Res. Commun.* 332(3): 845-52.

Shumilla JA, Wetterhahn KE, Barchowsky A. (1998) Inhibition of NF-kappa B binding to DNA by chromium, cadmium, mercury, zinc, and arsenite in vitro: evidence of a thiol mechanism. *Arch. Biochem. Biophys.* 349(2): 356-62.

Simons TJ. (1991) Intracellular free zinc and zinc buffering in human red blood cells. *J. Membr. Biol.* 123(1): 63-71.

Simpson M, Xu Z. (2006) Increased abundance of labile intracellular zinc during cell proliferation was due to increased retention of extracellular zinc in 3T3 cells. *J. Nutr. Biochem.* 17(8): 541-7.

Smirnova IV, Bittel DC, Ravindra R, Jiang H, Andrews GK. (2000) Zinc and cadmium can promote rapid nuclear translocation of metal response element-binding transcription factor-1. *J. Biol. Chem.* 275(13): 9377-84.

Smith JB, Smith L, Pijuan V, Zhuang Y, Chen YC. (1994) Transmembrane Signals and protooncogene induction evoked by carcinogenic metals and prevented by zinc. *Environ. Health Perspect.* 102(3): 181-9.

Smith PJ, Wiltshire M, Davies S, Chin SF, Campbell AK, Errington RJ. (2002) DNA damage-induced [Zn(2+)](i) transients: correlation with cell cycle arrest and apoptosis in lymphoma cells. *Am. J. Physiol. Cell. Physiol.* 283(2): C609-22.

Spahl DU, Berendji-Grün D, Suschek CV, Kolb-Bachofen V, Kröncke KD. (2003) Regulation of zinc homeostasis by inducible NO synthase-derived NO: nuclear metallothionein translocation and intranuclear Zn2+ release. *Proc. Natl. Acad. Sci. USA* 100(24): 13952-7.

St Croix CM, Wasserloos KJ, Dineley KE, Reynolds IJ, Levitan ES, Pitt BR. (2002) Nitric oxide-induced changes in intracellular zinc homeostasis are mediated by metallothionein/thionein. *Am. J. Physiol. Lung. Cell. Mol. Physiol.* 282(2): L185-92.

Stitt MS, Wasserloos KJ, Tang X, Liu X, Pitt BR, St Croix CM. (2006) Nitric oxide-induced nuclear translocation of the metal responsive transcription factor, MTF-1 is mediated by zinc release from metallothionein. *Vascul. Pharmacol.* 44(3): 149-55.

Szallasi Z, Bogi K, Gohari S, Biro T, Acs P, Blumberg PM. (1996) Non-equivalent roles for the first and second zinc fingers of protein kinase C delta. Effect of their mutation on phorbol ester-induced translocation in NIH 3T3 cells. *J. Biol. Chem.* 271(31): 18299-301.

Takahashi K, Akaishi E, Abe Y, Ishikawa R, Tanaka S, Hosaka K, Kubohara Y. (2003) Zinc inhibits calcineurin activity in vitro by competing with nickel. *Biochem. Biophys. Res. Commun.* 307(1): 64-8.

Tal TL, Graves LM, Silbajoris R, Bromberg PA, Wu W, Samet JM. (2006) Inhibition of protein tyrosine phosphatase activity mediates epidermal growth factor receptor signaling in human airway epithelial cells exposed to Zn(2+). *Toxicol. Appl. Pharmacol.* 214(1): 16-23.

Tanaka S, Akaishi E, Hosaka K, Okamura S, Kubohara Y. (2005) Zinc ions suppress mitogen-activated interleukin-2 production in Jurkat cells. *Biochem. Biophys. Res. Commun.* 335(1): 162-7.

Tang X, Shay NF. (2001) Zinc has an insulin-like effect on glucose transport mediated by phosphoinositol-3-kinase and Akt in 3T3-L1 fibroblasts and adipocytes. *J. Nutr.* 131(5): 1414-20.

Thompson, RB. (2005) Studying zinc biology with fluorescence: ain't we got fun? *Curr. Opin. Chem. Biol.* 9: 526-32.

Tom Dieck H, Doring F, Fuchs D, Roth HP, Daniel H. (2005) Transcriptome and proteome analysis identifies the pathways that increase hepatic lipid accumulation in zinc-deficient rats. *J. Nutr.* 135(2): 199-205.

Tom Dieck H, Doring F, Roth HP, Daniel H. (2003) Changes in rat hepatic gene expression in response to zinc deficiency as assessed by DNA arrays. *J. Nutr.* 133(4): 1004-10.

Tottey S, Harvie DR, Robinson NJ. (2005) Understanding how cells allocate metals using metal sensors and metallochaperones. *Acc. Chem. Res.* 38(10): 775-83.

Truong-Tran AQ, Carter J, Ruffin RE, Zalewski PD. (2001a) The role of zinc in caspase activation and apoptotic cell death. *BioMetals* 14(3-4): 315-30.

Truong-Tran AQ, Carter J, Ruffin R, Zalewski PD. (2001b) New insights into the role of zinc in the respiratory epithelium. *Immunol. Cell Biol.* 79(2): 170-7.

Turan B, Fliss H, Desilets M. (1997) Oxidants increase intracellular free Zn2+ concentration in rabbit ventricular myocytes. *Am. J. Physiol.* 272(5): H2095-106.

Udom AO, Brady FO. (1980) Reactivation in vitro of zinc-requiring apo-enzymes by rat liver zinc-thionein. *Biochem. J.* 187(2): 329-35.

Vallee BL, Falchuk KH. (1993) The biochemical basis of zinc physiology. *Physiol. Rev.* 73(1): 79-118.

Vasak M, Kägi JHR. (1994) Metallothioneins. In: Encyclopedia of inorganic chemistry, vol 4 (ed. RB King), 2229-41, Wiley, New York.

Verhaegh GW, Parat MO, Richard MJ, Hainaut P. (1998) Modulation of p53 protein conformation and DNA-binding activity by intracellular chelation of zinc. *Mol. Carcinog.* 21(3): 205-14.

Von Bülow V, Rink L, Haase H. (2005) Zinc-mediated inhibition of cyclic nucleotide phosphodiesterase activity and expression suppresses TNF-α and IL-1β production in monocytes by elevation of cGMP. *J. Immunol.* 175(7): 4697-705.

Wätjen W, Benters J, Haase H, Schwede F, Jastorff B, Beyersmann D. (2001) Zn^{2+} and Cd^{2+} increase the cyclic GMP level in PC 12 cells by inhibition of the cyclic nucleotide phosphodiesterase. *Toxicology* 157(3): 167-75.

Wätjen W, Haase H, Biagioli M, Beyersmann D. (2002) Induction of apoptosis in mammalian cells by cadmium and zinc. *Environ. Health Perspect.* 110(5): 865-7.

Weinberger RP, Rostas JA. (1991) Effect of zinc on calmodulin-stimulated protein kinase II and protein phosphorylation in rat cerebral cortex. *J. Neurochem.* 57(2): 605-14.

Wellinghausen N, Fischer A, Kirchner H, Rink L. (1996) Interaction of zinc ions with human peripheral blood mononuclear cells. *Cell. Immunol.* 171(2): 255-61.

Wellinghausen N, Kirchner H, Rink L. (1997a) The immunobiology of zinc. *Immunol. Today* 18(11): 519-21.

Wellinghausen N, Martin M, Rink L. (1997b) Zinc inhibits interleukin-1-dependent T cell stimulation. *Eur. J. Immunol.* 27(10): 2529-35.

Williams RJP. (1984) Zinc: what is its role in biology? *Endeavour* 8(2): 65-70.

Wu W, Graves LM, Gill GN, Parsons SJ, Samet JM. (2002a) Src-dependent phosphorylation of the epidermal growth factor receptor on tyrosine 845 is required for zinc-induced Ras activation. *J. Biol. Chem.* 277(27): 24252-7.

Wu W, Jaspers I, Zhang W, Graves LM, Samet JM. (2002b) Role of Ras in metal-induced EGF receptor signaling and NF-kappaB activation in human airway epithelial cells. *Am. J. Physiol. Lung Cell. Mol. Physiol.* 282(5): L1040-8.

Wu W, Silbajoris RA, Whang YE, Graves LM, Bromberg PA, Samet JM. (2005) p38 and EGF receptor kinase-mediated activation of the phosphatidylinositol 3-kinase/Akt pathway is required for Zn2+-induced cyclooxygenase-2 expression. *Am. J. Physiol. Lung Cell. Mol. Physiol.* 289(5): L883-9.

Wu W, Wang X, Zhang W, Reed W, Samet JM, Whang YE, Ghio AJ. (2003) Zinc-induced PTEN protein degradation through the proteasome pathway in human airway epithelial cells. *J. Biol. Chem.* 278(30): 28258-63.

Yamaguchi M. (1993) Regulatory effects of zinc and copper on the calcium transport system in rat liver nuclei. Relation to SH groups in the releasing mechanism. *Biochem. Pharmacol.* 45(4): 943-8.

Yamaguchi M, Fukagawa M. (2005) Role of zinc in regulation of protein tyrosine phosphatase activity in osteoblastic MC3T3-E1 cells: zinc modulation of insulin-like growth factor-I's effect. *Calcif. Tissue Int.* 76(1): 32-8.

Yang Y, Maret W, Vallee BL. (2001) Differential fluorescence labeling of cysteinyl clusters uncovers high tissue levels of thionein. *Proc. Natl. Acad. Sci. USA* 98(10): 5556-9.

Yang JP, Merin JP, Nakano T, Kato T, Kitade Y, Okamoto T. (1995) Inhibition of the DNA-binding activity of NF-kappa B by gold compounds in vitro. *FEBS Lett.* 361(1): 89-96.

Zabel U, Schreck R, Baeuerle PA. (1991) DNA binding of purified transcription factor NF-kappa B. Affinity, specificity, Zn2+ dependence, and differential half-site recognition. *J. Biol. Chem.* 266(1): 252-60.

Zalewski PD, Forbes IJ, Betts WH. (1993) Correlation of apoptosis with change in intracellular labile Zn(II) using zinquin [(2-methyl-8-p-toluenesulphonamido-6-quinolyloxy)acetic acid], a new specific fluorescent probe for Zn(II). *Biochem. J.* 296(2): 403-8.

Zalewski PD, Forbes IJ, Giannakis C, Betts WH. (1991) Regulation of protein kinase C by Zn(2+)-dependent interaction with actin. *Biochem. Int.* 24(6): 1103-10.

Zalewski PD, Forbes IJ, Giannakis C, Cowled PA, Betts WH. (1990) Synergy between zinc and phorbol ester in translocation of protein kinase C to cytoskeleton. *FEBS Lett.* 273(1-2): 131-4.

Zhang B, Georgiev O, Hagmann M, Günes C, Cramer M, Faller P, Vasak M, Schaffner W. (2003) Activity of metal-responsive transcription factor 1 by toxic heavy metals and H2O2 in vitro is modulated by metallothionein. *Mol. Cell. Biol.* 23(23): 8471-85.

Zhang Y, Wang H, Li J, Dong L, Xu P, Chen W, Neve RL, Volpe JJ, Rosenberg PA. (2006) Intracellular zinc release and ERK phosphorylation are required upstream of 12-lipoxygenase activation in peroxynitrite toxicity to mature rat oligodendrocytes. *J. Biol. Chem.* 281(14): 9460-70.

Zhou Z, Wang L, Song Z, Saari JT, McClain CJ, Kang YJ. (2004) Abrogation of nuclear factor-kappaB activation is involved in zinc inhibition of lipopolysaccharide-induced tumor necrosis factor-alpha production and liver injury. *Am. J. Pathol.* 164(5): 1547-56.

Zoli A, Altomonte L, Caricchio R, Galossi A, Mirone L, Ruffini MP, Magaro M. (1998) Serum zinc and copper in active rheumatoid arthritis: correlation with interleukin 1 beta and tumour necrosis factor alpha. *Clin. Rheumatol.* 17(5): 378-82.

In: Focus on Signal Transduction Research
Editor: Gavin McAlpine, pp. 37-66
ISBN 1-60021-376-6

Chapter 2

A ROLE FOR MAGNESIUM IN THE REGULATION OF RUMINAL SODIUM TRANSPORT

Friederike Stumpff* and Holger Martens
Department of Veterinary Physiology, Free University of Berlin, Germany

ABSTRACT

The ruminal epithelium is a stratified squamous epithelium that has evolved to display functions essential for the unique ability of cows and sheep to ferment dietary components like carbohydrates and protein, and to selectively absorb nutrients and minerals for the production of milk. A characteristic property of this tissue is its pronounced ability to transport magnesium against an electrochemical gradient. Absorption of magnesium is reduced by dietary elevation of ruminal potassium, leading to hypomagnesaemia that can reach clinical significance. Studies of the intact tissue and of isolated cells suggest that cellular magnesium uptake is decreased by apical depolarization of the ruminal membrane, resulting in both a lower cytosolic concentration and transepithelial transport of the element.

Another characteristic feature of this unusual epithelium is the expression of a sodium-conducting channel with functional properties that are clearly distinct from the epithelial sodium channel (ENaC) found in most mammalian epithelia. Thus, it has not been possible to demonstrate direct regulation of ruminal sodium transport by aldosterone, and the effects of amiloride are clearly limited to an inhibition of the sodium proton exchanger (NHE3) expressed by this tissue. Studies at the level of the animal and the tissue suggest that sodium conductance is enhanced by depolarization of the apical membrane. Recent *in vitro* studies have demonstrated that ruminal epithelial cells express non-selective cation channels in the apical membrane that are regulated by changes in cytosolic magnesium. We propose that the reduction in ruminal magnesium uptake observed after ingestion of high potassium fodder may be related to a role for magnesium

* Corresponding Author: Friederike Stumpff, M.D., Department of Veterinary Physiology, Free University of Berlin, Oertzenweg 19b, 14163 Berlin, Germany (0049 30 838 62595 (Office); 0049 30 838 62610 (Fax); E-Mail Address: stumpff@zedat.fu-berlin.de

in a signaling cascade that leads to an increase in the permeability of this non-selective cation channel for sodium, thus enhancing absorption of this ion from the rumen and restoring ruminal osmolarity, while contributing to the retention of potassium.

INTRODUCTION

Compared to the extensive research on various molecules involved in signal transduction, Mg^{2+} has not gotten much attention. One of the reasons for this may be that in man, hypomagnesaemia is a fairly rare condition and one that is easily treated by magnesium supplements when it does occur – as in pregnancy or lactation. Thus, the stimulus for research in the area has been limited. Conversely and due to the large amounts of magnesium lost in milk, hypomagnesaemia in cows and sheep continues to lead to deaths of afflicted animals or even part of entire herds [32, 47, 92, 104, 118]. As a result, there has been ongoing interest in the uptake of magnesium in this species.

In the following, the focus will be on the ruminant in general and on the very peculiar organ that gives this species its name. Thus, this topic may not appear to be of general interest. However, we feel that the story that is emerging from many decades of research on how the ruminant has adapted to a dramatically altered nutritional situation by utilizing the interactions of Mg^{2+} with uptake of other cations may have implications for other physiological situations.

THE RUMINANT SPECIES: ADAPTATION TO A ROUGHAGE DIET

The ruminant species and mankind evolved in parallel as survivors of a relatively short period of mass extinction during which the forests of the Oligocene perished and were replaced with vast grasslands [11]. During this era, the ancestral ruminant, a small and forest-dwelling species [60], perfected its ability to utilize poorly digestible fiber and adapted to uptake of grass and roughage for its nutritional requirements. This accomplishment has been essential for the survival of these animals, and probably also for that of the humanoid apes which left the dwindling forests and began to follow and hunt these herds of ruminants.

Central to the ability of ruminants to subsist on a diet of grass while excreting large amounts of milk is the development of an advanced digestive system with several specialized compartments (reticulum, rumen, and omasum) that precede the actual stomach (abomasum) and the lower digestive tract (Figure 1). The largest of these forestomaches, the rumen, fills almost half of the entire abdominal cavity of these animals with a volume reaching 60 to 100 l in cows [138]. Lined with squamous, stratified epithelium which is keratinized and non-glandular [48, 53], this pouch is probably of esophageal origin [98] and serves as both a water reservoir [138] and a fermentation chamber in which poorly digestible forage is broken down and reassembled into digestible nutrients by an abundant flora of microorganisms [8, 57, 58, 156]. Maintanance of a pH that should be slightly acidic (5.5 - 7.0) and an osmolarity that is marginally hypotonic before or slightly hypertonic after a meal (to blood) is central to this fermentation process. Ruminal homeostasis is ensured by a number of factors that include production of ample amounts of saliva, continuous mixing of saliva with ruminal content by ruminal contractions and rumination of roughage, and the ability of the rumen to serve as a barrier between blood and rumen content while absorbing certain minerals and organic compounds in a controlled manner.

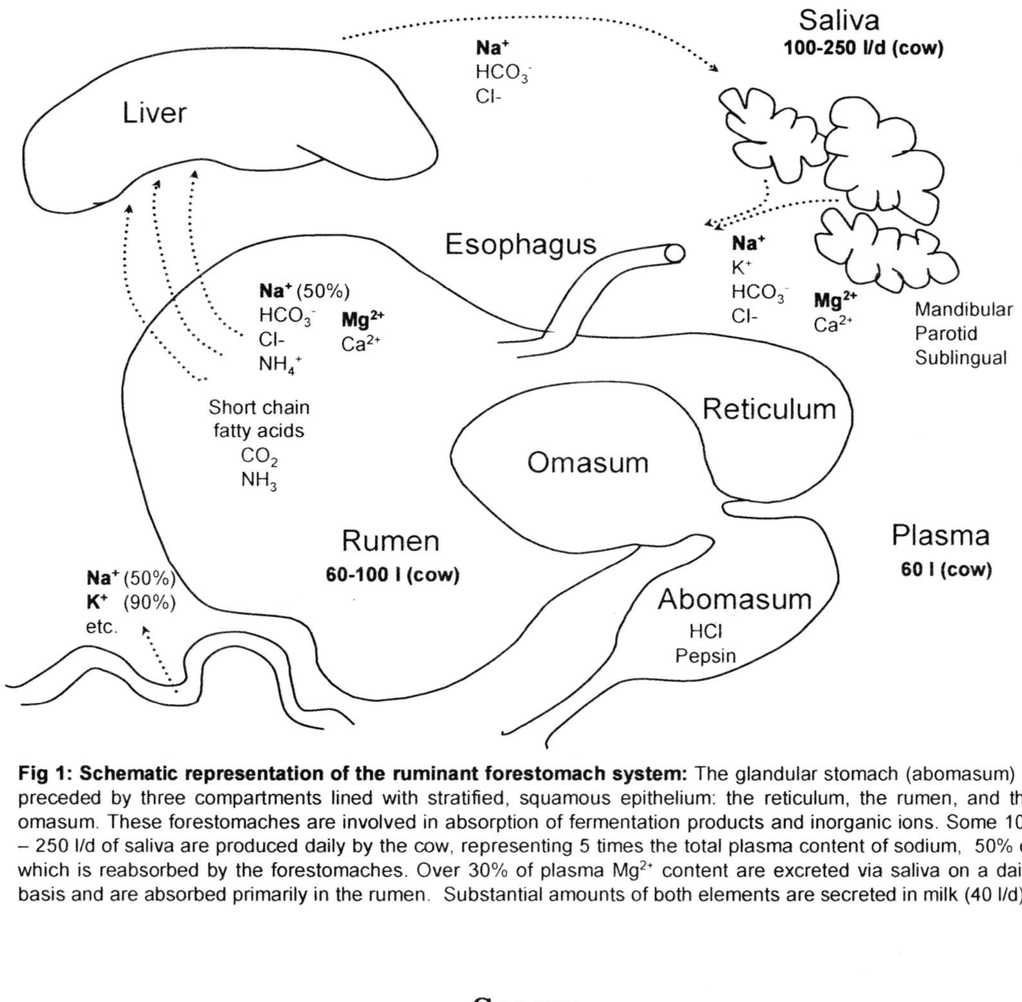

Fig 1: Schematic representation of the ruminant forestomach system: The glandular stomach (abomasum) is preceded by three compartments lined with stratified, squamous epithelium: the reticulum, the rumen, and the omasum. These forestomaches are involved in absorption of fermentation products and inorganic ions. Some 100 – 250 l/d of saliva are produced daily by the cow, representing 5 times the total plasma content of sodium, 50% of which is reabsorbed by the forestomaches. Over 30% of plasma Mg^{2+} content are excreted via saliva on a daily basis and are absorbed primarily in the rumen. Substantial amounts of both elements are secreted in milk (40 l/d).

SALIVA

A striking difference between ruminants and most monogastric species is the ability to produce an extremely large stream of well-buffered saliva that amounts to about 6-16 l/d in sheep [67] and 100 – 200 l/d in cattle [9, 15] with values of up to 250 l/d in lactating cows [23, 138] (Figure 1). This saliva is low in chloride (22 mmol/l) and contains large amounts of pH buffers such as HCO_3^- (120 mmol/l) and PO_4^- (30 mmol/l). After feeding, the flow rate but not the ionic composition of saliva increases, and can be so massive that the ADH and renin-angiotensin mechanism is activated [138]. Postprandial stimulation of salivary flow can be at least partially mimicked by the addition of volatile fatty acids to the rumen [15]. In contrast, in experiments in which ruminal pH was altered by addition of HCl, or ruminal osmolarity by addition of mannitol, no impact on the flow rate or composition of saliva could be observed in the sense of a feedback mechanism [15], although salivary secretion dropped when pH reached very low levels (<3.5) [15]. When strongly hypertonic solutions of sodium or potassium salts were infused into the rumen, or sodium salts or urea were infused into the blood, production of saliva decreased and the concentration of those substances increased in the saliva. Other treatments including physiological elevations of ruminal potassium content had little effect on salivary composition [157]. Thus, evidence for sensors in the gut or rumen are scarce and salivary flow appears to be regulated by the composition of the portal blood [21, 138], possibly via hepatic sensors [119, 151].

A major fact that emerged from these studies is that if experimentally obtained saliva is not returned to the rumen, flow rate decreases to 50% of the initial value within two hours [15]. This decrease can be prevented by administration of $NaHCO_3$ containing solution to the rumen [14] and points towards the importance of ruminal reabsorption of these elements.

With the saliva, the animal secretes an amount of Na^+ that is 1.2-1.5 mol/d in sheep and 15 – 30 mol/d in cows [82] and is thus five times higher than the sodium content of the body, and 15 times higher than the amount of sodium consumed with the fodder [139]. Salivary secretion of HCO_3^- reaches levels that can lead to acidaemia when salivary flow is high, as after a meal [111, 112, 129, 139], stressing the importance of a rapid reabsorption of HCO_3^-.

Sodium content of saliva is strongly regulated by aldosterone, which is of considerable importance given the fact that in the natural diet of the ruminant, sodium is a scarce element. In sodium depleted sheep, salivary content of this mineral can fall dramatically from concentrations of about 180 mmol/l to as low as 40 mmol/l. Potassium excretion rises concomitantly from a normal level of 4 to 6 mmol/l to values of over 130 mmol/l [14]. Conversely, ruminal reabsorption of Na^+ is not regulated by aldosterone [85].

Crisis situations can occur in spring, when animals are quickly transferred from a diet of hay, which is comparably low in potassium, to a diet of fresh spring grass which is rich in potassium. In these situations, potassium levels in the rumen can rapidly peak to over 100 mmol/l, while sodium levels drop to lows of 30 mmol/l. To restore the balance for sodium, animals thus have to be able to absorb sodium actively against this very high background of potassium, while any absorption of potassium has to be tightly coupled to its excretion in the kidney. Given the enormous surface area and volume of the rumen, and the huge quantities of potassium rich roughage, the demands on this organ are thus clear: it has to prevent potassium from entering the blood stream too quickly, and it has to be highly able to transport sodium, bicarbonate, magnesium and calcium from the rumen to the plasma against an electrochemical gradient.

The Rumen

Histology and Organogenesis

Interestingly, the esophagus of mammals including humans has been shown to posses transporting properties in that it forms a tight barrier against the efflux of K^+ [68] and has considerable potential for the active transport of Na^+ [109]. The physiological function of this pronounced transport of electrolytes across the esophagus is unclear and probably merely serves maintenance of cell equilibrium. However, it is tempting to speculate that these transporting properties may have facilitated the adaptation of the esophageal wall to the requirements of forming a chamber for fermentation in the evolutionary process.

Of course, dramatic differences between the two types of tissue predominate. From an electrophysiological point of view, one pronounced difference is the fact that while participation of the rumen in the total absorption of potassium is small [49, 113], the tissue is depolarized by high apical potassium [37, 38, 136]. This characteristic feature will be discussed in more detail further down.

The histology of ruminal epithelium has been studied in some detail [48] and shows a typical structure with stratum corneum, granulosum, spinosum and basale; tight junctions (claudin 1 and ZO1), gap junctions (connexin-43) and expression of increasing levels of the α subunit of the Na^+/K^+-ATPase towards the basal membrane of the multilayered epithelium. The surface area is enlarged by leaf-like papillae of 10 to 15 mm length in the cow. Conversely, the abomasum is lined with a glandular epithelium which secretes hydrochloric acid and pepsin and is comparable to the gastric mucosa of animal with simple stomachs.

Studies of organogenesis in ruminant embryos show that the forestomach system develops from an extension of the primordial esophagus to the dorsal side of the primordial stomach, and is subsequently separated from the proper esophagus by constriction [98].

Ruminal Osmoregulation

For short periods of time following the ingestion of a meal, an osmotic gradient between blood and rumen fluid may develop that can exceed 200 mosmol/kg [139], with prefeed values usually being reached within about 3 – 4 hours [129]. The survival both of the microorganisms that digest the roughage, and of the cells lining the rumen depends on a rapid restoration of ruminal osmolarity to the normal level and a number of mechanisms have been described that may facilitate this process. While saliva plays a major role in elevating ruminal osmolarity after drinking [139], hypertonicity cannot be restored by influx of saliva in a reasonable amount of time since saliva is isotonic with blood.

Hyperosmolarity after ingestion of a meal is based in part on an increase in fermentation products that can cross the membrane via lipid diffusion (Figure 1). These include short chain fatty acids (SCFA) [6, 75, 87, 108, 120, 132], ammonia, [1, 16, 17, 41, 54, 87, 93] and CO_2 [6, 87, 97]. Besides lowering ruminal osmolarity and preventing the influx of water and ruminal distention, absorption of SCFA is of essential importance for the energy metabolism of ruminants [12, 13].

Conversely, ruminal hyperosmolarity following ingestion of non-ionic, non-diffusible substances or fermentation of easily digestible carbohydrates such as glucose or starch can lead to problems such as loss of apetite or even ruminal damage in unadapted animals [22, 105]. *In vitro* and *in vivo* experiments have established that in these cases, osmolarity is restored by an increase in water influx into the rumen [31, 42, 78, 122, 129, 158, 160].

Of the electrolytes, potassium shows the biggest rise in concentration [129]. Surprisingly and in marked contrast to the problems that can occur when osmolarity rises after ingestion of easily digestible carbohydrates, the rumen is highly able to deal with the osmotic challenge of large amounts of fresh, potassium rich grass. As potassium in the rumen rises to values that can reach 100 mmol/l, sodium concentration falls so that osmolarity within the rumen remains constant (at about 130 mmol/l) (Figure 2). The inverse regulation of both ions seems to be a constant factor under many different feeding regimes [24, 50, 59, 61, 80, 96, 113, 130, 131, 136, 149, 159, 160]. However, we are only beginning to understand the mechanism by which this regulatory response occurs.

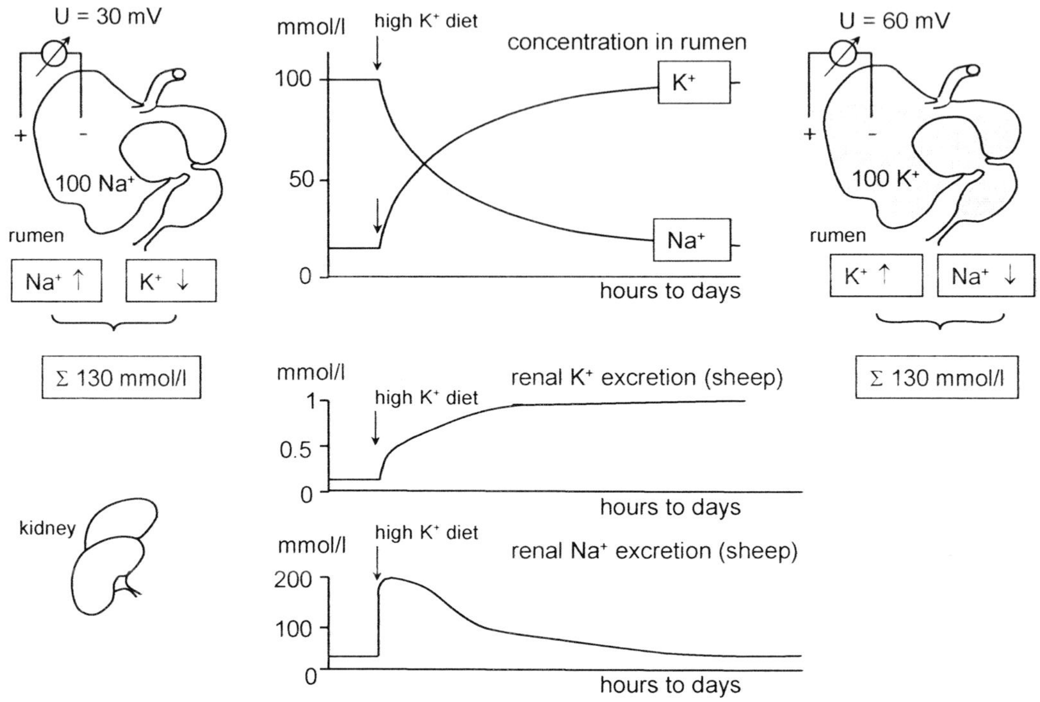

Fig 2: Ruminal osmoregulation: When sheep are transfered from a low potassium diet (i.e. hay) to a high potassium diet (i.e. grass), the concentration of potassium in the rumen continues to rise for a period of several hours to days. Concomittantly, ruminal concentration of sodium falls, so that the sum of both ions remains constant and ruminal osmolarity is restored. *In vivo* measurements show that the transepithelial potential increases with the concentration of potassium. Simultaneously, sodium excretion in urine peaks, while potassium excretion rises more slowly until maximum values are observed after about a week.

The Role of the Rumen in the Potassium Balance of Ruminants

Due to the large amounts of plant material that ruminants have to ingest to meet their energy requirements, daily intake of potassium can exceed 30 % of the total amount to be found in the body and is thus more than ten times higher than the dietary potassium intake of man (as related to body weight) [46, 69, 130]. As in man, about 90% of total potassium intake is absorbed [25, 46, 69, 112, 130, 149]. To prevent a rise to toxic concentrations in the plasma, potassium absorbed by the gut is rapidly redistributed to the cytosolic compartment of muscle, bone, liver and red blood cells, while the much slower renal corrections of disturbances in potassium balance require several hours [46, 70]. Thus, it should not come as a surprise that after a rapid transfer from a low potassium diet to one rich in potassium, the rumen is used as an additional storage space for this mineral [136] until urinary excretion of potassium has reached the level of intake. In the ruminant, this can take several hours or even days [149] (Figure 2).

Conversely, excretion of Na^+ sets in almost immediately after ingestion of a high potassium meal (1-3 hours) [25, 111, 112, 137, 149, 158, 160]. The sodium content of the diet does not appear to be related to this effect as natriuresis of sheep switched to a high potassium, low sodium diet is larger than that of sheep switched to a medium potassium, medium sodium diet [30]. The kidney of the sheep may retain sodium even when there are ample amounts in the diet, and lose sodium when intake is severely reduced [160], thus reflecting the replacement of Na^+ by K^+ due to ruminal osmoregulation.

From what is known about renal potassium excretion [46], an increase in the delivery of sodium to the distal nephron should stimulate uptake via ENaC, leading to a more lumen negative potential, and enhanced potassium excretion. Therefore, the sodium absorbed from the rumen should help to stimulate renal potassium excretion before ruminal K^+ reaches the gut and plasma potassium levels begin to rise. The ability of the rumen to absorb sodium thus appears central both for potassium balance and ruminal osmoregulation after ingestion of large amounts of potassium rich fodder.

Transport of Sodium and Chloride across the Rumen

Considering that the ability of ruminants to extract scarce sodium from grass and concentrate it in milk has been known for a long time, studies of Na^+ transport across the ruminal epithelium came relatively late. However, and perhaps for this reason, the ruminal epithelium was among the first tissues in which active transport of ions was investigated [143] following Hans Ussing's discovery of active sodium transport in frog skin [152]. The reasons for this interest are obvious: the ruminal epithelium is a moderately tight tissue [109] and thus, very suitable for determining active transport. An additional advantage of this organ is that due to its enormous size, experiments can easily be performed on animals *in vivo* by surgical fistulation of the rumen. Ruminal content can be removed by hand from such fistulas and exchanged for other solutions. It appears that animals adapt to this procedure reasonably well and will continue to ingest hay, if provided, during manipulations of the rumen. The solutions applied should not be dramatically out of the (considerable) physiological range and should contain appropriate amounts of Na^+, HCO_3^-, Ca^{2+} and Mg^{2+} if influx of saliva into the rumen is prevented in the technique of the isolated rumen.

The first study of ruminal transport was performed on goat rumen pouches in situ and left no doubt that chloride was being transported against a chemical gradient from rumen to blood [143]. In the same paper, and five years prior to the discovery of the Na^+/K^+-ATPase [142], the authors also noted absorption of Na^+ against an electrochemical gradient, and the secretion of a small amount of K^+ into the ruminal pouch. Subsequently, Dobson [27-29] published results from *in vivo* experiments in the sheep demonstrating that sodium is actively transported across the tissue. In these studies, electrodes were used to measure the potential difference, PD_t, generated across the ruminal epithelium. A mean PD_t of 30 mV with the blood side positive was observed, and many subsequent studies have confirmed the hypothesis that this potential is generated by active transport of sodium across the ruminal epithelium, with chloride following along the electrochemical gradient generated by this transport [36-39, 53, 81, 83, 85, 86, 128, 131, 136, 160]. Half of the sodium that is secreted by saliva is thus absorbed by the rumen [27], corresponding to a recirculation of total body sodium 2-3 times a day across this tissue [138]. Thus, water and electrolyte balance in the ruminant cannot be understood without taking the ruminal-portal-salivary exchanges into account (Figure 1).

It has since been demonstrated that the major part of ruminal Na^+ transport takes place by apical uptake of Na^+ via NHE3 exchange [42, 82, 83, 145], blockable by amiloride (1 mmol/l) (Figure 3). Conversely, bumetanide and furosemide have no effect on Na^+ transport across the intact ruminal epithelium, and there is no other indication of Na^+-K^+-Cl^- cotransport in the rumen [83, 87].

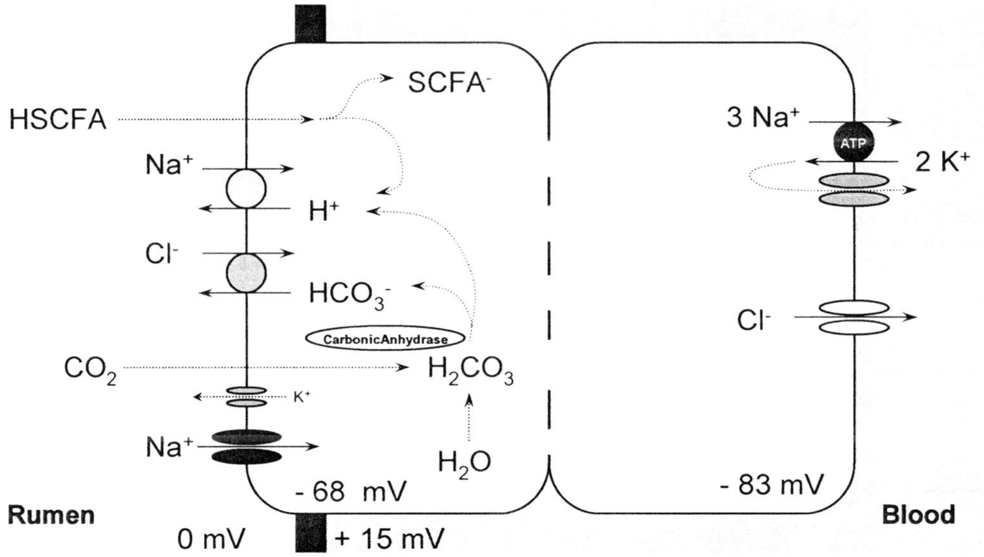

Fig 3: Current model of Na^+ transport across the ruminal epithelium: Na^+ transport from the rumen can occur both via sodium-proton exchange (NHE) coupled to chloride-bicarbonate exchange and via an electrogenic pathway. On the basolateral side, active transport of sodium occurs via the Na^+/K^+-ATPase which generates a potential (blood side positive) that drives basolateral efflux of chloride through a channel. This, in turn, stimulates apical uptake of Cl^- via Cl^-/HCO^-_3 exchange and Na^+ via NHE. Carbondioxide and short chain fatty acids (HSCFA) are thought to diffuse freely through the lipid membrane and stimulate electroneutral sodium transport. (It should be noted that the ruminal epithelium consists of more than two layers. Potential levels are for orientation and can vary considerably from tissue to tissue.)

Basolateral extrusion of Na^+ takes place via the Na^+/K^+-ATPase mentioned above. Substantial evidence suggests that K^+ is mainly recycled via basolateral potassium channels, since apical potassium secretion is small [16, 39, 53, 73, 76]. The resulting negative cytosolic potential drives Cl^- out of the cell through a basolateral Cl^- channel [1, 74] and into the blood. This, in turn, may generate a chemical gradient for apical uptake of Cl^- via Cl^-/HCO_3^- exchange [27, 82], utilizing HCO_3^- that is continuously produced from CO_2 diffusing into the cell across the apical membrane. Accordingly, Cl^- transport is sensitive to acetazolamide in the manner known from many other transporting epithelia [36]. It should be noted that electroneutral coupling of chloride with bicarbonate exchange ensures that the potential gradient built up by the Na^+/K^+-ATPase does not break down. In line with this model, changes in chloride concentration do not lead to large changes of PD_t [52], although smaller effects of exchanging chloride for sulfate have been noted [37, 80]. However, any apical or paracellular conductance for chloride appears to be relatively small, or blocked by other ions under most circumstances.

Apical Na^+/H^+ exchange is stimulated both by CO_2 ($\Leftrightarrow$ H_2CO_3) and by the large amounts of short chain fatty acids (SCFA) found in the rumen. Both diffuse through the apical membrane in their lipophylic form, and donate a proton for exchange with Na^+ after dissociation [40, 43, 45, 75, 133-135].

In contrast, the response of ruminal epithelium to the third major fermentation product of the rumen, ammonia, depends on ruminal pH [1, 2]. At an (unphysiologically high) pH of 7.4, addition of ammonia to the ruminal epithelium *in vitro* inhibits the uptake of Na^+ via NHE due to the influx of NH_3 with subsequent reconstitution to NH_4^+ [1] and an increase of cellular pH [97]. At the more acidic values to be found physiologically (5.5-7.0), the amount of NH_3 is reduced greatly due to protonation, and the dominant observation is a stimulation of Na^+ uptake due to uptake of NH_4^+ with subsequent cytosolic release of protons for exchange in NHE [1]. Thus, in the healthy rumen with physiological pH values, the major products of fermentation, short chain fatty acids, bicarbonate and ammonium, all stimulate sodium reabsorption as a requirement for the postprandial stimulation of salivary flow.

Electrogenic Na^+ Conductance in the Rumen

The epithelial sodium channel (ENaC) is the channel that is responsible for the conductance of sodium across most mammalian transporting epithelia in the lung, gut and kidney [71, 115]. Characteristic features are block by low doses of amiloride in the micromolar range, stimulation by aldosterone and lack of voltage dependence (although a slightly higher open probability after *hyperpolarization* of the cell has been observed [114]).

In contrast, amiloride given in a dose as high as 1 mmol/l does not significantly affect the PD_t and the short circuit current across the ruminal epithelium, although Na^+ transport via NHE is significantly reduced. In addition and contrary to expectations, extensive *in vivo* [85] and *in vitro* experiments (unpublished results) by our laboratory failed to show any regulation of ruminal sodium transport by aldosterone. Likewise, adrenalectomy did not affect the PD_t of the ruminal epithelium *in vivo* [131]. In micropuncture experiments on the intact epithelium, it was found that *depolarization* of the apical membrane decreases the fractional apical resistance [72, 122] and increases the conductance of the ruminal epithelium for sodium. Thus, the electrogenic Na^+ conductance of ruminal epithelium has none of the properties associated with the expression of epithelial sodium channels (ENaC).

In an attempt to find further characteristics of the electrogenic sodium conductance of forestomach epithelia, it was observed in Ussing chamber experiments that removal of mucosal Ca^{2+}, Mg^{2+} or both increased the PD_t and the short circuit current across the omasal [121] and ruminal epithelium [73]. At physiological concentrations of both divalent cations, the conductance saturates at a Na^+ concentration of about 30 mmo/l [116]. In light of the great physiological variability of ruminal Na^+ content (which varies from 25 mmol/l to above 100 mmol/l), this property is essential for maintaining sodium conductance at low levels of Na^+, while preventing a massive Na^+ influx at higher concentrations which might damage cell homeostasis.

It should be noted that these experiments were performed in absence of an electrochemical gradient for sodium and can thus not be explained by opening of a paracellular shunt pathway due to removal of divalent cations. Since both short circuit current and potential difference rose almost immediately after removal of divalent cations, it can also be concluded that the Na^+/K^+-ATPase is not the rate limiting step in sodium transport across the ruminal epithelium.

In microelectrode experiments on intact epithelia, removal of mucosal divalent cations led to a depolarization of the apical membrane [72], while paracellular conductance was not affected. Maneuvers that have been shown to reduce cytosolic Mg^{2+} concentration in isolated cells such as the elevation of cAMP [77, 126], magnesium deprivation [73, 124], or reduction of serosal Na^+ [121], can also increase Na^+ conductance through forestomach epithelia.

In patch clamp experiments on isolated cells of the ruminal epithelium, we have observed similar effects: conductance for both Na^+ and K^+ increases when external divalents cations are removed. Interestingly, a further opening could be achieved by hyperpolarization, and it was found that the voltage dependence of this conductance was altered by the cytosolic level of Mg^{2+} [77, 147]. A residual sodium conductance can also be observed in physiological solutions with normal concentrations of external divalent cations (Figure 4 A and B). However, when the Mg^{2+} concentration in the pipette is set to values at the upper end of the range reported for intact ruminal epithelial cells [123, 124, 126], no significant conductance for Na^+ can be seen (Figure 4 C). Under these high Mg^{2+} conditions, sodium conductance could be stimulated by elevation of cytosolic cAMP, either by direct application to the cytosol via the patch pipette, or via PGE2 [146]. The latter effect may rest both on a reduction of cytosolic magnesium [126], and/or upon direct effects on the channel protein. It is noteworthy that an endogenous production of prostaglandins has been observed in intact ruminal epithelium [40].

Although the precise nature of the Na^+ conducting channel in ruminal epithelium has yet to be identified, our experiments suggest that electrogenic uptake of Na^+ by the sheep rumen is regulated by voltage dependent binding of external and internal magnesium to sites in the pore of a non-selective cation channel.

Mg^{2+} Gating of Ion Channels

Regulation of channel conductance by intracellular or extracellular magnesium is known from a large number of ion channels, such as the inwardly rectifying potassium channel, the NMDA receptor, a number of non-selective cation channels from the TRP family [3, 100, 154] or the epithelial sodium channel, ENaC [26].

While voltage gated channels respond to a change in membrane voltage by changes in the conformation of the channel protein, in magnesium gated ion channels, removal of Mg^{2+} in the solution eliminates the voltage dependence. Inwardly rectifying potassium channels have been studied extensively for many decades [10, 56, 117], and it appears that voltage dependence is related to binding of Mg^{2+} to a binding site on the intracellular side (Figure 5 A). When the interior of the cell is hyperpolarized versus the outside, the resulting potential gradient will tend to push positively charged ions into the cell and away from the cell membrane. If the binding of Mg^{2+} to the internal mouth of the channel is weak, the probability that Mg^{2+} will dissociate and move into the cell lumen rises, and thus, the number of channels that are open at any given moment will increase. A characteristic and distinctive property of the inward rectifier is that the conductance change is a function not of the absolute potential E, but of the potential difference $E - E(K^+)$ [56, 66] where $E(K^+)$ is the reversal potential for potassium. The reason for this is that as the external potassium concentration rises, influx of potassium ions will push the blocking Mg^{2+} ion out of internal mouth of the channel pore [55] and the channel will open at voltages where it would otherwise have been closed. Thus, paradoxically, a channel that normally only opens when it is hyperpolarized to low levels will also open when the external side is exposed to a high concentration of potassium. The relatively simple mechanism of Mg^{2+} block allows depolarization of the cell when extracellular potassium concentration rises and allows influx of potassium when the membrane voltage is below the Nernst potential for this ion, but prevents efflux of potassium out of the cell at higher voltage levels.

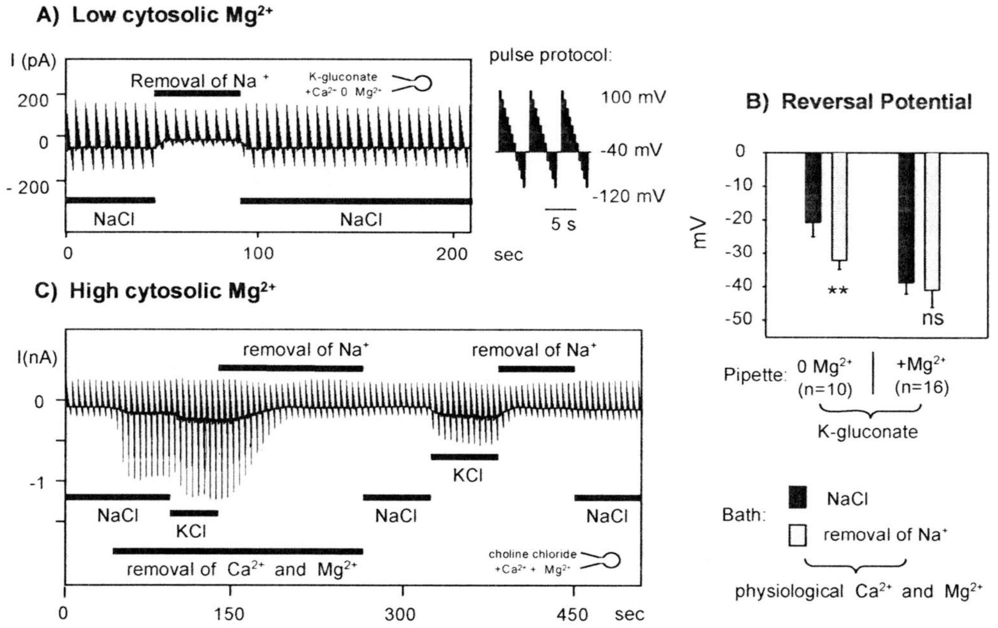

Fig 4: Sodium conductance of ruminal epithelial cells is blocked by external and internal Mg^{2+}: Original patch clamp recordings. Cells were filled with different pipette solutions, and currents in response to various voltages (pulse protocol) were measured. Removal of Na^+ from a bath solution with physiological concentrations of Ca^{2+} and Mg^{2+} resulted in a visable reduction of inward currents (A) and significant depolarization of the cells (B) when filled with a pipette solution that contained no Mg^{2+}. When Mg^{2+} was added to the cytosolic medium in a concentration of 0.9 mmol/l, effects of removing sodium from the bath solution were not significant. Removal of external Ca^{2+}, Mg^{2+}, or both increased both sodium and potassium conductance of the cells.

NMDA receptors of neurons are channels with a high permeability for calcium and have been studied extensively, facilitated undoubtedly in part by the low resting conductance of the neuron's cell membrane. These channels are opened by certain agonists, but conduct poorly in normal bathing medium with physiological concentrations of magnesium, even in the presence of an agonist [55]. However, when the cell is depolarized in the presence of an agonist, their conductance increases [4, 62]. In these channels, the binding sites for Mg^{2+} are thought to be inside the channel pore, within the region where transmembrane voltage drops from the external level to the level found on the internal side of the cell membrane. As above, it has been established that the affinity to Mg^{2+} changes as a function of membrane voltage (Figure 5 B). However, at depolarized membrane voltages, the probability of Mg^{2+} occupying these *external* sites is lower, the block is less pronounced, and the conductance of the NMDA receptor increases. Interestingly, this block appears to be modulated by binding of permeant monovalent cations to the external channel vestibule of NMDA receptors [110]. A further feature is an additional block with different binding affinity by *internal* Mg^{2+} [63]. It has been suggested that Mg^{2+} can permeate the channel at a low, voltage dependent rate and would thus impede the flow of other ions through the pore, with the number of channels blocked in this manner rising with the concentration of Mg^{2+}. The different blocking properties of external or internal Mg^{2+} can be explained by channel models in which several ions move in single file through a long narrow pore with asymmetrical energy barriers [63], [56]. Note that the electrical field within the channel pore and thus, the forces that bind and repel the blocking Mg^{2+} ion, depend on the other ions within the pore. It has been proposed that the combination of block by internal and external Mg^{2+} make the NMDA-activated channel a bidirectional rectifier with an optimal voltage range for conductance [63] (Figure 5 C).

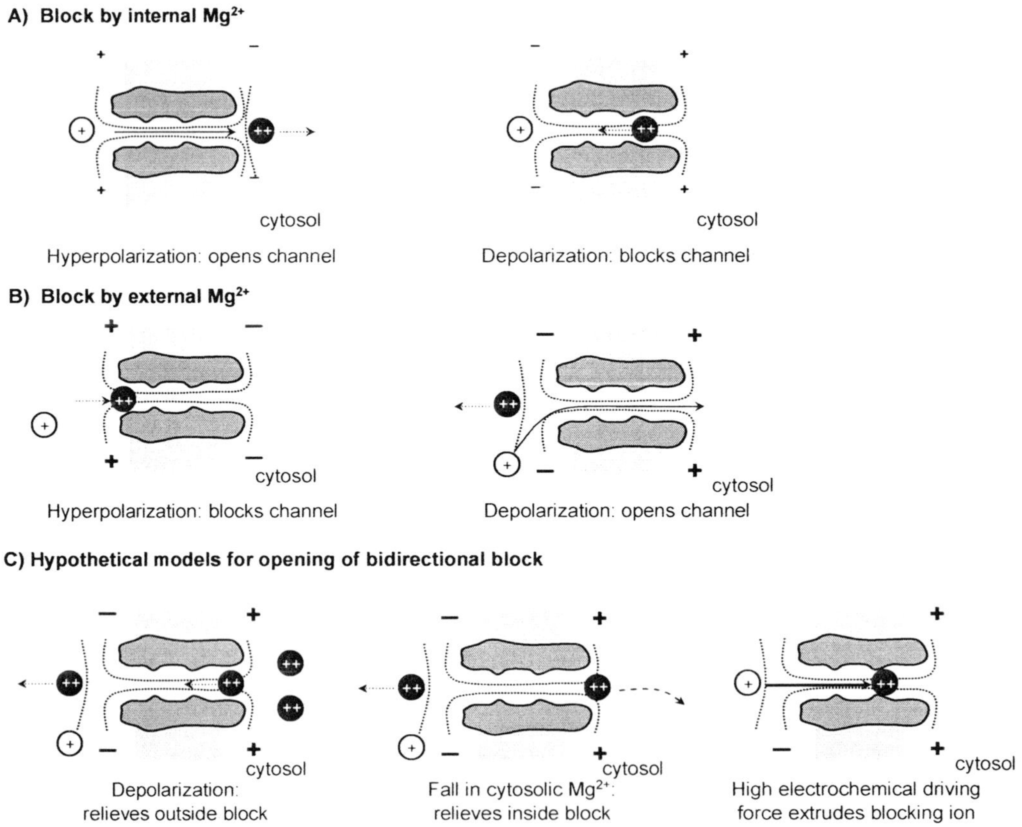

Fig 5: Voltage dependent block of ion channels by Mg^{2+}: some models (see text)

The epithelial calcium channels ECaC 1 (TRPV6) and ECaC 2 (TRPV5) are known for their extremely high Ca^{2+} selectivity (P_{Ca}/P_{Na} >100) *in vitro*, and appear to be involved in Ca^{2+} reabsorption in kidney and intestine [106, 154]. When divalent cations are removed, they become permeable to monovalent cations and open at negative potentials with strong inward rectification. As with the inward rectifier, this rectification depends at least in part on the presence of intracellular Mg^{2+}. Block by external divalent cations is decreased when intracellular Mg^{2+} levels drop. As with the NMDA receptor, interaction with permeant cations has been observed. At high potentials, Mg^{2+} block is relieved, indicating that Mg^{2+} may be able to permeate the selectivity filter of the channel if the driving force is high enough [154].

In terms of conductance for sodium and block by divalent cations, the ruminal conductance for sodium has striking similarity to channels belonging to the TRP family [106]. However, in isolated cells of the ruminal epithelium, the blocking effects of divalent cations are much larger than any currents that might be attributed to a conductance for Ca^{2+} and Mg^{2+}, even if the electrochemical driving force is set to very high levels in patch clamp experiments [148]. This is in line with the *in vitro* and *in vivo* observation that influx of Mg^{2+} into the cells and across the rumen saturates [85, 86, 123], although initial uptake rates are high and fast. Thus, while it cannot and should not be excluded that Mg^{2+} uptake by the rumen may involve channels from this group, these channels do not display the high rates of Mg^{2+} conductance reported for channels such as TRPM7 [95, 99].

Interestingly, low levels of messenger RNA of both TRPV5 and TRPV6 have been found in the rumen of sheep [162] where functionally, they may be involved in the transport of Ca^{2+} [120]. In the same study, a higher expression of both channels not just on the mRNA but also on the protein level was found in other parts of the gut of the sheep where absorption of Ca^{2+} is thought to play a minor role, and thus, the physiological role of these channels in the gastrointestinal tract needs to be clarified.

In the case of TRPV5, both high Ca^{2+} selectivity and sensitivity to extracellular divalent cations appear to depend on a single aspartate residue in the pore region between transmembrane domains 5 and 6 [101]. Point mutations can be expected to dramatically alter the conductance of ion channels, and the TRP family is no exception. Associated proteins may lead to further differences [153]. It should be noted that buffering of Ca^{2+} and Mg^{2+} in physiological solutions is by no means trivial [51, 79]. Since the cytosolic concentration of these ions has impacts on the selectivity of these channels, a prediction of their conductance *in vivo* from *in vitro* experiments has to be done with care. In most patch clamp studies [101] including our initial experiments with ruminal epithelial cells, [1, 101], the concentration of Mg^{2+} in the pipette solution is higher than the free concentration of the ion that is found in resting intact cells [123, 127], and the situation in the intact epithelium *in vivo* may be even more complex.

A recent study has shown that the esophageal epithelium of the rabbit possesses a sodium conductance with similar properties, and it has been suggested that in this tissue, ENaC subunits assemble to form a non-selective cation channel that is not blockable by amiloride [7]. Data from RT-PCR experiments in our laboratory suggest that alpha subunits of ENaC are expressed in ruminal epithelium (personal communication, Jatti Priesnitz). As mentioned, regulation of ENaC by intracellular Mg^{2+} has been demonstrated [26]. However, attempts to demonstrate other subunits have not been convincing so far, and while this is certainly an intriguing possibility, we do not feel that at this point that the involvement of other cation channels (i.e. TRP channels [100]) in ruminal sodium transport should be ruled out in light of the failure of aldosterone or amiloride to show any effects on electrogenic sodium conductance in the forestomach.

It may be argued that since the greater amount of sodium absorbed by the rumen is transported by the NHE as outlined above, any interest in a further mechanism is purely academic. However, the apical electrogenic entry of sodium determines the magnitude of the transepithelial potential (blood side positive), [27, 37-39, 82], and is thus an essential contributor to the total driving force for the basolateral efflux of choride and other anions such as bicarbonate [27]. Basolateral efflux of chloride drives apical uptake of chloride, which, in turn, should stimulate Na^{+} uptake via coupling of NHE with Cl^{-}/HCO_3^{-} exchange.

More importantly, perhaps, stimulation of electrogenic sodium transport should have a negative impact on the electrochemical driving force for potassium efflux from the rumen. The central importance of electrogenic sodium transport for transruminal potential thus makes it a gate-keeper for the transport of ions across the rumen, and since the electrogenic Na^+ conductance of ruminal epithelium is regulated by changes in cytosolic magnesium, the uptake of this ion by the ruminal epithelium should be of some interest.

Absorption of Magnesium

Interest in the uptake of magnesium by the ruminant is old and related to a surge of cases of an affliction known as "grass tetany" that led to widespread loss of cattle in the last century [90, 125]. After sudden transfer from normal winter rations to young spring grass, animals became afflicted with excitability, salivation, ataxia, recumbency, and tetanic muscle spasms [140, 141]. Despite normal magnesium content of the fodder, the animals were apparently unable to absorb this mineral and thus suffered from hypomagnesaemia. Sjollema's hypothesis that the surge in frequency was related to the introduction of fertilizers containing potash has since been confirmed by a solid body of evidence which has been reviewed elsewhere [34, 90]. It should be noted that a number of other factors are known to contribute to malabsorption of magnesium that include sodium deficiency [118] and high protein content of the diet with subsequent high levels of ruminal ammonia [41].

Absorption of magnesium from the rumen has been studied extensively at the level of the animal, the tissue, and the cell. It has been established that the main site for absorption of magnesium in the ruminant is the forestomach [107]. When the rumen is bypassed, decreased plasma Mg^{2+} concentrations, decreased urinary Mg^{2+} excretion and increased fecal Mg^{2+} excretion is observed. This negative magnesium balance cannot be compensated for by post-ruminal infusion of Mg^{2+} [150].

Given that the serosal side of the rumen is positive versus the mucosal side (30 mV), any paracellular transport will move Mg^{2+} from blood into the rumen [20]. Transport must thus be transcellular and active and was first demonstrated by Martens and Harmeyer in 1978 [86]. In the same study, it was shown that the transport is ouabain sensitive and saturable. The presence of basolateral Na^+/ Mg^{2+} exchange coupled to Na^+/K^+-ATPase has since been established [77, 126] as the mechanism underlying the active step of absorption (Figure 6 A).

In contrast, it has been shown that apical uptake depends on the electrochemical driving force for Mg^{2+} across the apical membrane (Figure 6 B). Depolarization of this membrane by potassium was found to reduce transport of Mg^{2+} across the rumen *in vivo* using various approaches that include the temporary isolation of the rumen from influx of saliva [19, 50, 80, 85, 87, 89, 94, 164]. Experiments across isolated ruminal epithelium in the Ussing chamber established that the potential difference, and not potassium concentration, were central to the reduction in the uptake of Mg^{2+} [76, 84]. Likewise, exposure of isolated cells of the ruminal epithelium in culture to high potassium solution resulted in a reduced uptake of Mg^{2+}[123, 124]. It has been shown that this maneuver depolarizes isolated cells of the ruminal epithelium via potassium channels [1]. In contrast, plasma magnesium concentration had no influence on the rate of Mg^{2+} absorption [91], and the effects of K^+ on Mg^{2+} absorption are not attributable to high levels of absorbed K^+, but rather to the concentration of K^+ present in the rumen [113]. From these experiments, a model has emerged that resembles that developed for the distal convoluted tubule, where absorption occurs via an apical Mg^{2+} channel in conjunction with a basolateral Na^{+}/ Mg^{2+} exchanger [65, 161]. In the context of this review, it is tempting to point out that in humans, a fall in plasma magnesium is known to lead to kaliuresis [161] [64] [105], and that reduction in plasma Mg^{2+} levels should thus help with excretion of potassium.

It should be noted that there is evidence for a second transporting process that is not potential dependent and which facilitates uptake of magnesium into depolarized cells at higher ruminal potassium concentrations of this mineral [61, 76, 124]. In addition, a number of factors are known to stimulate absorption of magnesium from the rumen, including a rise in bicarbonate, short chain fatty acids, and a low but physiological pH of around 6.4 [90].

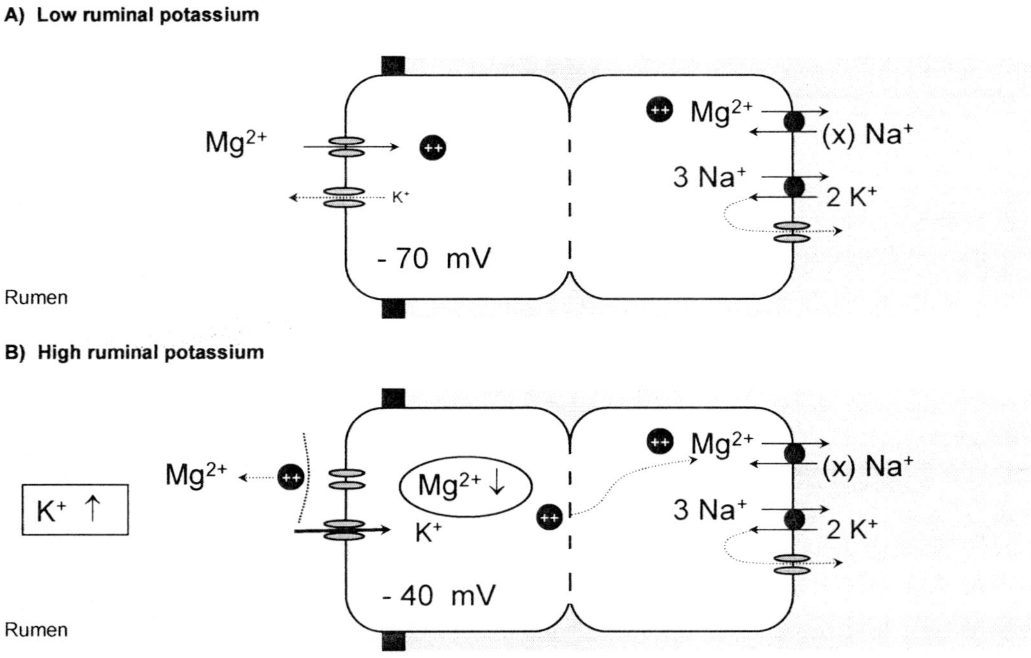

Fig 6: Reduction of Mg^{2+} uptake by elevation of ruminal potassium: Apical Mg^{2+} influx occurs down the electrochemical gradient, driven by the negative potential across the apical membrane (A). Basolateral efflux occurs in exchange for sodium. The stochiometry has not been identified.

After elevation of ruminal potassium (B), the apical membrane is depolarized and the gradient for uptake of Mg^{2+} decreases. Cytosolic Mg^{2+} concentration drops, and thus, also basolateral extrusion. Note that the relative contribution of the potassium conductance to total membrane permeability may rise as ruminal potassium levels increase.

Interestingly, reductions in the absorption of Mg^{2+} were also observed after acute exposure to ammonia [18, 41, 54, 87, 89]. For reasons that are not clear at present, compensatory mechanisms set in after a few days [41] in this case, so that hypomagnesaemia is transient. It is possible to speculate about the role that the electrogenic vacuolar H-ATPase might play in this regulatory response [124] or about up-regulation of enzymes which are involved in ammonia metabolism in the rumen epithelium [102, 103].

The practical implications of these physiological background facts are clear and have led to a reduction in the frequency of the affliction. Adjusting cattle to a high potassium and high protein diet with production of ammonia must be done slowly. Mg^{2+} concentration in the rumen must be elevated by feeding Mg^{2+} salts to activate the non-potential dependent magnesium uptake mechanism. Fertilization techniques which greatly raise potassium intake should be avoided. In addition, care has to be taken that animals are not sodium depleted, since this increases aldosterone levels, salivary excretion of potassium and thus, elevates ruminal levels of potassium [90] [118] with the detrimental effect on Mg^{2+} absorption from the forestomachs described above [88].

RUMINAL TRANSPORT OF POTASSIUM

Given the high concentration of potassium in the ruminant diet and that the volume of the rumen can exceed that of the plasma, it is clear that any absorption of potassium from this organ has to be slow. When total potassium absorption is related to feed intake and recirculation of saliva is not considered, the contribution of the rumen to total potassium absorption is marginal, and most studies regard the small intestine as the major site for the absorption of dietary potassium [69, 164].

However, when the rate of absorption is related to the concentration found in the rumen, a certain permeability of the ruminal wall for potassium becomes apparent. Secretion of potassium under experimental conditions with very low ruminal potassium concentration was first noted by Sperber and Hyden [143], and this finding has been confirmed many times [35, 59, 80, 130, 160]. The simplest explanation for this observation is to assume a leak of potassium into the rumen paracellularly from the blood side (4-5 mmol/l K^+), driven by the positive PD_t of about 30 mV (blood side positive). In this case, an increase in ruminal potassium should lead to an increase in the potential across the ruminal wall, and such a potential was, indeed, measured in feeding experiments [136]. Subsequent parallel *in vivo* and *in vitro* investigations [37, 38] confirmed that the potential rose strongly with the luminal concentration of potassium both in the completely isolated rumen in anaesthetized sheep, and in isolated ruminal epithelium in the Ussing chamber. Conversely, raising Na^+ from a basal level of 20 mmol/l had little (*in vivo*) or no (*in vitro*) effect [37, 38]. Later *in vivo* studies (under more physiological conditions with HCO_3^- and acetate instead of Cl^- as the major anions [18, 80]) confirmed the finding that K^+ concentration contributes more to overall transepithelial potential than concentrations of Na^+ exceeding 20 mmol/l. However, it should be noted that the increasing transepithelial potential with falling ruminal sodium concentration in these experiments suggests that any paracellular leak of sodium into the rumen has to be small. Thus, the simple model of a hypothetic "leak" pathway postulated above cannot be quite so simple after all and must, at the very least, be highly selective for potassium over sodium.

At first glance, these results seem to suggest that the efflux of potassium has to be considerable since the transruminal potential has to be continuously maintained by a corresponding flux of ions under *in vivo* non-equilibrium conditions. This is very clearly not the case. The ability of the rumen to concentrate potassium has already been outlined above (Figure 2). When sheep are transferred from a low potassium diet of hay and meals to a diet of high potassium grass from a heavily fertilized meadow, ruminal potassium concentrations can rise from levels of about 30 mmol/l to peaks of 100 mmol/l [136]. Concomitantly, the concentration of sodium falls from levels as high as 100 mmol/l to values of 25 mmol/l. Thus, while there can be no doubt about the coupling of potassium and transruminal potential [37, 38, 61, 76, 80, 84, 131, 155], and while a small passive flux of potassium down its electrochemical gradient has been demonstrated [59,160], the major ion being absorped under these conditions is sodium [59, 130, 144, 158, 160]. In line with this, a complete removal of sodium (or application of ouabain) abolished the potential and the short circuit current across the ruminal epithelium [53], establishing Na^+ as the major ion responsible for the potential across the ruminal wall. An explanation for the failure to observe changes in transepithelial potential in response to the elevation of ruminal sodium above 20 mmol/l [37, 38] has since emerged: *in vitro*, electrogenic sodium conductance saturates at 30 mmol/l of Na^+ [116]. Thus, raising Na^+ concentration from 20 mmol/l to higher values should only have marginal effects.

These data do not support the originally proposed model of a major paracellular leak pathway for potassium and suggest a transcellular electrogenic uptake pathway highly selective for potassium over sodium, with potassium channels being ideal candidates. In line with this, a high concentration of potassium was found to depolarize the apical membrane of ruminal epithelium and decrease the fractional apical resistance [76]. *In vitro* studies showing that active, ouabain sensitive potassium secretion can take place under equilibrium conditions [39, 53, 76, 163], with evidence for the existence of additional basolateral potassium channels [76], support the model of channel mediated transcellular uptake.

A similar notion of apical potassium channels evolved independently from *in vivo* experiments on sheep, demonstrating that ruminal uptake of ammonia cannot be explained exclusively by assumption of lipophillic diffusion of NH_3 [17]. *In vitro* experiments in the Ussing chamber followed, that postulated electrogenic uptake of NH_4^+ via a quinidine sensitive potassium channel [16]. This suggestion is supported by a recent study demonstrating stimulating effects of NH_4^+ on sodium transport, thus establishing that transport of NH_4^+ has to be transcellular [1]. Using the patch clamp technique, it was possible to directly demonstrate corresponding quinidine sensitive potassium channels with a conductance for both potassium and NH_4^+ in isolated ruminal epithelial cells [1]. Interestingly and typically [5, 33, 55, 56, 165], the open probability of these channels [1] rose with the concentration of the permeant ion. This property of potassium channels may help to explain why the cell is depolarized with increasing levels of potassium and falling levels of sodium [39, 53, 76] despite the fact that in low potassium ringer, the conductance for sodium exceeds that for potassium.

The assumption of transcellular flux via potassium channels integrates certain observations in a satisfactory way, since uptake of NH_4^+ is favored by the electrochemical gradient across the apical membrane of the rumen, whereas passive influx of K^+ into the cytosol is not possible except at high ruminal potassium concentrations due to the high cytosolic concentration of potassium. This is very much in line with the *in vivo* observations of ammonia transport at low pH and low concentrations of ammonia [17], while very high concentrations of potassium are necessary in order to observe sizable transport rates for potassium [160]. However, the failure to observe a clear effect of quinidine on potassium transport across the rumen *in vivo* does leave some room for a debate that should include the very wide range of quinidine effects on cellular metabolism [35]. Despite this caveat, the *in vivo* and *in vitro* evidence for the existence of potassium channels with a substantial contribution to total conductance of the apical and basolateral membrane of the ruminal epithelium is solid, and a viable model is needed to explain why nevertheless, potassium movements across the ruminal wall remain small [69, 136, 160].

An attempt to integrate the various observations can be made by considering the interaction of the electrogenic sodium conductance with potassium transport (Figure 7). As stated before, this pathway is stimulated by the depolarization of the apical membrane [72] and part of the potential generated by an increase of potassium in the rumen should be due to stimulation of this sodium conductance. Now, the paradoxical observations fall into place: potassium depolarizes the apical membrane, opens the sodium conductance [72, 77], the transepithelial potential increases, the electrochemical driving force for the efflux of potassium from the rumen falls. Therefore, we see an increase in sodium transport, a rise in transepithelial potential, but potassium efflux from the rumen remains limited.

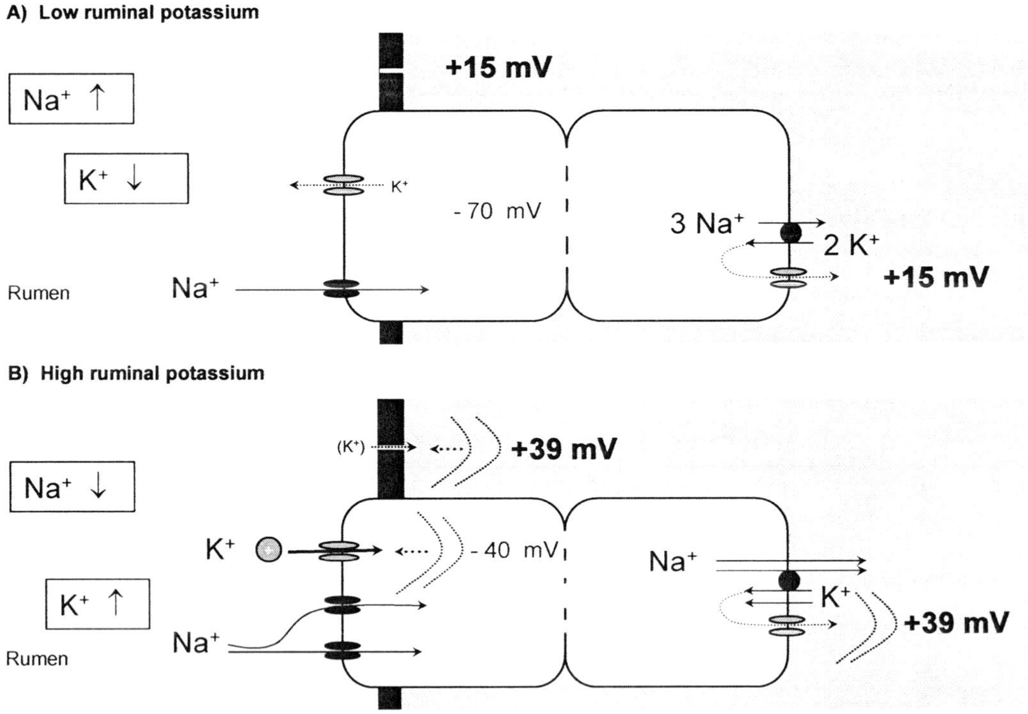

Fig 7: Uptake of K^+ by the ruminal epithelium: Flux of K^+ across the ruminal epithelium is passive and proportional to the transepithelial electrochemical driving force. With increasing concentration of ruminal K^+, the efflux of K^+ from the rumen into the plasma increases. However, stimulation of electrogenic Na^+ transport with depolarization of the apical membrane and generation of a strong, lumen positive transepithelial potential limits the amount of K^+ that leaks out of the rumen. In vivo, transepithelial potential can reach values of 60 – 70 mV.

A Role for Magnesium in Ruminal Osmoregulation

Numerous studies demonstrate that rising levels of potassium stimulate the absorption of sodium across the rumen *in vivo* [59, 130, 144, 158, 160]. Harrison notes that *in vitro* stimulation of Na^+ transport by elevation of K^+ only succeeded under open circuit conditions [52], and thus, a potential dependent mechanism appears to be involved. This effect cannot be attributed to a rise in osmotic pressure, since the effects of urea or mannitol on sodium absorption were discrete [160]. As has been outlined above, a high increase in osmotic pressure due to administration of non-ionic osmotic agents can even reduce Na^+ absorption, and induce water influx into the rumen [44, 78, 122]. Conversely, potassium stimulates absorption of sodium even in the absence of an osmotic gradient [85].

In light of the information we have at this point, it is possible to develop a fairly consistent model that may describe the events that occur when sheep or cows are transferred from a low potassium diet of hay to a diet higher in potassium, such as fresh spring grass (Figure 8). As the concentration of potassium in the rumen increases, the conductance of the apical membrane for potassium rises [1, 55] and the apical membrane is depolarized [76], the binding affinity of divalent cations at the external mouth of the non-selective cation channel is loosened, and simultaneously, uptake of magnesium decreases. Now, the non-selective cation channel begins to open [77, 147].

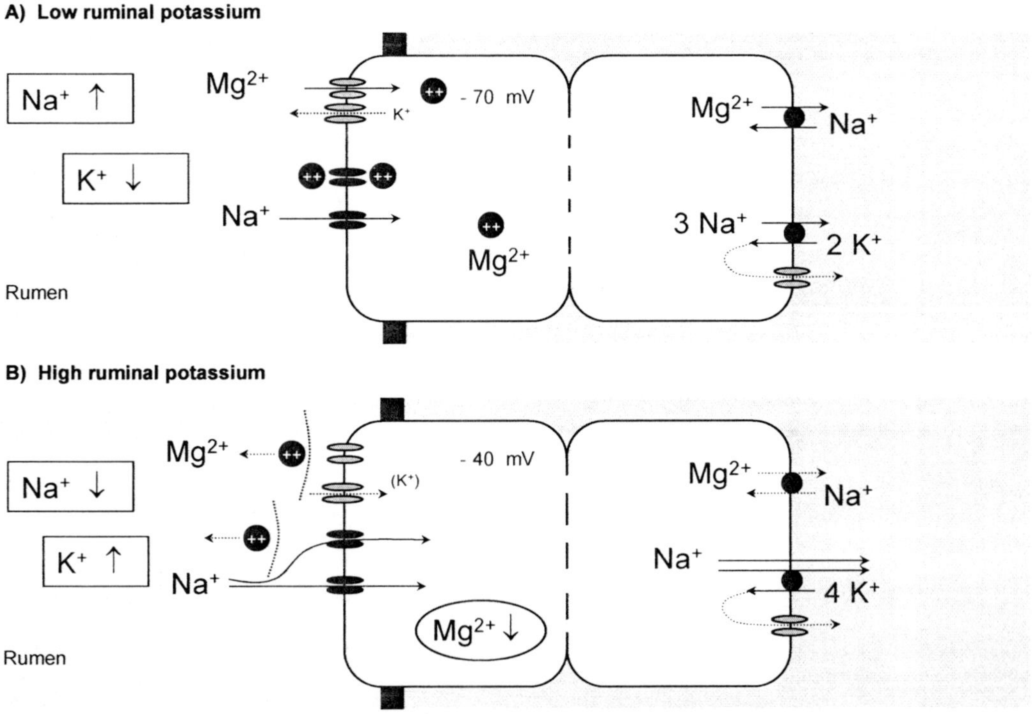

Fig 8: Stimulation of ruminal sodium transport by elevation of ruminal potassium: A) Divalent cations block many of the non-selective cation channels in the apical membrane, limiting electrogenic sodium influx. Despite a large concentration of apical Na^+, total sodium flux is low. After elevation of the ruminal K^+ concentration (B), the apical membrane is depolarized, uptake of Mg^{2+} decreases and cytosolic Mg^{2+} concentration drops. In addition, depolarization relieves the block of the non-selective cation from the external side. Electrogenic Na^+ transport rises creating an additional depolarizing stimulus and reducing the electrochemical gradient for the uptake of K^+.

As Na^+ rushes in, the apical membrane is depolarized even further and a catastrophic cascade of events would unfold but for the fact that at a certain potential level, residual internal Mg^{2+} begins to block the pore from the inside, thus limiting a further depolarization. It appears conceivable that at elevated cytosolic concentrations of sodium, basolateral Na^+/ Mg^{2+} exchange slows down so that internal Mg^{2+} levels begin to rise again despite reduced uptake. And thus, this model can predict that at medium levels of potassium, sodium transport is stimulated, while at very high levels of potassium, sodium transport is depressed [160].

An intriguing feature of this model is that as the transepithelial potential depolarizes due to the electrogenic efflux of sodium, the driving force for the flux of potassium through leaks from the rumen such as the paracellular pathway is decreased. Again, we can predict what happens when potassium levels rise: as sodium transport is depressed, potassium conductance begins to rise and this is, indeed, what is observed [160]. Thus, the introduction of potassium channels in the apical membrane of the rumen does make "sense" from an evolutionary point of view – or did, until potash fertilizers came up [140].

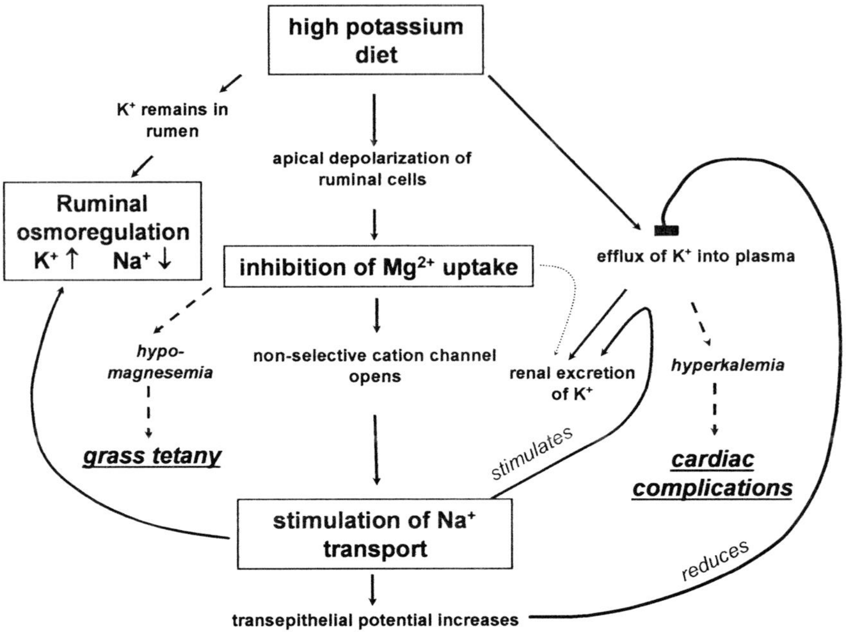

Fig 9: A role for magnesium in ruminal osmoregulation: The adverse effects of an inhibition of transcellular ruminal Mg^{2+} transport are balanced by the positive effects of a stimulation of Na^+ efflux from the rumen, namely, inhibition of K^+ efflux into the plasma, stimulation of renal K^+ excretion, and regulation of ruminal osmolarity.

Conclusion

Ruminants can deal with abrupt and extremely large rises in potassium intake due to their ability to utilize the rumen as an additional reservoir for buffering K^+ until renal excretion of K^+ has set in. Ruminal concentrations of potassium can rise to values of 100 mmol/l in these situations, with sodium content decreasing in proportion so that osmolarity remains constant throughout. This is essential since rising osmolarity would interfere with ruminal fermentation processes and food uptake by the animal.

In contrast to the pronounced ability to maintain physiological ruminal osmolarity and plasma potassium levels, hypomagnesemia is a known complication with economic significance after transfer of ruminants to a high potassium diet. The pathogenesis of this condition has been shown to involve a decrease in Mg^{2+} absorption by the rumen after exposure to high ruminal potassium.

We feel that the two phenomena may be causally linked via a signaling cascade that involves depolarization of the apical membrane by high potassium, reduced uptake of magnesium, and opening of a sodium conductance with subsequent rise in sodium transport out of the ruminal pouch (Figure 9). In evolutionary terms, the potassium-induced reduction in digestibility of Mg^{2+} may be balanced by the increased ability to quickly ingest very large amounts of fresh grass after a longer period of famine.

REFERENCES

[1] Abdoun K, Stumpff F, Wolf K, and Martens H. Modulation of electroneutral Na transport in sheep rumen epithelium by luminal ammonia. *Am J Physiol Gastrointest Liver Physiol* 289: G508-520, 2005.

[2] Abdoun K, Wolf K, Arndt G, and Martens H. Effect of ammonia on Na^+ transport across isolated rumen epithelium of sheep is diet dependent. *Br J Nutr* 90: 751-758, 2003.

[3] Albitz R, Magyar J, and Nilius B. Block of single cardiac sodium channels by intracellular magnesium. *Eur Biophys J* 19: 19-23, 1990.

[4] Antonov SM and Johnson JW. Permeant ion regulation of N-methyl-D-aspartate receptor channel block by Mg(2+). *Proc Natl Acad Sci U S A* 96: 14571-14576, 1999.

[5] Aqvist J and Luzhkov V. Ion permeation mechanism of the potassium channel. *Nature* 404: 881-884, 2000.

[6] Ash RW and Dobson A. The Effect of Absorption on the Acidity of Rumen Contents. *J Physiol* 169: 39-61, 1963.

[7] Awayda MS, Bengrine A, Tobey NA, Stockand JD, and Orlando RC. Nonselective cation transport in native esophageal epithelia. *Am J Physiol Cell Physiol* 287: C395-402, 2004.

[8] Bach A, Calsamiglia S, and Stern MD. Nitrogen metabolism in the rumen. *J Dairy Sci* 88 Suppl 1: E9-21, 2005.

[9] Bailey CB. Saliva secretion and its relation to feeding in cattle. 4. The relationship between the concentrations of sodium, potassium, chloride and inorganic phosphate in mixed saliva and rumen fluid. *Br J Nutr* 15: 489-498, 1961.

[10] Balser JR, Roden DM, and Bennett PB. Single inward rectifier potassium channels in guinea pig ventricular myocytes. Effects of quinidine. *Biophys J* 59: 150-161, 1991.

[11] Beja-Pereira A, Luikart G, England PR, Bradley DG, Jann OC, Bertorelle G, Chamberlain AT, Nunes TP, Metodiev S, Ferrand N, and Erhardt G. Gene-culture coevolution between cattle milk protein genes and human lactase genes. *Nat Genet* 35: 311-313, 2003.

[12] Bergman EN. Energy contributions of volatile fatty acids from the gastrointestinal tract in various species. *Physiol Rev* 70: 567-590, 1990.

[13] Bergman EN, Reid RS, Murray MG, Brockway JM, and Whitelaw FG. Interconversions and production of volatile fatty acids in the sheep rumen. *Biochem J* 97: 53-58, 1965.

[14] Blair-West JR, Coghlan JP, Denton DA, Goding JR, and Wright RD. The effect of aldosterone, cortisol, and corticosterone upon the sodium and potassium content of sheep's parotid saliva. *J Clin Invest* 42: 484-496, 1963.

[15] Boda JM, McDonald PG, and Walker JJ. Effects of the Addition of Fluids to the Empty Rumen on the Flow Rate and Chemical Composition of Bovine Mixed Saliva. *J Physiol* 177: 323-336, 1965.

[16] Bödeker D and Kemkowski J. Participation of NH_4^+ in total ammonia absorption across the rumen epithelium of sheep (Ovis aries). *Comp Biochem Physiol A Physiol* 114: 305-310, 1996.

[17] Bödeker D, Winkler A, and Holler H. Ammonia absorption from the isolated reticulo-rumen of sheep. *Exp Physiol* 75: 587-595, 1990.

[18] Care AD, Brown RC, Farrar AR, and Pickard DW. Magnesium absorption from the digestive tract of sheep. *Q J Exp Physiol* 69: 577-587, 1984.

[19] Care AD, Farrar AR, and Pickard DW. Factors affecting the absorption of magnesium from the rumen in sheep. *Journal of Physiology* 325, 1982.

[20] Care AD and van't Klooster AT. In Vivo Transport of Magnesium and Other Cations across the Wall of the Gastro-Intestinal Tract of Sheep. *J Physiol* 177: 174-191, 1965.

[21] Carr DH and Titchen DA. Post prandial changes in parotid salivary secretion and plasma osmolality and the effects of intravenous of saline solutions. *Q J Exp Physiol Cogn Med Sci* 63: 1-21, 1978.

[22] Carter RR, Allen OB, and Grovum WL. The effect of feeding frequency and meal size on amounts of total and parotid saliva secreted by sheep. *Br J Nutr* 63: 305-318, 1990.

[23] Cassida KA and Stokes MR. Eating and resting salivation in early lactation dairy cows. *J Dairy Sci* 69: 1282-1292, 1986.

[24] Denton DA. The effect of variation in the concentration of individual extracellular electrolytes on the response of the sheep's parotid gland to Na^+ depletion. *J Physiol* 140: 129-147, 1958.

[25] Dewhurst JK, Harrison FA, and Keynes RD. Renal excretion of potassium in the sheep. *J Physiol* 195: 609-621, 1968.

[26] Diakov A and Korbmacher C. A novel pathway of epithelial sodium channel activation involves a serum- and glucocorticoid-inducible kinase consensus motif in the C terminus of the channel's alpha-subunit. *J Biol Chem* 279: 38134-38142, 2004.

[27] Dobson A. Active transport through the epithelium of the reticulo-rumen sac. *J Physiol* 146: 235-251, 1959.

[28] Dobson A. The forces moving sodium ions through rumen epithelium. *J Physiol* 128: 39-40P, 1955.

[29] Dobson A and Phillipson AT. The forces moving chloride ions through rumen epithelium. *J Physiol* 125: 26-27P, 1954.

[30] Dobson A, Scott D, and Bruce JB. Changes in sodium requirement of the sheep associated with changes of diet. *Q J Exp Physiol Cogn Med Sci* 51: 311-323, 1966.

[31] Dobson A, Sellers AF, and Gatewood VH. Dependence of Cr-EDTA absorption from the rumen on luminal osmotic pressure. *Am J Physiol* 231: 1595-1600, 1976.

[32] Donovan GA, Steenholdt C, McGehee K, and Lundquist R. Hypomagnesemia among cows in a confinement-housed dairy herd. *J Am Vet Med Assoc* 224: 96-99, 54, 2004.

[33] Doyle DA. Structural themes in ion channels. *Eur Biophys J* 33: 175-179, 2004.

[34] Dua K and Care AD. Impaired absorption of magnesium in the aetiology of grass tetany. *Br Vet J* 151: 413-426, 1995.

[35] Dua K and Care AD. Lack of effect of quinidine on divalent mineral absorption from the reticulorumen of sheep. *Res Vet Sci* 56: 114-115, 1994.

[36] Emanovic D, Harrison FA, Keynes RD, and Rankin JC. The effect of acetazolamide on ion transport across isolated sheep rumen epithelium. *Journal of Physiology* 254: 803-812, 1976.

[37] Ferreira HG, Harrison FA, and Keynes RD. The potential and short-circuit current across isolated rumen epithelium of the sheep. *J Physiol* 187: 631-644, 1966.

[38] Ferreira HG, Harrison FA, Keynes RD, and Nauss AH. Observations on the potential across the rumen of sheep. *J Physiol* 187: 615-630, 1966.

[39] Ferreira HG, Harrison FA, Keynes RD, and Zurich L. Ion transport across an isolated preparation of sheep rumen epithelium. *J Physiol* 222: 77-93, 1972.

[40] Gäbel G, Butter H, and Martens H. Regulatory role of cAMP in transport of Na^+, Cl^- and short-chain fatty acids across sheep ruminal epithelium. *Exp Physiol* 84: 333-345, 1999.

[41] Gäbel G and Martens H. The effect of ammonia on magnesium metabolism in sheep. *J Anim Physiol Anim Nutr* 55: 278-287, 1986.

[42] Gäbel G, Martens H, Suendermann M, and Galfi P. The effect of diet, intraruminal pH and osmolarity on sodium, chloride and magnesium absorption from the temporarily isolated and washed reticulo-rumen of sheep. *Q J Exp Physiol* 72: 501-511, 1987.

[43] Gäbel G and Sehested J. SCFA transport in the forestomach of ruminants. *Comp Biochem Physiol A Physiol* 118: 367-374, 1997.

[44] Gäbel G, Suendermann M, and Martens H. The influence of osmotic pressure, lactic acid and pH on ion and fluid absorption from the washed and temporarily isolated reticulo-rumen of sheep. *J Vet Med A* 34: 220-226, 1987.

[45] Gäbel G, Vogler S, and Martens H. Short-chain fatty acids and CO_2 as regulators of Na^+ and Cl^- absorption in isolated sheep rumen mucosa. *J Comp Physiol [B]* 161: 419-426, 1991.

[46] Giebisch G. Renal potassium transport: mechanisms and regulation. *Am J Physiol* 274: F817-833, 1998.

[47] Goff JP. Macromineral disorders of the transition cow. *Vet Clin North Am Food Anim Pract* 20: 471-494, v, 2004.

[48] Graham C and Simmons NL. Functional organization of the bovine rumen epithelium. *Am J Physiol Regul Integr Comp Physiol* 288: R173-181, 2005.

[49] Greene LW, Webb KE, Jr., and Fontenot JP. Effect of potassium level on site of absorption of magnesium and other macroelements in sheep. *J Anim Sci* 56: 1214-1221, 1983.

[50] Grings EE and Males JR. Effects of potassium on macromineral absorption in sheep fed wheat straw-based diets. *J Anim Sci* 64: 872-879, 1987.

[51] Günzel D, McGuigan JA, and Schlue WR. Use of Mg^{2+} and Ca^{2+} macroelectrodes to measure binding in extracellular-like physiological solutions. *Front Biosci* 10: 905-918, 2005.

[52] Harrison FA. Ion transport across rumen and omasum epithelium. *Philos Trans Royal Soc Lond B Biol Sci* 262: 301-305, 1971.

[53] Harrison FA, Keynes RD, Rankin JC, and Zurich L. The effect of ouabain on ion transport across isolated sheep rumen epithelium. *J Physiol* 249: 669-677, 1975.

[54] Head MJ and Rook JA. Hypomagnesaemia in dairy cattle and its possible relationship to ruminal ammonia production. *Nature* 176: 262-263, 1955.

[55] Hille B. Ion Channels of Excitable Membranes. *Sinauer Associates, Inc*: Chapter 15, pg. 482, 2001.

[56] Hille B and Schwarz W. Potassium channels as multi-ion single-file pores. *J Gen Physiol* 72: 409-442, 1978.

[57] Hoover WH and Miller TK. Rumen digestive physiology and microbial ecology. *Vet Clin North Am Food Anim Pract* 7: 311-325, 1991.

[58] Hungate RE, Bryant MP, and Mah RA. The rumen bacteria and protozoa. *Annu Rev Microbiol* 18: 131-166, 1964.

[59] Hyden S. Observations on the absorption of inorganic ions from the reticulo-rumen of the sheep. *Kugl Lantbrukshögkolans Annaler* 27: 273-285, 1961.

[60] Janis CM, Damuth J, and Theodor JM. Miocene ungulates and terrestrial primary productivity: Where have all the browsers gone? *Proc Natl Acad Sci U S A* 97: 7899-7904, 2000.

[61] Jittakhot S, Schonewille JT, Wouterse HS, Yuangklang C, and Beynen AC. The relationships between potassium intakes, transmural potential difference of the rumen epithelium and magnesium absorption in whethers. *Br J Nutr* 91: 183-189, 2004.

[62] Johnson JW and Ascher P. Glycine potentiates the NMDA response in cultured mouse brain neurons. *Nature* 325: 529-531, 1987.

[63] Johnson JW and Ascher P. Voltage-dependent block by intracellular Mg^{2+} of N-methyl-D-aspartate-activated channels. *Biophys J* 57: 1085-1090, 1990.

[64] Kamel KS, Harvey E, Douek K, Parmar MS, and Halperin ML. Studies on the pathogenesis of hypokalemia in Gitelman's syndrome: role of bicarbonaturia and hypomagnesemia. *Am J Nephrol* 18: 42-49, 1998.

[65] Kamel KS, Oh MS, and Halperin ML. Bartter's, Gitelman's, and Gordon's syndromes. From physiology to molecular biology and back, yet still some unanswered questions. *Nephron* 92 Suppl 1: 18-27, 2002.

[66] Katz B. Les constantes électriques de la membrane du muscle. *Arch Sci Physiol* 2: 285-299, 1949.

[67] Kay RN. The rate of flow and composition of various salivary secretions in sheep and calves. *J Physiol* 150: 515-537, 1960.

[68] Khalbuss WE, Alkiek R, Marousis CG, and Orlando RC. Potassium conductance in rabbit esophageal epithelium. *Am J Physiol* 265: G28-34, 1993.

[69] Khorasani GR, Janzen RA, McGill WB, and Kennelly JJ. Site and extent of mineral absorption in lactating cows fed whole-crop cereal grain silage of alfalfa silage. *J Anim Sci* 75: 239-248, 1997.

[70] Koeppen BM and Stanton BA. Renal Physiology. (Third edition, Mosby Inc, St. Louis, Missouri, ISBN 0-323-01242-01246). 2001.

[71] Kunzelmann K and Mall M. Electrolyte transport in the mammalian colon: mechanisms and implications for disease. *Physiol Rev* 82: 245-289, 2002.

[72] Lang I and Martens H. Na^+ transport in sheep rumen is modulated by voltage-dependent cation conductance in apical membrane. *Am J Physiol* 277: G609-618, 1999.

[73] Leonhard-Marek S. Divalent cations reduce the electrogenic transport of monovalent cations across rumen epithelium. *J Comp Physiol [B]* 172: 635-641, 2002.

[74] Leonhard-Marek S, Breves G, and Busche R. Effect of chloride on pH microclimate and electrogenic Na^+ absorption across the rumen epithelium of goat and sheep. *Am J Physiol Gastrointest Liver Physiol*, 2006.

[75] Leonhard-Marek S, Gäbel G, and Martens H. Effects of short chain fatty acids and carbon dioxide on magnesium transport across sheep rumen epithelium. *Exp Physiol* 83: 155-164, 1998.

[76] Leonhard-Marek S and Martens H. Effects of potassium on magnesium transport across rumen epithelium. *Am J Physiol* 271: G1034-1038, 1996.

[77] Leonhard-Marek S, Stumpff F, Brinkmann I, Breves G, and Martens H. Basolateral Mg^{2+}/Na^+ exchange regulates apical nonselective cation channel in sheep rumen epithelium via cytosolic Mg^{2+}. *Am J Physiol Gastrointest Liver Physiol* 288: G630-645, 2005.

[78] Lodemann U and Martens H. Effects of diet and osmotic pressure on Na^+ transport and tissue conductance of sheep isolated rumen epithelium. *Exp Physiol* 91: 539-550, 2006.

[79] Luthi D, Günzel D, and McGuigan JA. Mg-ATP binding: its modification by spermine, the relevance to cytosolic Mg^{2+} buffering, changes in the intracellular ionized Mg^{2+} concentration and the estimation of Mg^{2+} by 31P-NMR. *Exp Physiol* 84: 231-252, 1999.

[80] Martens H and Blume I. Effect of intraruminal sodium and potassium concentrations and of the transmural potential difference on magnesium absorption from the temporarily isolated rumen of sheep. *Q J Exp Physiol* 71: 409-415, 1986.

[81] Martens H and Blume I. Studies on the absorption of sodium and chloride from the rumen of sheep. *Comp Biochem Physiol A* 86: 653-656, 1987.

[82] Martens H and Gäbel G. Transport of Na and Cl across the epithelium of ruminant forestomachs: rumen and omasum. A review. *Comp Biochem Physiol A* 90: 569-575, 1988.

[83] Martens H, Gäbel G, and Strozyk B. Mechanism of electrically silent Na and Cl transport across the rumen epithelium of sheep. *Exp Physiol* 76, 1991.

[84] Martens H, Gäbel G, and Strozyk H. The effect of potassium and the transmural potential difference on magnesium transport across an isolated preparation of sheep rumen epithelium. *Q J Exp Physiol* 72: 181-188, 1987.

[85] Martens H and Hammer U. [Magnesium and sodium absorption from the isolated sheep rumen during intravenous aldosterone infusion (author's transl]. *Dtsch Tierarztl Wochenschr* 88: 404-407, 1981.

[86] Martens H and Harmeyer J. Magnesium transport by isolated rumen epithelium of sheep. *Res Vet Sci* 24: 161-168, 1978.

[87] Martens H, Heggemann G, and Regier K. Studies on the effect of K, Na, NH_4^+, VFA and CO_2 on the net absorption of magnesium from the temporarily isolated rumen of heifers. *Zentralbl Veterinarmed A* 35: 73-80, 1988.

[88] Martens H, Kubel OW, Gäbel G, and Honig H. Effects of low sodium intake on magnesium metabolism in sheep. *J agric Sci*: 237-243, 1987.

[89] Martens H and Rayssiguier Y. Magnesium metabolism and hypomagnesaemia. In: Proceedings of the 5th International Symposium on Ruminant Physiology (ed. Ruckebush, Y and Thivend, P.). 447-466, 1980.

[90] Martens H and Schweigel M. Pathophysiology of grass tetany and other hypomagnesemias. Implications for clinical management. *Vet Clin North Am Food Anim Pract* 16: 339-368, 2000.

[91] Martens H and Stossel EM. Magnesium absorption from the temporarily isolated rumen of sheep: no effect of hyper- or hypomagnesaemia. *Q J Exp Physiol* 73: 217-223, 1988.

[92] McCoy MA. Hypomagnesaemia and new data on vitreous humour magnesium concentration as a post-mortem marker in ruminants. *Magnes Res* 17: 137-145, 2004.

[93] McDonald IW. The absorption of ammonia from the rumen of the sheep. *Biochem J* 42: 584-587, 1948.

[94] McLean AF, Buchan W, and Scott D. Magnesium absorption in mature ewes infused intrarumenally with magnesium chloride. *Br J Nutr* 52: 523-527, 1984.

[95] Monteilh-Zoller MK, Hermosura MC, Nadler MJ, Scharenberg AM, Penner R, and Fleig A. TRPM7 provides an ion channel mechanism for cellular entry of trace metal ions. *J Gen Physiol* 121: 49-60, 2003.

[96] Morris JG and Gartner RW. The effect of potassium on the sodium requirements of growing steers with and without alpha-tocopherol supplementation. *Br J Nutr* 34: 1-14, 1975.

[97] Muller F, Aschenbach JR, and Gäbel G. Role of Na^+/H^+ exchange and HCO_3^- transport in pH_i recovery from intracellular acid load in cultured epithelial cells of sheep rumen. *J Comp Physiol [B]* 170: 337-343, 2000.

[98] Mutoh K and Wakuri H. Early organogenesis of the caprine stomach. *Nippon Juigaku Zasshi* 51: 474-484, 1989.

[99] Nadler MJ, Hermosura MC, Inabe K, Perraud AL, Zhu Q, Stokes AJ, Kurosaki T, Kinet JP, Penner R, Scharenberg AM, and Fleig A. LTRPC7 is a Mg.ATP-regulated divalent cation channel required for cell viability. *Nature* 411: 590-595, 2001.

[100] Nilius B, Vennekens R, Prenen J, Hoenderop JG, Bindels RJ, and Droogmans G. Whole-cell and single channel monovalent cation currents through the novel rabbit epithelial Ca^{2+} channel ECaC. *J Physiol* 527 Pt 2: 239-248, 2000.

[101] Nilius B, Vennekens R, Prenen J, Hoenderop JG, Droogmans G, and Bindels RJ. The single pore residue Asp542 determines Ca^{2+} permeation and Mg^{2+} block of the epithelial Ca^{2+} channel. *J Biol Chem* 276: 1020-1025, 2001.

[102] Nocek JE and Polan CE. Influence of ration form and nitrogen availability on ruminal fermentation patterns and plasma of growing bull calves. *J Dairy Sci* 67: 1038-1042, 1984.

[103] Oba M, Baldwin RLt, Owens SL, and Bequette BJ. Metabolic fates of ammonia-N in ruminal epithelial and duodenal mucosal cells isolated from growing sheep. *J Dairy Sci* 88: 3963-3970, 2005.

[104] Odette O. Grass tetany in a herd of beef cows. *Can Vet J* 46: 732-734, 2005.

[105] Owens FN, Secrist DS, Hill WJ, and Gill DR. Acidosis in cattle: a review. *J Anim Sci* 76: 275-286, 1998.

[106] Owsianik G, Talavera K, Voets T, and Nilius B. Permeation and selectivity of TRP channels. *Annu Rev Physiol* 68: 685-717, 2006.

[107] Pfeffer E, Thompson A, and Armstrong DG. Studies on intestinal digestion in the sheep. 3. Net movement of certain inorganic elements in the digestive tract on rations containing different proportions of hay and rolled barley. *Br J Nutr* 24: 197-204, 1970.

[108] Phillipson AT. Absorption of acetate from the rumen of sheep. *J Physiol* 109: Proc, 31, 1949.

[109] Powell DW, Morris SM, and Boyd DD. Water and electrolyte transport by rabbit esophagus. *Am J Physiol* 229: 438-443, 1975.

[110] Qian A, Antonov SM, and Johnson JW. Modulation by permeant ions of Mg(2+) inhibition of NMDA-activated whole-cell currents in rat cortical neurons. *J Physiol* 538: 65-77, 2002.

[111] Rabinowitz L. Aldosterone and potassium homeostasis. *Kidney Int* 49: 1738-1742, 1996.

[112] Rabinowitz L, Green DM, Sarason RL, and Yamauchi H. Homeostatic potassium excretion in fed and fasted sheep. *Am J Physiol* 254: R357-380, 1988.

[113] Rahnema SH and Fontenot JP. Effects of intravenous infusion of high levels of potassium and sodium on mineral metabolism in sheep. *J Anim Sci* 68: 2833-2838, 1990.

[114] Rossier BC. The epithelial sodium channel: activation by membrane-bound serine proteases. *Proc Am Thorac Soc* 1: 4-9, 2004.

[115] Rossier BC, Pradervand S, Schild L, and Hummler E. Epithelial sodium channel and the control of sodium balance: interaction between genetic and environmental factors. *Annu Rev Physiol* 64: 877-897, 2002.

[116] Rübbelke MK. In vitro Untersuchungen des Pansenepithels von Schafen zur Characterisierung eines elektrogenen, calcium-sensitiven Natriumtransportes (Inaugural-Dissertation, Institut für Veterinär-Physiologie, Freie Universität Berlin). Berlin, 1998 (Journal-Nr: 2169), 1998.

[117] Sakmann B and Trube G. Conductance properties of single inwardly rectifying potassium channels in ventricular cells from guinea-pig heart. *J Physiol* 347: 641-657, 1984.

[118] Sargison ND, Macrae AI, and Scott PR. Hypomagnesaemic tetany in lactating Cheviot gimmers associated with pasture sodium deficiency. *Vet Rec* 155: 674-676, 2004.

[119] Sawchenko PE and Friedman MI. Sensory functions of the liver--a review. *Am J Physiol* 236: R5-20, 1979.

[120] Schröder B, Vossing S, and Breves G. In vitro studies on active calcium absorption from ovine rumen. *J Comp Physiol [B]* 169: 487-494, 1999.

[121] Schultheiss G and Martens H. Ca-sensitive Na transport in sheep omasum. *Am J Physiol* 276: G1331-1344, 1999.

[122] Schweigel M, Freyer M, Leclercq S, Etschmann B, Lodemann U, Bottcher A, and Martens H. Luminal hyperosmolarity decreases Na transport and impairs barrier function of sheep rumen epithelium. *J Comp Physiol [B]* 175: 575-591, 2005.

[123] Schweigel M, Lang I, and Martens H. Mg(2+) transport in sheep rumen epithelium: evidence for an electrodiffusive uptake mechanism. *Am J Physiol* 277: G976-982, 1999.

[124] Schweigel M and Martens H. Anion-dependent Mg^{2+} influx and a role for a vacuolar H^+-ATPase in sheep ruminal epithelial cells. *Am J Physiol Gastrointest Liver Physiol* 285: G45-53, 2003.

[125] Schweigel M and Martens H. [Electrophysiologic changes in rumen epithelium in their effect on magnesium transport--a review]. *Berl Munch Tierarztl Wochenschr* 113: 97-102, 2000.

[126] Schweigel M, Park HS, Etschmann B, and Martens H. Characterization of the Na^+-dependent Mg^{2+} transport in sheep ruminal epithelial cells. *Am J Physiol Gastrointest Liver Physiol* 290: G56-65, 2006.

[127] Schweigel M, Vormann J, and Martens H. Mechanisms of Mg(2+) transport in cultured ruminal epithelial cells. *Am J Physiol Gastrointest Liver Physiol* 278: G400-408, 2000.

[128] Scott D. Absorption of chloride from the rumen of the sheep. *Res Vet Sci* 11: 291-293, 1970.

[129] Scott D. Changes in mineral, water and acid base balance associated with feeding and diet. Digestion and metabolism in the ruminant: Proceedings of the IV International Symposium on Ruminant Physiology (1974). ed: I.W. McDonald and A.C.I. Warner. The University of New England Publishing Unit (1975), 1975.

[130] Scott D. The effects of potassium supplements upon the absorption of potassium and sodium from the sheep rumen. *Q J Exp Physiol Cogn Med Sci* 52: 382-391, 1967.

[131] Scott D. The effects of sodium depletion and potassium supplements upon electrical potentials in the rumen of the sheep. *Q J Exp Physiol Cogn Med Sci* 51: 60-69, 1966.

[132] Sehested J, Diernaes L, Moller PD, and Skadhauge E. Interaction between absorption of sodium and acetate across the rumen epithelium of cattle. *Acta Vet Scand Suppl* 89: 107-108, 1993.

[133] Sehested J, Diernaes L, Moller PD, and Skadhauge E. Ruminal transport and metabolism of short-chain fatty acids (SCFA) in vitro: effect of SCFA chain length and pH. *Comp Biochem Physiol A Mol Integr Physiol* 123: 359-368, 1999.

[134] Sehested J, Diernaes L, Moller PD, and Skadhauge E. Transport of butyrate across the isolated bovine rumen epithelium--interaction with sodium, chloride and bicarbonate. *Comp Biochem Physiol A Mol Integr Physiol* 123: 399-408, 1999.

[135] Sehested J, Diernaes L, Moller PD, and Skadhauge E. Transport of sodium across the isolated bovine rumen epithelium: interaction with short-chain fatty acids, chloride and bicarbonate. *Exp Physiol* 81: 79-94, 1996.

[136] Sellers AF and Dobson A. Studies on reticulo-rumen sodium and potassium concentration and electrical potentials in sheep. *Res Vet Sci* 1: 95, 1960.

[137] Sellers AF, Gitis TL, and Roepke MH. Studies of electrolytes in body fluids of dairy cattle. III. Effects of potassium on electrolyte levels in body fluids in midlactation. *Am J Vet Res* 12: 296-301, 1951.

[138] Silanikove N. The struggle to maintain hydration and osmoregulation in animals experiencing severe dehydration and rapid rehydration: the story of ruminants. *Exp Physiol* 79: 281-300, 1994.

[139] Silanikove N and Tadmor A. Rumen volume, saliva flow rate, and systemic fluid homeostasis in dehydrated cattle. *Am J Physiol* 256: R809-815, 1989.

[140] Sjollema B. On the nature and therapy of grass staggers. *Vet Rec* 10: 425, 1930.

[141] Sjollema B. Untersuchungen über die Ursache der Grastetanie und der großen Frequenzzunahme dieser Krankheit. *Deutsche Tierärztliche Wochenzeitschrift* 15, 1932.

[142] Skou JC. The influence of some cations on an adenosine triphosphatase from peripheral nerves. *Biochim Biophys Acta* 23: 394-401, 1957.

[143] Sperber I and Hyden S. Transport of chloride through the ruminal mucosa. *Nature* 169: 587, 1952.

[144] Stacy BD and Warner AC. Balances of water and sodium in the rumen during feeding: osmotic stimulation of sodium absorption in the sheep. *Q J Exp Physiol Cogn Med Sci* 51: 79-93, 1966.

[145] Stevens CE. Transport of sodium and chloride by the isolated rumen epithelium. *Am J Physiol* 206: 1099-1105, 1964.

[146] Stumpff F, Brinkmann I, and Martens H. Regulation of a non-selective cation channel in cells of the ovine rumen epithelium by PGE2 and Forskolin. *Pflügers Arch* 449 (Suppl.): S116, 2005.

[147] Stumpff F, Brinkmann I, Schweigel M, and Martens H. High potassium diet, sodium and magnesium in ruminants: the story is not over. *Production diseases in farm animals*: 284-285, 2004.

[148] Stumpff F, Martens H, and Schweigel M. A patch-clamp and furaptra study of electrodiffusive Mg uptake into cells of the ruminal epithelium. *Acta Physiol* 186 (Suppl): 232 (PW203A-217), 2006.

[149] Suttle NF and Field AC. Studies on magnesium in ruminant nutrition. 8. Effect of increased intakes of potassium and water on the metabolism of magnesium, phosphorus, sodium, potassium and calcium in sheep. *Br J Nutr* 21: 819-831, 1967.

[150] Tomas FM and Potter BJ. The site of magnesium absorption from the ruminant stomach. *Br J Nutr* 36: 37-45, 1976.

[151] Tsuchiya Y, Nakashima S, Banno Y, Suzuki Y, and Morita H. Effect of high-NaCl or high-KCl diet on hepatic Na^+- and K^+-receptor sensitivity and NKCC1 expression in rats. *Am J Physiol Regul Integr Comp Physiol* 286: R591-596, 2004.

[152] Ussing HH and Zerahn K. Active transport of sodium as the source of electric current in the short-circuited isolated frog skin. *Acta Physiol Scand* 23: 110-127, 1951.

[153] van de Graaf SF, Hoenderop JG, and Bindels RJ. Regulation of TRPV5 and TRPV6 by associated proteins. *Am J Physiol Renal Physiol* 290: F1295-1302, 2006.

[154] Voets T, Janssens A, Prenen J, Droogmans G, and Nilius B. Mg^{2+}-dependent gating and strong inward rectification of the cation channel TRPV6. *J Gen Physiol* 121: 245-260, 2003.

[155] Wadhwa DR and Care AD. The absorption of calcium ions from the ovine reticulo-rumen. *J Comp Physiol [B]* 170: 581-588, 2000.

[156] Wallace RJ. Ruminal microbial metabolism of peptides and amino acids. *J Nutr* 126: 1326S-1334S, 1996.

[157] Warner AC and Stacy BD. Influence of ruminal and plasma osmotic pressure on salivary secretion in sheep. *Q J Exp Physiol Cogn Med Sci* 62: 133-142, 1977.

[158] Warner AC and Stacy BD. Intraruminal and systemic responses to variations in intake of sodium and potassium by sheep. *Q J Exp Physiol Cogn Med Sci* 57: 89-102, 1972.

[159] Warner AC and Stacy BD. Solutes in the Rumen of the Sheep. *Q J Exp Physiol Cogn Med Sci* 50: 169-184, 1965.

[160] Warner AC and Stacy BD. Water, sodium and potassium movements across the rumen wall of sheep. *Q J Exp Physiol Cogn Med Sci* 57: 103-119, 1972.

[161] Warnock DG. Renal genetic disorders related to K^+ and Mg^{2+}. *Annu Rev Physiol* 64: 845-876, 2002.

[162] Wilkens M, Kunert-Keil C, and Schröder B. Expression of the epithelial Ca^{2+} channels TRPV5 and TRPV6 in ovine Ca^{2+} transporting tissues. *Acta Physiol* 186(Suppl): 232 (PW203A-215), 2006.

[163] Wolffram S, Frischknecht R, and Scharrer E. Influence of theophylline on the electrical potential difference and ion fluxes (Na, Cl, K) across the isolated rumen epithelium of sheep. *Zentralbl Veterinarmed A* 36: 755-762, 1989.

[164] Wylie MJ, Fontenot JP, and Greene LW. Absorption of magnesium and other macrominerals in sheep infused with potassium in different parts of the digestive tract. *J Anim Sci* 61: 1219-1229, 1985.

[165] Yellen G. The voltage-gated potassium channels and their relatives. *Nature* 419: 35-42, 2002.

In: Focus on Signal Transduction Research
Editor: Gavin McAlpine, pp. 67-102

ISBN 1-60021-376-6

Chapter 3

WW45 Expression is Upregulated Upon T Cell Activation and its Targeted Downregulation in Activated T Cells Decreases T Cell-Mediated Bone Resorption in Experimental Periodontal Disease in Rats

Paloma Valverde*

Department of General Dentistry, Division of Oral Biology. Tufts University School of Dental Medicine. One Kneeland Street. Boston, MA

Abstract

The protein Salvador (also named Shar-pei) has been shown to limit cell proliferation and stimulate apoptosis in *Drosophila melanogaster* [1, 2]. In vertebrates the function and expression pattern of the Salvador homolog WW45 [3] under different physiological and pathological situations have not been characterized in detail. WW45 transcripts have been previously shown to be ubiquitously distributed in a variety of tissues of mouse and human origin [3] and to be undetectable in several human cancer cell lines [1]. In this chapter changes in the expression of WW45 upon T cell activation and the putative role of this protein in T cell mediated-bone resorption were investigated. WW45 transcript was significantly increased in human naïve $CD3^+$ T lymphocytes when activated with anti-CD3, anti-CD28 or the combination of both antibodies at 24 h, but not after 72 h of stimulation. WW45 mRNA expression was also high in mononuclear gingival infiltrates of patients with periodontal disease, whose loss of alveolar bone has been associated with the high content of activated/memory T cells in gingival tissues [4, 5] and the elevated expression of receptor activator of NFkB ligand (RANKL) [5-9] that increases osteoclast differentiation and bone resorption [10-12]. To explore the putative involvement of WW45 in T cell-induced inflammatory bone resorption, a rat model for

* email: paloma.valverde@tufts.edu

experimental periodontal disease was used [13, 14], in which local gingival stimulation of Th1-type memory T cells with antigen presenting cells (APC) and specific antigen increases alveolar bone resorption [13,14]. When activation of these memory T cells with APC and their specific antigen was performed in the presence of antisense ($WW45^{ANT}$) or control scrambled oligonucleotides ($WW45^{SC}$) for WW45, the decreased WW45 expression in the former cells correlated with the decreased production of RANKL. Rats adoptively transferred with $WW45^{ANT}$-treated T cells exhibited and average of 90% lower experimental bone resorption and gingival RANKL expression than rats injected with $WW45^{SC}$-treated T cells. To summarize, WW45 expression appears to be transiently upregulated during T cell activation of human naïve T cells whereas exhibits a sustained upregulation upon antigen-specific stimulation of memory T cells. Furthermore, downregulation of WW45 in antigen-stimulated Th1-type memory T cells is associated with the reduction of RANKL expression and T cell-induced bone resorption *in vitro* and *in vivo*. Therefore, careful manipulation of WW45 expression in diseases characterized by disrupted T cell homeostasis and excessive RANKL function may lead to a novel therapeutic approach to inhibit T cell-induced bone loss.

Key Words: WW45/Salvador homologue 1 protein, T cell activation, Naïve, Memory, Proliferation, Differentiation, Apoptosis, RANKL, Bone Resorption, Periodontal Disease.

ABBREVIATIONS

AICD:	Activation-induced cell death;
AL:	Alveolar ledge;
APC:	Antigen presenting cells;
CRAC:	Ca^{2+} release-activated Ca^{2+};
CEJ:	Cemento-enamel junction;
CPM:	Counts per minute;
CTLA4:	Cytotoxic T-lymphocyte-associated antigen 4;
DC:	Dendritic cells;
ELISA:	Enzyme-linked immunosorbent assay;
EST:	Expressed sequence tag;
FBS:	Fetal bovine serum;
GAPDH:	Glyceraldehyde-3-phosphate dehydrogenase;
IL2:	Interleukin 2;
IP_3:	Inositol (1,4,5)-triphosphate;
LPS:	Lipopolysaccharide;
JNK:	c-Jun N-terminal kinase;
Lats:	Large tumor suppressor;
MAP:	Multiantigenic peptides;
M-CSF:	Macrophage-colony stimulating factor;
MHC:	Major histocompatibility complex;
MOCP-5:	Mouse osteoclast precursor cell line;
Mst:	Mammalian STE20-like kinase;
NFAT:	Nuclear factor of activated T cells;

PBMC:	Peripheral blood mononuclear cells;
PLC:	Phospholipase C;
PMA:	Phorbol 12-myristate 13-acetate;
PKC:	Protein kinase C;
RACE:	Rapid amplification of complementary DNA ends;
RT-PCR:	Reverse transcriptase-polymerase chain reaction;
RANK(L):	Receptor activator of NFkB (ligand);
TRACP^{+} MNC:	Tartrate resistant acid phosphatase positive multinucleated cells;
TCR:	T cell-receptor;
TNF:	Tumor necrosis factor;
WW45^{ANT}:	Antisense oligonucleotides for WW45;
WW45^{SC}:	Control scrambled oligonucleotides.

INTRODUCTION

Regulation of cell number and organ size during development by the Salvador/Warts/Hippo signaling complex in Drosophila melanogaster: Tight coupling of cell proliferation, differentiation and apoptosis are required to determine the optimum cell number and organ size during embryogenesis and to maintain homeostasis during adulthood in multicellular organisms. Among the different genes involved in this regulation, the *Drosophila melanogaste*r gene *salvador* (also named *shar-pei*) appears to play an important role during organ development by limiting cell proliferation and stimulating apoptosis [1, 2] Salvador contains different domains for putative interaction with other proteins including 2 WW domains and a SARAH domain. WW domains [15,16] are small globular domains of 38 to 40 amino acids that are characterized by the presence of two conserved tryptophans separated by 20 to 22 amino acids, aromatic residues in the middle of the two tryptophans and a proline residue at position +2 in relation to the second tryptophan. Several WW domain-binding motifs have been identified including, PPXY (where X signifies any amino acid) [17], PPLP [18], PGM [19] and phosphoserine or phosphothreonine-binding modules [20]. The WW domains present in Salvador have been reported to interact with the PPXY motifs of the serine/threonine protein kinase Warts [1]. In addition to WW domains, Salvador contains a protein-protein interaction motif named SARAH that spans approximately 50 amino acids and contains a carboxy-terminal portion with moderate to high coiled-coil propensity [21]. In fact, Salvador has been shown to interact with Hippo, an Ste20-family protein kinase, that also contains a SARAH domain [22-25]. In addition, Hippo was reported to phosphorylate both Warts and Salvador, and the phosphorylation of Warts by Hippo was enhanced by the presence of Salvador [23, 25]. These data suggested that Warts, Salvador and Hippo might form a ternary complex in which Salvador would function as an scaffold protein to bring Warts in close proximity to Hippo [22]. Importantly, mutations at either *salvador*, *warts* or *hippo* have been associated to overgrowth phenotypes in developing *Drosophila* tissues that resulted from increased proliferation and inhibition of apoptosis [1, 2, 22-28]. In addition, defective cell differentiation was also reported for some mutations of *warts* and *hippo* [23, 27].

Known functions of WW45, Lats and Mst, the vertebrate homologues of Salvador, Warts and Hippo respectively: The putative vertebrate homologues for Salvador, Warts and Hippo are known as WW45 [3] or Salvador homolog 1 protein [1], large tumor suppressor (Lats) 1 and 2 for Warts [29-32] and mammalian STE20-like kinase (Mst) 1 and 2 for Hippo [33, 34] respectively. WW45 was initially described as a WW domain-containing protein with a predicted molecular weight of 45 kDa [3]. WW45 transcripts were detected during early stages of mouse embryonic development and were ubiquitously distributed in adult tissues of mouse and human origin [3]. WW45 protein sequence was also shown to contain a high degree of similarity among different species including *Drosophila*, along the WW and the coiled-coil domains [3]. Human WW45 gene was mapped to a region of chromosome 14 normally affected by deletions and translocations in a variety of cancers [3], and indeed several cancer cells lines [1] but not primary human tumors [35] were later reported to contain mutations in this gene that abolished its transcription. Thus, very little is known about the function of WW45 in vertebrates or its expression distribution during physiological or pathological processes other than cancer. More information is available about the functions of Lats1/2 and Mst1/2 in vertebrates. Thus, human Lats1 and Lats2 kinases have been implicated in regulating cell cycle progression and apoptosis [30, 36, 37]. In addition, Lats1 exhibits properties of tumor suppressor gene in mouse and human. Thus, mice with a disrupted Lats1 gene have been shown to exhibit ovarian stromal cell tumors and an increased incidence of soft tissue sarcomas [38], whereas Lats1 expression was found either reduced or absent in a number of human soft tissue sarcomas [39]. On the other hand, the vertebrate homologues of Hippo Mst1 and Mst2 protein kinases, have been implicated in apoptosis both upstream and downstream of caspase activation [33, 34, 40-42]. Indeed, either Mst1 or Mst2 have been shown to be activated by several pro-apoptotic inducers including staurosporin, Fas ligand, the pharmacological drug for osteoporosis bisphosphonate or the anti-tumor drug cytotrienin A [33, 34, 40-42]. Upon caspase cleavage of cytoplasmic Mst1/Mst2, the truncated forms can be rapidly translocated into the nuclei [43] and induce apoptosis partly by phosphorylating histone H2B and facilitating DNA condensation [44]. However, caspases in general and caspase-3 in particular are capable of regulating not only apoptosis but also nonapoptotic functions such as inducing nuclear extrusion in differentiating lens epithelium [45] or keratinocytes [46], inducing differentiation of megakaryocytes [47], skeletal myocytes [48] or lens epithelial cells [49] and inducing proliferation during early stages of T cell activation [50]. Therefore activation of Mst kinase activity by caspases may be expected to induce apoptosis, differentiation or proliferation depending on the cell type. Indeed, Mst1, either intact or cleaved by caspases, has been implicated in non-apoptotic functions such as in the induction of megakaryocyte differentiation [47] and skeletal myocyte differentiation [48]. Therefore, interaction between WW45, Lats and Mst is likely to be different depending on the cell type and the physiological or pathological context in which these proteins are being expressed.

Activation of Salvador/Warts/Hippo and WW45/Lats/Mst signaling complexes may occur differently in Drosophila melanogaster and vertebrates: Despite the high sequence similarity between the components of the vertebrate and *Drosophila* signaling complexes described in the introduction of this chapter, recent reports suggest that their activation may occur by different mechanisms. Indeed, experiments in which WW45, Lats and Mst were overexpressed in human cells were not able to confirm that Lats1, Mst2 and WW45 could form stable ternary complexes [51], or that WW45 could interact with Lats1, although Mst2 was able to bind to either Lats1 or WW45 [51]. On the other hand, whereas inactivation of *hippo* has been associated with overgrowth phenotypes due to decreased apoptosis in *Drosophila* [22-25, 28], the cardiac-specific overexpression of the Hippo homologue Mst1 has been shown to increase apoptosis of terminally differentiated cardiac myocytes as well as the size of adult heart [52]. Furthermore, activation of Mst1 by bisphosphonates increases osteoclast apoptosis [40] that would be expected to increase bone mass by inhibiting bone resorption. These results suggest that *Drosophila* Salvador/Warts/Hippo and vertebrate WW45/Lats/Mst signaling complexes may be subjected to different regulatory mechanisms depending on the context in which the components of the pathway are expressed. If that is the case, activation of this signaling module is expected to regulate cell number and organ size by different mechanisms in developing and adult tissues and during physiological and pathological processes. From the studies performed in the *Drosophila* model, it is tempting to speculate that changes in the expression of WW45, that seems to be required for the activation of the ternary signaling module, would occur differently during embryogenesis and adulthood, and under different physiological and pathological situations in adult cells of vertebrates. Therefore, the expression of WW45 and its adaptor function would be expected to be different during normal and pathological T cell-mediated immunoregulation that involve different coupling of the processes of proliferation and apoptosis during T cell activation, as will be described in more detail in the following section.

T cell activation and regulation of activated/memory T cell number: Naïve peripheral T cells circulate through the blood and peripheral lymphoid organs in a proliferatively and transcriptionally quiescent state until they encounter dendritic cells (DC) that act as antigen-presenting cell (APC) by bearing a processed antigenic peptide bound to major histocompatibility complex (MHC) molecules [53]. Interaction of T cells with APC leads to the formation of synapses at the T cell/APC contact site. Synapse formation is associated with rapid clustering of T cells receptor (TCR) molecules binding to peptide/MHC complexes on APC and activation of phospholipase C (PLC), which leads to the activation of protein kinase C (PKC) and the generation of inositol (1,4,5)-triphosphate (IP_3) [54]. IP_3 then induces the release of Ca^{2+} from intracellular stores, which in turn, leads to the opening of Ca^{2+} release-activated Ca^{2+} (CRAC) channels and results in a sustained increase in cytosolic Ca^{2+} [54, 55]. Simultaneous activation of CD28 by the costimulatory molecule CD80 or CD86 further amplifies the resulting Ca^{2+} signaling. The sustained increase in the cytosolic Ca^{2+} concentration activates the phosphatase calcineurin, which then dephosphorylates the cytoplasmic transcription factor NFAT (nuclear factor of activated T cells), enabling it to translocate into the nucleus to bind to NFAT-response elements of several genes [56]. In a separate pathway, activation of PKC leads to activation of the Ras and JNK (c-Jun N-terminal kinase) pathways [57], which results in the assembly of Fos/Jun (AP1) transcription factors. Coordinated activity of Ca^{2+}- and PKC-dependent signaling pathways culminates in the cooperative interaction of NFAT and AP1 that is essential for a productive immune response, by stimulating the expression of a growing number of cytokines, chemokines and cell surface receptors including IL2, and by inducing T cell proliferation [58, 59]. In the absence of AP1 cooperative binding, activation of NFAT (by elevation of intracellular calcium levels with ionomycin or anti-CD3 antibody treatment) leads to the expression of anergy associated genes [60] involved in T cell hyporesponsiveness to antigen stimulation, that constitutes one of the mechanism imposing peripheral tolerance.

As the foreign antigen is eliminated and the productive immune response winds down, the majority of activated T cells are eliminated by an apoptotic process termed activation-induced cell death (AICD) that involves death factors such as tumor necrosis factor (TNF) and Fas ligand [61-63]. A small percentage of antigen-specific T cells escapes cell death and enters a pool of long-lived memory T cells with an apoptosis resistant phenotype, that can be distinguished from naïve T cells by their different expression of cell-surface molecules [64]. Such memory cells play important roles in generating secondary immune responses upon subsequent reexposure to antigen [64]. Although it is not known whether the signaling complex WW45/Lats/Mst plays a role in controlling the differential outcome of a proliferative response or AICD by activated T cells, several reports have proven that the cytokine IL2 is a key player in coupling these two events. Indeed, IL2 has been shown to be mandatory for the initial clonal expansion after TCR stimulation, and once the cells are proliferating, IL2 also primes mature T cells for AICD [65-67] by driving T cells into the S phase of the cell cycle, where they are sensitive to TCR-triggered AICD [68]. It was reported that IL2 enhances AICD by upregulation of Fas ligand and downregulation of inhibitors of the Fas ligand signaling pathway [69]. In support to this dual role of IL2, mice deficient in IL2 or its receptor exhibit an abnormal accumulation of activated T cells and autoimmunity [70, 71]. Therefore, uncoupling of AICD and proliferation of activated T cells by alterations in the IL2-mediated signaling, the calcineurin-NFAT-mediated cellular reactions and hypothetically by an aberrant WW45/Mst/Lats signaling, could lead to pathological processes of the immune system characterized by an abnormal number of activated/memory T cells. On one hand, reduced rates of cell death are often associated with cancer, autoimmunity and inflammatory diseases, whereas too much cell death may result in immunodeficiency and leukopenia [62, 63]. Interestingly several of the pathologies associated with an excessive number of activated/memory T cells are also characterized by a disrupted bone homeostasis and excessive bone loss as will be described in more detail in the following section.

Involvement of activated/memory T cells in regulating bone homeostasis: Maintenance of adult bone homeostasis is possible due to the process of bone remodeling that involves the tight coupling of bone resorption by osteoclasts [72] and bone formation by osteoblasts [73]. The key molecular regulators of this process include receptor activator of (NF)-kB ligand (RANKL) [6,7, 10-12], its cellular receptor, receptor activator of NF-kB (RANK), and the decoy receptor, osteoprotegerin (OPG) [74]. RANKL (also known as TRANCE or tumor necrosis factor-related activation-induced cytokine) is a member of the tumor necrosis factor (TNF) family of ligands that is expressed in osteoblasts and stromal cells under physiological conditions [6,7, 10-12]. In the presence of macrophage colony-stimulating factor (M-CSF), RANKL is both necessary and sufficient for the differentiation of osteoclast precursors into mature activated osteoclasts that can resorb bone [8-10]. RANKL can also induce or inhibit osteoclast apoptosis in osteoclast precursor cells or mature osteoclasts respectively [8, 9, 75, 76], and inhibits apoptosis of dendritic cells (DC) that indirectly increases DC-mediated T-cell proliferation [7]. Effects of RANKL are counteracted by OPG that is secreted by various tissues and prevents the binding of RANKL to its receptor RANK on osteoclasts and dendritic cells [77, 78]. Naïve T cells do not express significant levels of RANKL and hence do not alter bone homeostasis [77, 78]. However, activated/memory T cells and certain cancer cell types that metastasize to bone express high levels of this cytokine and can therefore disrupt bone homeostasis by increasing osteoclast differentiation and bone resorption [77-80]. For example, the alveolar bone destruction observed in periodontal infections and in animal models of periodontal disease, has been shown to be partly mediated by microorganism-triggered induction of RANKL expression on T cells and the subsequent activation of osteoclasts [4, 5, 8, 9, 81, 82]. In addition, the inflammation and cartilage destruction observed in patients with arthritis and animal models of autoimmune arthritis, have been correlated with the presence of T cells in synovial fluid that express high levels of RANKL [77, 78, 83]. The contribution of RANKL to the pathological bone resorption observed in certain autoimmune diseases, such as diabetes mellitus and lupus erythematosus, in metastatic breast and lung cancers or even in allergic diseases is also beginning to be elucidated [77, 78]. Exploration of signaling pathways coordinating proliferation, differentiation and apoptosis during T cell activation, may help develop novel therapeutic approaches to control aberrant T cell activation and T cell-induced bone loss. In this chapter, the pattern of WW45 expression upon T cell activation is described for the first time in either human naïve T cells or a rat Th1 memory cell clone that can induce bone resorption *in vitro* and *in vivo* upon antigen-specific stimulation. Further experimental evidence is shown to support a correlation between downregulation of WW45 expression in this antigen-stimulated rat memory T cell clone, downregulation of RANKL expression and decreased severity of the associated T cell-mediated bone resorption *in vitro* and *in vivo*.

Materials and Methods

Isolation of normal human peripheral blood T lymphocytes and mononuclear infiltrates from healthy patients and individuals with chronic periodontitis: Peripheral blood mononuclear cells (PBMC) were collected from healthy individuals under informed consent agreement (n=5, 28-45 year-old, 4 males and 1 female). The mononuclear cell fraction was separated from blood by gradient centrifugation using Histopaque-1077 (Sigma) and incubated in RPMI-1640 medium supplemented with 10% FBS, 2-mercapto-ethanol, L-glutamine, penicillin and streptomycin. $CD3^+$ enriched-T cell fraction was obtained by glass-wool and nylon-wool column purification as described by others [84] followed by selection with magnetic beads coated with anti-CD3 Mab (OKT3, ATCC, Rockville, MD; UCTH1, R&D systems). Mononuclear cells isolated from gingival tissues of healthy individuals and from those diagnosed with chronic periodontitis were kindly provided by Drs. Taubman and Kawai (The Forsyth Institute, Boston, MA). Isolation of mononuclear cells was performed as previously described [85].

Activation of naïve human T cells: The $CD3^+$ enriched-T cell fraction from peripheral blood of healthy individuals was stimulated with immobilized antibodies to CD3 (OKT3; 5 μg/ml), CD28 (CD28.2; 2 μg/ml) or a mix of both antibodies on 96-well plates (5 x 10^4 cells per well) or 24-well plates (5 x 10^5 cells per well) for the times indicated in the figures or figure legends. Alternatively, the $CD3^+$ enriched-T cell fraction was stimulated with 20 mM of phorbol 12-myristate 13-acetate (PMA) and 1 μM ionomycin in the presence and absence of 100 ng/ml cyclosporine.

Preparation of specific antigen and activation of rat memory Th1-clone cells: Actinobacillus actinomycetemcomitans, American Type Culture Collection no. 43718 (strain Y4; Manassas, VA) was grown in pleuro-pneumonia-like organism broth (Difco Laboratories, Detroit, MI) with glucose (3 g/L) and sodium bicarbonate (1 g/L) for 72 h at 37°C under increased CO_2 (candle jar). The cultured bacteria at log-growth phase were killed with formalin (5%), washed 5 times in PBS and resuspended in RPMI to serve as T cell antigen. The *Actinobacillus actinomycetemcomitans* Omp29 and *Actinobacillus actinomycetem comitans* lipopolysaccharide (LPS) were prepared as previously described [86, 87]. Purified Omp29 protein was kindly provided by Dr. Hitoshi Komatsuzawa (Hiroshima University, Japan). G23 Th1-type clone cells specific for *Actinobacillus actinomycetemcomitans* Omp29 [88] were stimulated with sterile formalin-killed *Actinobacillus actinomycetemcomitans* and irradiated (3,300 rad) syngeneic rat spleen cells (APC) for 1 to 3 days on 24-well plates. In some experiments, G23 Th1-type clone cells were treated with PBS, kaliotoxin (50 nM) or CTLA4Ig (20 μg/ml) for 15 min prior to the addition of the APC and the specific antigen. Kaliotoxin is a blocker of the voltage-gated potassium channel Kv1.3 [89-92]. In the first few hours after mitogenic stimulation, specifically when Ca^{2+} elevation is necessary [54], blockers for Kv1.3 have been demonstrated to inhibit Ca^{2+} influx and T cell proliferation [89-92] as well as the production of a variety of cytokines, including IL2, TNFα, IFN-γ, RANKL and OPG [8, 14, 93, 94] in human or rat memory T cell clones. Kaliotoxin has also been shown to decrease bone resorption induced by activated G23 Th1-type clone cells in the rat model of experimental periodontal disease used in this chapter [8, 13, 14]. CTLA4Ig is a human fusion protein consisting of the binding domain of human cytotoxic T-lymphocyte-associated antigen 4 (CTLA4) and the Fc fragment of human IgG1 that inhibits antigen-specific T cell response by blocking B7/CD28 costimulatory signaling [95]. Preventing this binding blocks the engagement of CD28 that is important in the progression of autoimmune and inflammatory diseases such as rheumatoid arthritis. Either local of systemic administration of CTLA4Ig, can also abrogate the bone resorption induced by activated G23 Th1 clone cells in experimental periodontal disease in rats [13].

Targeted downregulation of WW45 by antisense treatments in human naïve T cells and rat memory Th1 clone cells: WW45 antisense ($WW45^{ANT}$) or scrambled control ($WW45^{SC}$) oligonucleotides were synthesized by Integrated DNA Technologies (Coralville, IA) with a phosphorothioate backbone to improve resistance to endonucleases as described by others [96] and then purified by high performance liquid chromatography. Prior to their use in transfection experiments with Oligofectamine™ reagent (Invitrogen), optimization experiments were performed with an unrelated control phosphorothioate oligonucleotide (5'-TCCGAAGGCAGAAAGGCTTCA-3') described by others [97] at a concentration range of 0.1 to 3 μM delivered with 0.6 μl/well or 2 μl/well of Oligofectamine™ reagent for experiments performed on 96- or 24-well plates respectively (data not shown). Concentrations of the control oligonucleotide higher than 1.5 μM led to some toxicity in experiments performed with human naïve T cells (data not shown). Therefore, all experiments described in this chapter were performed with concentrations of oligonucleotides equal or lower than 1 μM.

The sequences of phosphorothioate oligonucleotides used in experiments with human cells were as follows: 5'-TCGGGACAGCATCCTTCTAG-3' and 5'-CTGGAGACCGTA CCTTTCGA-3' for $WW45^{ANT}$ and $WW45^{SC}$ respectively. The sequences used in experiments with rat cells were as follows: 5'-TTGCGGGACAGCATCCTCGCG-3' and 5'-GATGCGGGACAGCTCGGCTCGTCG-3' for $WW45^{ANT}$ and $WW45^{SC}$ respectively. $WW45^{ANT}$ or $WW45^{SC}$ were delivered into the cells with Oligofectamine™ reagent (Invitrogen). At the time of transfection, Oligofectamine™ reagent (0.6 μl/well or 2 μl/well for experiments performed on 96 or 24-well plates) and oligonucleotide solutions (0.25-1 μM) were diluted in serum-containing media without antibiotics, incubated for 10 min at room temperature in separate and then mixed gently and incubated at room temperature for an additional 30 min period. In the experiments with human naïve T cells, oligofectamine-oligonucleotide complexes were then added slowly to cell suspensions and incubated for 2 h at 37 °C prior to their stimulation on plate-bound antibodies [CD3 (OKT3; 5 μg/ml), CD28 (CD28.2; 2 μg/ml) or a mix of both antibodies] for up to 72 h. In the experiments with rat T clone cells, oligofectamine-oligonucleotide complexes were added slowly to cell suspensions plated on either 96-well or 24-well plates and incubated for 2 h at 37 °C. Then, cell suspensions were stimulated in the presence of APC and specific antigen for a maximum of 72 h.

Gingival challenge and adoptive transfer of Th1-type clone cells of Rowett rats: Two to three-month old female Rowett rats were maintained under pathogen-free conditions in laminar flow cabinets as previously described [13, 14]. Experiments using these animals were performed under the approval of the Forsyth Institute's Internal Animal Care and Use Committee (Boston, MA).

Rowett strain female rats received three palatal gingival injections (1 μl/site) on the mesial of the first molar and in the papillae between first and second and third molars on the right and left sides of the maxilla. Gingival injections consisted of 500 μg/ml *Actinobacillus actinomycetemcomitans* Omp29 plus 500 μg/ml *Actinobacillus actinomycetemcomitans* Omp29 LPS (left, experimental side) and PBS as a control (right, control side) on day 0 as previously described [13, 14]. G23 T cells were stimulated with sterile formalin-killed *Actinobacillus actinomycetemcomitans* and irradiated (3,300 rad) syngeneic rat spleen cells (APC) for 3 days on 24-well plates in the presence of 0.25 μM of $WW45^{ANT}$, $WW45^{SC}$ or PBS. Cells were then isolated by gradient centrifugation and transferred intravenously on day 0 (4×10^6 cells/animal) as previously described [13, 14]. Animals were sacrificed at different time points depending on the experimental procedure to be followed, either on day 0 (defined as immediately after the gingival challenge and adoptive transfer) or on days 2 or 10 (defined as two or ten days after the gingival challenge and adoptive transfer).

Isolation of T lymphocytes from rat gingival tissue: In the experimental periodontal disease model used herein, mononuclear cells isolated from gingival tissue of the experimental side demonstrated the highest Omp29-specific T cell proliferation on the 2nd and 3rd days after the T cell adoptive transfer [13, 14]. However, bone resorption could only begin to be detected after 5 days [13] with the maximum bone resorption observed 10 days after the T cell transfer [13]. Therefore, gingival tissues were collected on days 0 and 2 for RT-PCR analysis. $CD3^+$ T cells infiltrating into the gingival tissue were isolated as previously described [14]. To obtain an average of 10^5 $CD3^+$ T cells for RT-PCR analyses, gingival samples were pooled from the experimental sides of 2 rats on day 2. Recovery of $CD3^+$ T cells was negligible from the experimental side on day 0 (n = 2 gingival samples, data not shown) or the control side on days 0 or 2 (n = 2 gingival samples, data not shown). Gingival tissue was cut into small pieces and digested in RPMI/10% FBS containing 2 mg/ml collagenase (Worthington Biochemical Corporation, Freehold, NJ) for 1 h at 37°C. Cell debris was excluded by passing cells through a steel mesh. Mononuclear cells were isolated by density gradient centrifugation using Histopaque (density 1.077 ± 0.001; Sigma Diagnostics, Inc, MO). The lymphocyte-containing interface layer was collected and washed twice with RPMI/10% FCS. $CD3^+$ T cells were then enriched by positive selection with mouse monoclonal anti-rat αβ TCR antibody (R73) followed by Dynabeads (Dynal Biotech, NY) coated with anti-mouse IgG as described [13,14]. The αβ TCR antibody R73 was kindly provided by Dr. Thomas Hünig (University of Wurzberg, Germany).

Reverse transcriptase polymerase-chain reaction (RT-PCR) analysis: Total RNA was extracted by using an RNeasy Mini Kit (Qiagen, Valencia, CA) from approximately 10^5 T cells when isolated from tissue or peripheral blood and 10^6 cultured G23 T clone cells essentially as previously described [14]. Freshly isolated RNA was reverse transcribed with a SuperScript™ first-strand synthesis system (Invitrogen, Carlsbad, CA) following the manufacturer's recommendations to increased sensitivity. The resulting cDNA was then amplified by PCR with HotStartTaq DNA polymerase (Qiagen, Valencia, CA) as recommended by the manufacturer. The sequences of the primers used in the amplification of human RANKL were: 5'-TTCAGAGCGCAGATGGATCCTAATAGAATA-3' and 5'-CTTCCTCCTTTCATCAGGGTATGAGAAC-3' and for rat RANKL were: 5'-CCAA CCGAGACTACGGCAAGTACCT-3' and 5'-TCAGTCTATGTCTTGAACTTTGAAAG CC-3'. The sequences of the primers used in the amplification of human glyceraldehyde-3-phosphate dehydrogenase (GAPDH) were: 5'-GCTCAGACACCATGGGGAAGGTGAA GG-3' and 5'-GGTGGTCCAGGGGTCTTACTCCTTGGA-3' and for rat GAPDH were: 5'-CATAGACAAGATGGTGAAGGTCGGTGTC-3' and 5'-CTTACTCCTTGGAGGCCA TGTAGGCCA-3'. Human WW45 was amplified with the primers previously described [3]. Rat WW45 was amplified with the primers: 5'-CGACGTCCGCGAGGATGCTG TCCCGCA-3' and 5'-CCTGGCTTGTGGAAAGTTAACTTAA-3'. All primers used in this study were purchased from Integrated DNA Technologies (Coralville, IA) at the 25 nM scale. Linearity of the PCR conditions (in terms of cDNA amount and number of cycles) was determined for each primer pair in preliminary experiments (data not shown). PCR amplifications were performed with 200 pg of RNA derived from human T cells or cultured G23 T clone cells and an estimated 150 pg of RNA for samples derived from gingival tissues (either total tissue, $CD3^+$ T cells or mononuclear cells). Total RNA was isolated from whole gingival tissue of the experimental side (injected with LPS and antigen) on day 0 and from the control side (injected with PBS) on days 0 and 2, since recovery of $CD3^+$ T cells was negligible. Images of the amplified products in 1% agarose gels were processed with Molecular Analyst Software (BioRad, Hercules, CA) and analyzed by Adobe Photoshop 6.0 (Adobe Systems Incorporated, San Jose, CA) and Scion Image Beta 4.02 software (Scion Image, Frederick, Maryland).

Isolation of rat WW45 cDNA: To obtain the full-length cDNA for rat WW45, RT-PCR was used to amplify the open reading frame using as starting material adult heart first stranded cDNA. Primer sequences, based on available information of overlapping rat expressed sequence tags (EST) highly homologues to the open reading frame mouse WW45 cDNA [3], were 5'-CGACGTCCGCGAGGATGCTGTCCCGCA-3' and 5'-CCTGGCTTGTGGAAAGTTAACTTAA-3'. The resulting PCR product was subcloned into a TOPO-TA cloning vector (Invitrogene) and sequenced on both strands with vector specific primers. A Marathon™ cDNA amplification kit (Clontech) was then used to prepare an adult rat heart RACE library following the manufacturer's recommendations and perform 5' and 3' RACE to confirm the sequences on the untranslated regions. RACE reactions were performed with Advantage 2 Polymerase Mix (Clontech) using specific primers (see below) and the adaptor primers AP1 and AP2 (Clontech) as recommended by the manufacturer. The sequence of the specific primers for the 5' and 3' rapid amplification of complementary DNA ends (RACE) reactions were 5'-CTCCAAGTGGCCTCCTTCTCTGA-3' and 5'-TGGTATGCCCAGCACCATGGCAAG-3' respectively. The resulting 5' and 3' RACE reactions were subcloned into a TOPO-TA-cloning vector (Invitrogene) and the longest clones were sequenced. Sequencing reactions were performed at the University of Maine DNA Sequencing Facility by using the ABI model 373A Stretch DNA Sequencer with the XL Upgrade.

Anti-WW45 antibody preparation: Polyclonal antibodies to WW45 were obtained by immunizing BALB/c mice with the multiantigenic peptides (MAP, n=8) DRAQEDYRYYEYNHD or RQALLTELENRKQRQ (synthesized by Research Genetics) corresponding to the amino acids 151 to 165 and 359 to 373 respectively of murine WW45 [3]. Mice immunizations with the WW45 peptides were performed with ImmunEasy™ mouse Adjuvant (Qiagen). A mix of both antibodies (1:1) was used in subsequent experiments.

Western blot analysis: Total cell lysates from cultured Omp29-specific T cell clones (1-$2x10^6$) were prepared with the Mammalian Cell-PE LB™ buffer system supplemented with the Protease-Arrest™ protease inhibitor cocktail from GenoTechnology Inc. (St Louis, MO). Protein extracts were resolved on 4-12% Novagen SDS-PAGE gels (Invitrogen, Carlsbad, CA) and transferred to Invitrolon™ PVDF membranes (Invitrogen, Carlsbad, CA). Western blots were performed and analyzed essentially as described with minor modifications [14, 98-100]. Briefly, membranes were incubated in blocking solution (2.5% non-fat milk in PBS) for 1h at room temperature. Primary and secondary antibodies required in the detection of RANKL and GAPDH were purchased from Santa Cruz Biotechnology (Santa Cruz, CA). Primary antibodies (1:400 dilution) were incubated in blocking solution containing 0.02% sodium azide at 4 ^{o}C overnight. Pretreatment of the PVDF membranes with 1X strong stripping solution from Chemicon for 30 min at room temperature was performed before adding the anti-WW45 antibody mix, because it was found to improve the immunoreactivity of the antibody mix (data not shown) for endogenous rat and human WW45. Secondary horseradish peroxidase linked anti-IgG antibodies were incubated in 1% non-fat milk in PBS for 1h at room temperature. Blots were visualized using an enhanced chemiluminescence kit (Pierce, Rockford, IL) as recommended by the manufacturer. Band intensities on the X-rays (Pierce, Rockford, IL) were analyzed by Adobe Photoshop 6.0 (Adobe Systems Incorporated, San Jose, CA) and Scion Image Beta 4.02 software (Scion Image, Frederick, Maryland) essentially as previously described [99, 100].

Assessment of tartrate-resistant acid phosphatase (TRACP)-positive multinucleated cells ($TRACP^+$ MNC) and bone resorption pit formation in coculture experiments of fixed activated T cells and osteoclast-precursor cells: Activated G23 T cells were fixed with a cold solution of 50% acetone and 35% ethanol for 5 min. Fixed T cells (10^5 cells/well) were then co-cultured with a mouse osteoclast precursor cell line MOCP-5 [101] (5 x 10^2 cells/well) in α-MEM supplemented with 10 ng/ml macrophage-colony stimulating factor (M-CSF) and 15% FBS on 96 well-plates for 7 days. The MOCP-5 osteoclast precursor cell line was kindly provided by Dr. Yi-Ping Li (The Forsyth Institute, Boston, MA). Differentiated osteoclasts were then identified as red-stained cells with 3 or more nuclei by cytochemical staining for TRACP as described previously [14]. To evaluate bone resorption pit formation, co-cultures of fixed T cells and MOCP-5 were incubated on BD BioCoat Osteologic discs (BD Biosciences, Ontario, Canada) for 7 days. Areas of resorption lacunae were determined essentially as previously described [14, 100]. Briefly, cells were removed with bleach as recommended by the manufacturer and digital images of the osteologic discs were taken under dark field microscopy with a Nikon Eclipse E600 microscope and Spot Advanced software. Resorbed areas on the digital images were measured with the Scion Image Beta 4.02 software (Scion Corporation, Frederick, Maryland) and expressed as % resorbed area=(Resorbed area per disc/total disc area) x 100.

Proliferation assays of rat splenic T cells and human naïve T cells: For the *ex vivo* proliferation assays of rat splenic T cells, T cells (>95% $CD3^+$ T cells) were recovered from rat spleens 10 days after adoptive transfer as described [13,14] and cultured in 96-well plates (5×10^3 cells/well) in the presence of irradiated normal splenic APC (5×10^5 cells/well) with or without killed *A. actinomycetemcomitans* bacteria (10^7/well) as antigen. [^{3}H] Thymidine (0.5 μCi/well) was applied for the last 16 h of a total of 3 days in culture. Samples were harvested onto glass fiber filters, and radioactivity (counts per minute, cpm) was measured in a β-scintillation Spectrometer (Beckman Coulter, CA). Antigen-specific proliferation of the splenic T cells was calculated and expressed as the antigen-specific stimulation index = (cpm of cells cultured in the presence of antigen / cpm of cells cultured in the absence of antigen). Proliferation assays for human naïve T cells were performed essentially as above. Changes in proliferation were expressed as stimulation index = (cpm of cells cultured in the presence of activating antibodies / cpm of cells cultured in the absence of activating antibodies).

Enzyme-Linked Immunosorbent Assay (ELISA) determinations: ELISA using a DuoSet ELISA development System to detect human IL2 from R&D systems (Minneapolis, MN) was used to measure the levels of IL2 in supernatants of human activated naïve T cell cultures, as recommended by the manufacturer. ELISA using horseradish peroxidase conjugated anti-rat IgG2a (sheep, Binding site, San Diego, CA) and o-phenylenendiamine (Sigma) as substrate was employed to detect IgG2a antibody levels to Omp29 in rat serum at day 10 as previously described [13,14]. Hyperimmune serum from rats immunized *sc* with Omp29 was used as a reference standard. The OD at 490 nm of this serum diluted at 1:8500 was chosen as 100 ELISA units, and all rat serum IgG2a reactions to Omp29 were evaluated based on a reference curve provided by dilution of this hyperimmune serum.

Measurement of alveolar bone resorption in vivo: Ten days after adoptive transfer, rats were sacrificed and the jaws were defleshed. Periodontal bone resorption was measured on the palatal surface of the maxillary molars essentially as previously described [13,14]. The distances from the cemento-enamel junction (CEJ) to the alveolar ledge (AL) of experimental injected sites (upper left palatal side injected with LPS and Omp29) and saline injected control sites (upper right palatal side) were measured using a reticule eyepiece at 25x magnification. A total of 5 measurements were evaluated in each quadrant, including 2 roots of the first molar, both roots of the second molar and one root of the third molar. The evaluation of bone resorption was calculated and expressed as % bone resorption = [(total CEJ-AL distance of 5 points on the left experimental side) – (total CEJ-AL distance of 5 points on the right control side)] / (total CEJ-AL distance on the 5 points of right control side) x 100.

Statistical analysis: Image analyses results obtained from RT-PCR or western blot experiments are shown as the mean of two independent experiments at the bottom of the picture of a representative experiment or in histograms as the mean ± SE of 3 or more experiments. Other results are shown as the mean ± SE of 3 or more experiments. One-way

ANOVA was used to test significance using the software package Origin 6.1 (OriginLab, Northampton, MA). Values of p lower than 0.05 were considered statistically significant.

RESULTS

WW45 expression is transiently upregulated upon T cell activation of naïve T cells: Human and mouse WW45 have been shown to be ubiquitously distributed in a variety of adult tissues by RT-PCR and northern blot analyses [3]. However, the pattern of expression of WW45 upon TCR stimulation in the presence and absence of co-stimulation has never been reported. Full activation of $CD3^+$ T cells requires engagement of both the TCR-CD3 complex and a costimulatory receptor, such as CD28 [102, 103]. In some situations, activations of the TCR-CD3 complex alone in the absence of CD28 co-stimulation can result in T cell anergy [104]. To determine whether the expression of WW45 was subjected to any change upon conditions that would lead to full T cell activation or T cell anergy, human $CD3^+$ T cells were isolated from peripheral blood of healthy individuals. Then, $CD3^+$ T cells were activated with plate-bound CD3, CD28 or both antibodies for different times *in vitro*, because *in vitro* co-stimulation by antibodies against CD3 and CD28 can mimic the normal antigen presenting cell stimulation and lead to full activation of $CD3^+$ T cells [105], whereas treatment with CD3 antibody alone increases intracellular calcium concentration and leads to the induction of the NFAT-induced anergy program [60]. Then, expression pattern of WW45 mRNA was analyzed by semiquantitative RT-PCR. Amplification of the mRNA for the remodeling cytokine RANKL, was also performed for comparison purposes, since its expression pattern upon T cell activation has been widely described in the literature. Thus, RANKL has been previously reported to be upregulated upon T cell activation of human naïve T cells up to 96 h of stimulation [77-80]. As shown in Figure 1A to 1C, mRNA expression of both WW45 and RANKL was found to be significantly upregulated upon 24 h of treatment with either one of the antibody treatments. Importantly, either WW45 or RANKL induction required both CD3 stimulation and CD28 costimulation for maximal expression of mRNA at 24 h. However the pattern of expression of WW45 was more transient than that of RANKL and was only observed within the first 24 h of activation (Figure 1A to 1C and data not shown). Indeed, levels of expression of RANKL remained higher in activated T cells than in non-activated cells at 72 h, whereas levels of expression of WW45 were not significantly different before and after stimulation with any of the antibody treatments at such time point (Figure 1A to 1C). Similar experiments performed with mouse peripheral blood T cells, indicated that WW45 upregulation upon TCR stimulation with or without co-stimulation, was only detectable up to 24 h (data not shown). This transient pattern of upregulation of WW45 suggests that this protein may play a role during the early stages of the immune response of naïve T cells.

Additional experiments were performed to determine the involvement of the calcium signaling and PKC activation in the upregulation of WW45, by treating human naïve T cells with the calcium ionophore ionomycin and the activator of PKC phorbol 12-myristate 13-acetate (PMA) respectively. These treatments resulted in comparable levels of upregulation of WW45 than those obtained with CD3 and CD28 antibodies (Figure 1D). Similarly, the upregulation RANKL was also found to be calcium- and PKC-dependent (Figure 1D) as described by others [6, 7, 77-80]. Importantly, the potent immunosuppressant cyclosporine, known to inhibit the calcium/calmodulin dependent phosphatase calcineurin and NFAT-dependent transcriptional events such as IL-2 production and T cell proliferation [106], significantly decreased the upregulation of either WW45 or RANKL expression (Figure 1D). These results indicated that, as previously shown for RANKL and other factors of the TNF family of ligands [6, 77-80], the TCR stimulation of WW45 expression was calcineurin-dependent and required increased cytoplasmic calcium levels and activation of PKC for its maximal level of upregulation.

Downregulation of WW45 expression decreases IL-2 production and the number of activated T cells: Since WW45 appeared to be transiently upregulated upon activation of the TCR-CD3 with or without co-stimulation with CD28, the next set of experiments was performed to determine whether its downregulation could have any effect in controlling the total number of activated T cells. To that end, $CD3^+$ T cells isolated from human peripheral blood, were transfected with $WW45^{ANT}$ to downregulate WW45 expression and then stimulated with plate-bound CD3 or the combination of CD3 and CD28 antibodies. Proliferation assays were then performed 72 h later to assess the effect of downregulating WW45 in the total number of activated T cells. As shown in Figure 2, treatment with $WW45^{ANT}$ decreased the stimulation index of cells stimulated with CD3 or the combination of CD3 and CD28 antibodies, and therefore the total number of activated T cells resulting from either treatment was decreased upon WW45 downregulation. In contrast, when similar experiments were performed in cells that were transfected with $WW45^{SC}$, the stimulation index of the activated T cells was similar than in untransfected cells (Figure 2A). Since IL2 is a cytokine that is upregulated during early stages of T cell activation (like WW45), and is involved in controlling the decision between a proliferative response and a death response by activated T cells [58, 59, 65-67], levels of IL2 were determined by ELISA in supernatants of activated T cells treated with $WW45^{ANT}$ or $WW45^{SC}$. As shown in Figure 2B, levels of IL2 at 24 h or even 16 h (data not shown) were significantly decreased in T cells activated by CD3 antibodies in the presence of $WW45^{ANT}$ but not when activated in the presence of $WW45^{SC}$. These results together with those shown in the previous section, suggested that WW45 upregulation during early stages of the immune response, may be involved in controlling the number of activated T cells by a mechanism involving Ca^{2+}/calcineurin/NFAT/AP1-dependent signaling, that controls IL2 levels as well as those of Fas ligand.

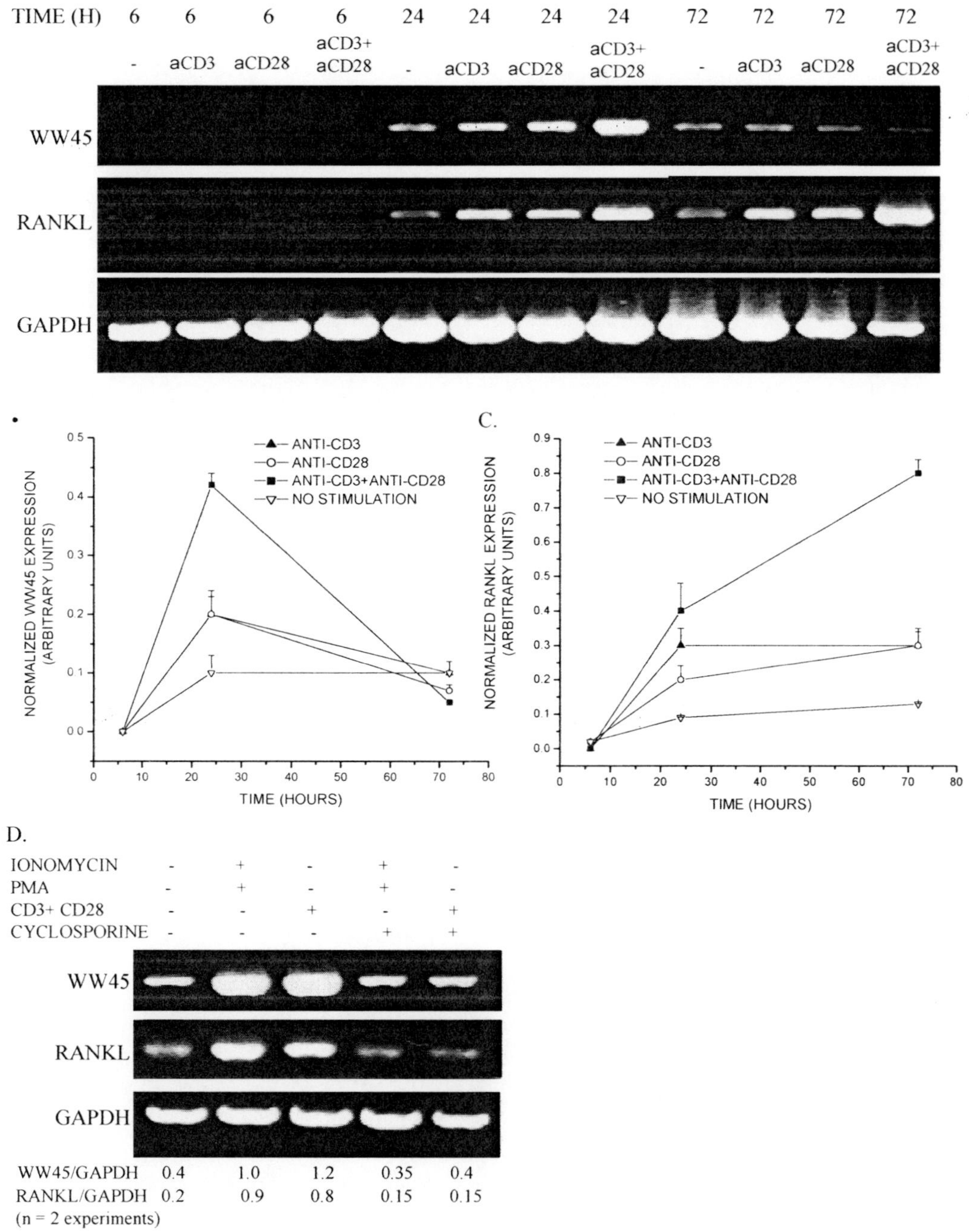

Figure 1. WW45 expression increases transiently upon T cell activation of human naïve T cells *in vitro*. (A) Representative example of WW45 expression pattern based upon semiquantitative RT-PCR analysis. Normal human $CD3^+$ T cells were activated in the presence and absence of αCD3, αCD28 or the combination of both antibodies for different times. RT-PCR analysis was performed to determine the expression pattern of WW45 expression upon T cell activation. The pattern of expression of RANKL was analyzed for comparison purposes. Expression of ubiquitously expressed GAPDH was used as a loading control. (B) Normalized expression of WW45 in human naïve T cells before and after T cell activation. The intensity of the amplified bands for WW45 was normalized with that of GAPDH. Data are shown as the mean ± SE of three independent experiments. C) Normalized expression of

RANKL in human naïve T cells upon T cell activation. The intensity of the amplified bands for RANKL was normalized with that of GAPDH. Data are shown as the mean ± SE of three independent experiments. (D) Effects of cyclosporine in the upregulation of WW45 or RANKL upon T cell activation *in vitro*. Normal human CD3$^+$ T cells were activated with antibodies αCD3 and αCD28 or with ionomycin and PMA for 24 h in the presence and absence of cyclosporine. RT-PCR analysis was performed to determine the expression pattern of WW45 and RANKL. GAPDH was used as a loading control, and normalized expression values of WW45 and RANKL from two different experiments are shown at the bottom of a representative example.

Mononuclear infiltrates of patients with periodontal disease exhibit high WW45 and RANKL expression levels: Inflammatory bone-resorptive diseases such as periodontal disease and rheumatoid arthritis are characterized by the presence of abundant mononuclear infiltrates including activated/memory T cells in gingival tissues [4, 5, 8] and synovial fluid [77, 78] respectively. Both inflammation and local bone resorption are important events playing a critical role in the progression of these diseases [5, 77, 78]. RANKL, expressed by either B cells or T cells, has been shown to be higher in patients with these diseases than in healthy individuals and to play a key role in inducing excessive bone resorption [8, 9, 77, 78]. As shown in Figure 3, mononuclear cells from patients with chronic periodontal disease expressed an average of 6-fold higher levels of RANKL than samples obtained from healthy individuals (Figure 3 and data not shown). Interestingly, the expression of WW45 was also about 4-fold higher in patients with periodontal disease than in healthy individuals (Figure 3 and data not shown). These results suggested for the first time that levels of expression of WW45 could be upregulated in a pathological situation. So far, only downregulation of WW45 expression has been reported in some cancer cell lines [1] as previously mentioned in the introduction. Thus, upregulation of WW45 could play a role in the pathogenesis of periodontal disease and most likely in other pathologies in which excessive activated/memory T cells are present and are actively involved in disrupting homeostasis of the immune and bone systems.

Isolation of rat WW45 cDNA and detection of endogenous rat/human WW45 proteins by western blot analysis: To further elucidate the involvement of WW45 in T cell-mediated bone resorption associated with periodontal disease, a rat memory Th1 clone cell line that expresses RANKL when activated by its specific antigen and APC [5, 13, 14, 76] and a rat model for periodontal disease [13, 14] were available for this study. Before undertaking this analysis, the sequence of rat WW45 cDNA was determined and antibodies to recognize endogenous levels of WW45 were generated. To obtain the full-length rat WW45 cDNA, a combination of RT-PCR and RACE techniques were used as described in the materials and methods section of this chapter. The sequence of rat WW45 cDNA isolated through these studies, predicted an open reading frame of 387 amino acids (Figure 4), containing two WW domains for its putative interaction with Lats1/2 and a SARAH domain for its putative interaction with Mst1/2 (Figure 5A, 5B). Furthermore, the overall protein sequence of rat WW45 exhibited high identity to that previously published for mouse (96%) or human (93%) (Figure 5B; [3]). Due to the high sequence identity of WW45 in mouse, rat and human,

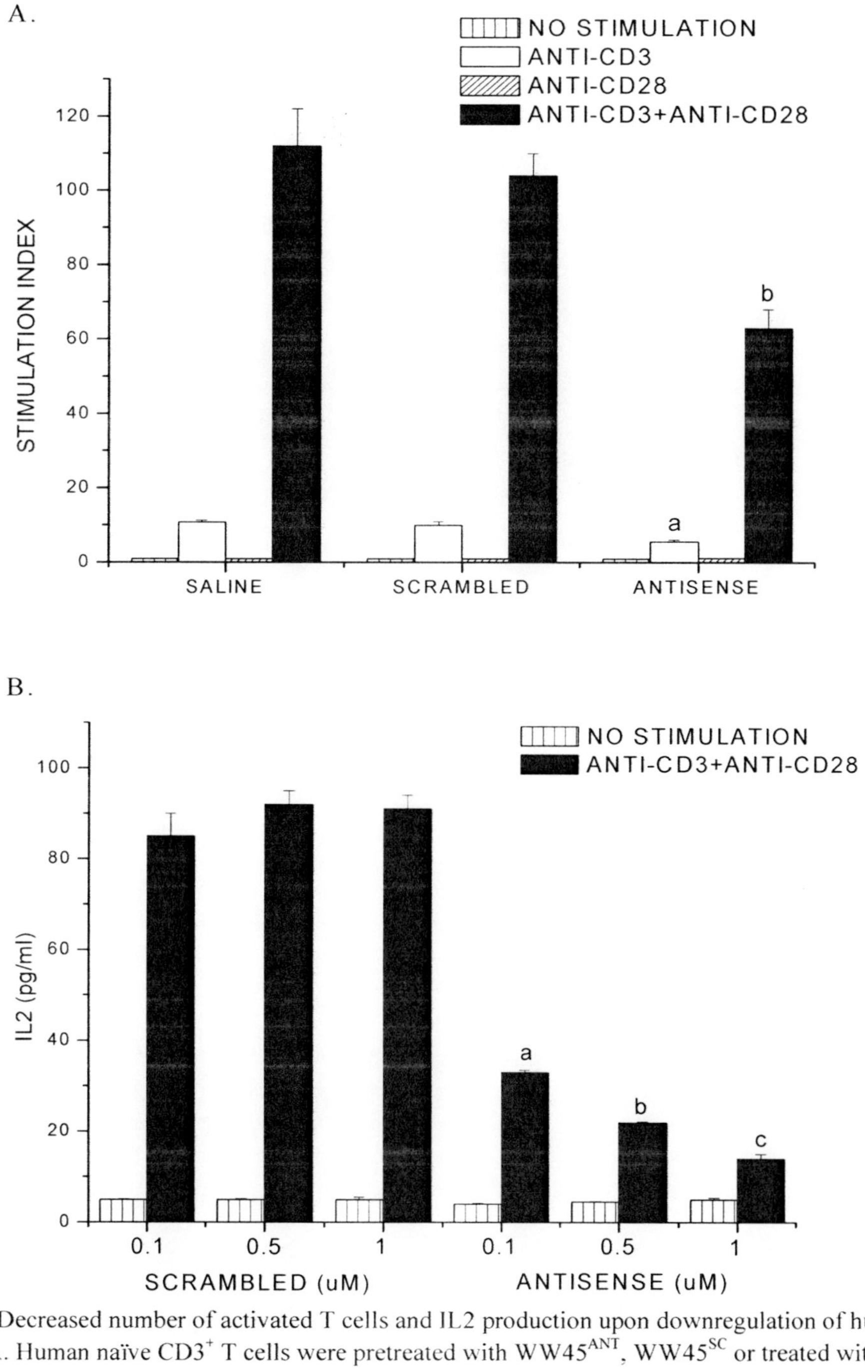

Figure 2. Decreased number of activated T cells and IL2 production upon downregulation of human WW45. A. Human naïve $CD3^+$ T cells were pretreated with $WW45^{ANT}$, $WW45^{SC}$ or treated with an equivalent volume of PBS for 2 h prior to their activation with CD3 or a mix of CD3 and CD28 plate-bound antibodies for 3 days. The proliferation of the resulting cells was measured by $[^3H]$Thymidine incorporation (cpm). The stimulation index was expressed as a ratio of the cpm obtained with cells cultured in the presence and absence of antibodies (mean ± SE). [a]significantly different respect to the cells activated with CD3 antibody ($p<0.05$); [b]significantly different respect to the cells activated with CD3 and CD28 antibodies ($p<0.01$). B. IL2 production measured by ELISA was detected at 24 h (or 16 h, data not shown) in supernatants of unstimulated cells or those activated with CD3 and CD28 plate-bound antibodies that were transfected with $WW45^{ANT}$or $WW45^{SC}$ at different concentrations (0.1-1

μM). [a, b, c]significantly different respect to cells transfected with 0.1, 0.5 or 1 μM of $WW45^{SC}$ respectively ($p<0.05$).

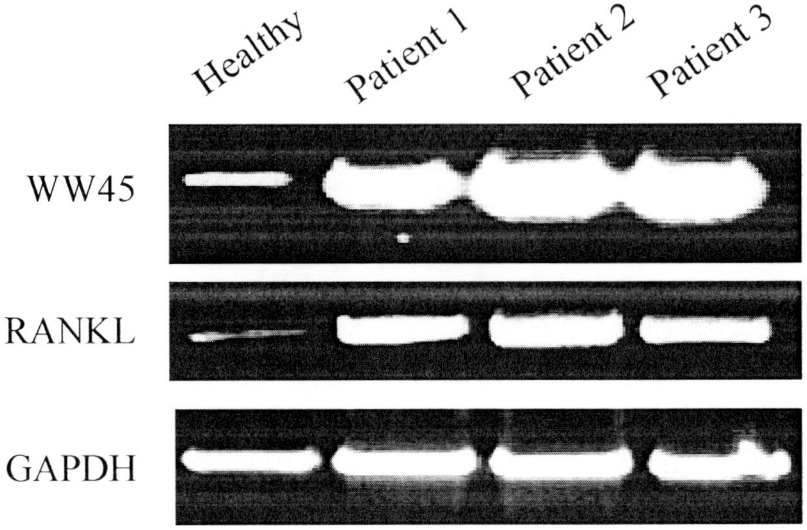

	Healthy	Patient 1	Patient 2	Patient 3
WW45/GAPDH	0.4	1.4	1.8	1.8
RANKL/GAPDH	0.2	1.2	1.4	1.1

(n = 2 experiments)

Figure 3. Human WW45 is upregulated in patients with periodontal disease. Semiquantitative RT-PCR analysis was performed to compare the expression pattern of WW45 and RANKL in gingival mononuclear infiltrates of patients with periodontal disease and one healthy subject. Expression of the ubiquitously expressed GAPDH was used as a loading control for normalization purposes. Normalized WW45 and RANKL expression levels are shown at the bottom of the figure of a representative experiment.

```
1      gtgagcggcggcgggcgcgcttagcgccgccaacatgcggcggagcctgcgggcgggcgc
61     gggcggcggagacccgaggcgaaggcggcgccggcccgggaggcggcggcgggcggcggg
121    ctcgcggccctagcgaggctgctggggcagcggcgggcgacgtccgcgaggatgctgtcc
                                                          M  L  S
181    cgcaagaaaaccaaaaacgaggtgtctaagccggccgaggttcagggcaagtacgtgaag
       R  K  K  T  K  N  E  V  S  K  P  A  E  V  Q  G  K  Y  V  K
241    aaggagacgtcgcccctgctgcggaatctcatgccttcattcattcggcacggtccaaca
       K  E  T  S  P  L  L  R  N  L  M  P  S  F  I  R  H  G  P  T
301    attccaagacggactgacctctgtcttccagagtcaagcgctagtgctttctcagcttct
       I  P  R  R  T  D  L  C  L  P  E  S  S  A  S  A  F  S  A  S
361    ggagatggcgtagtttcaagaaaccagagtttcctgagaactccagttcaaaggacgcct
       G  D  G  V  V  S  R  N  Q  S  F  L  R  T  P  V  Q  R  T  P
421    catgaagtaatgagaagagaaagcaacagactgtccgcaccgtcttcttaccttgtcagg
       H  E  V  M  R  R  E  S  N  R  L  S  A  P  S  S  Y  L  V  R
481    agcctagcggatgtccctcgggaatacggctcatcacagtcattttaacagaagttaat
       S  L  A  D  V  P  R  E  Y  G  S  S  Q  S  F  L  T  E  V  N
541    tttgctgttgaaaatggagactctggttcccgatattttattcagataactttttgat
```

```
      F  A  V  E  N  G  D  S  G  S  R  Y  F  Y  S  D  N  F  F  D
601   ggtcagagaaggaggccacttggagatcgtgcacaagaagactatagatattatgaatac
      G  Q  R  R  R  P  L  G  D  R  A  Q  E  D  Y  R  Y  Y  E  Y
661   aaccatgatctcttccaaagaatgccacagaatcagggaggcacacatcaggtattggg
      N  H  D  L  F  Q  R  M  P  Q  N  Q  G  R  H  T  S  G  I  G
721   agagtcactgctacatctttaggaaatttaactaaccatggatctgaagatttaccctt
      R  V  T  A  T  S  L  G  N  L  T  N  H  G  S  E  D  L  P  L
781   cctcctggctggtctgtggactggacaatgagagggagaaaatactacatagatcataac
      P  P  G  W  S  V  D  W  T  M  R  G  R  K  Y  Y  I  D  H  N
841   acaaacacaactcattggagtcatcctcttgaacgagaaggacttcctcctgggtgggag
      T  N  T  T  H  W  S  H  P  L  E  R  E  G  L  P  P  G  W  E
901   cgagtggaatcatcagaatttggaacctattatgtggatcacacaaataaaaggctcag
      R  V  E  S  S  E  F  G  T  Y  Y  V  D  H  T  N  K  R  A  Q
961   tacaggcacccctgtgctcccagtgtgcctcggtatgatcagcctccacccatcacgtat
      Y  R  H  P  C  A  P  S  V  P  R  Y  D  Q  P  P  P  I  T  Y
1021  cagccacagcaaactgaaagaaatcagtctctcctggtccctgcaaacccctaccatact
      Q  P  Q  Q  T  E  R  N  Q  S  L  L  V  P  A  N  P  Y  H  T
1081  gcagaaattcctgactggcttcaggtttatgcccgagcccctgtgaaatacgaccacatt
      A  E  I  P  D  W  L  Q  V  Y  A  R  A  P  V  K  Y  D  H  I
1141  ctgaagtgggagctcttccagctggctgacctggacacgtaccagggaatgctgaagctg
      L  K  W  E  L  F  Q  L  A  D  L  D  T  Y  Q  G  M  L  K  L
1201  ctcttcatgaaggagctggagcagattgtcaagttgtatgaggcctacagacaggctctt
      L  F  M  K  E  L  E  Q  I  V  K  L  Y  E  A  Y  R  Q  A  L
1261  gtcaccgagctggaaaaccgcaagcagaggcagcagtggtatgcccagcaccatggcaag
      V  T  E  L  E  N  R  K  Q  R  Q  Q  W  Y  A  Q  H  H  G  K
1321  aagttttaagttaactttccacaagccaggttttgtaagagctttaaaatattttcaga
      K  F  L  S  *
1381  taaatgactgcaaacaagttctttggttaataaaggcaacttactatttggaattacaaa
1441  attctagttttacatatattatcaactttttcttttattgaaccaagaaagttttatcac
1501  ttaaaaagccaatatggcatatactaccaaatgctgttaggaaactgctttggagactgt
1561  ggatggaaaggagtacaggaagagactgtcataaatggcattaaacaggagtggacacac
1621  ccttcttgttccgaaaggaagcggacgagagcacactgtgaacagcggctccttgtggt
1681  gtcaggtgttttctggaatgagatggtgtgaccattgccatgggtatagttcagaataaa
1741  aacagggcatggagcctttatggggacaaacgtgtatatggtgtgtgtggtcatctactc
1801  ccacctcctccctatgtgagggacacctttgaaggagggagataaagcctttcaggaata
1861  gagtcaggaatacgggaatataataaatacgggcctcagttggtgcgaaactgttgctaa
1921  tactgcttccttccttcctttgagccagagcacatgagtacagtcagcaaggctctggtg
1981  agacacagcacgggcttcctgaccacagggaatactgtcttccatcaggctgggctacgt
2041  gccaaaa
```

Figure 4. Structure of rat WW45 full length cDNA and deduced amino acid sequence. Full length cDNA was obtained by a combination of RT-PCR (to amplify the cDNA corresponding to the open reading frame) and RACE (to obtain sequence information about the untranslated regions). The RT-PCR product contained the sequence spanning the nucleotides 166 and 1351. The longest 5' RACE and 3' RACE products contained the sequence spanning nucleotides 1 to 625 and 1297 to 2047 respectively. Numbers on the left indicate the nucleotide number. The first ATG corresponding to the first methionine is shown in bold, polyadenylation sites are double underlined and one in-frame stop codon in the 5' untranslated region is single underlined. The nucleotide sequence has been deposited with EMBL under Accession number AM261215.

antibodies for WW45 were generated against MAP-modified peptides, that are known to exhibit a good level of immunogenicity. To increase specificity for endogenous WW45, two different anti-MAP peptide antibodies for WW45 were used in combination in subsequent western blot analysis. As shown in Figure 5C, this anti-WW45 antibody mix recognized a protein band migrating at approximately 47 kDa, which is similar to the predicted molecular mass of 45 kDa, in western blot experiments performed with human naïve T cells or adult rat heart or rat brain lysates. Endogenous protein levels of WW45 in rat and human lysates (Figure 5C and data not shown) were in agreement with the mRNA expression pattern previously reported for mouse and human tissues [3], and for human naïve T cells (Figure 1)

A

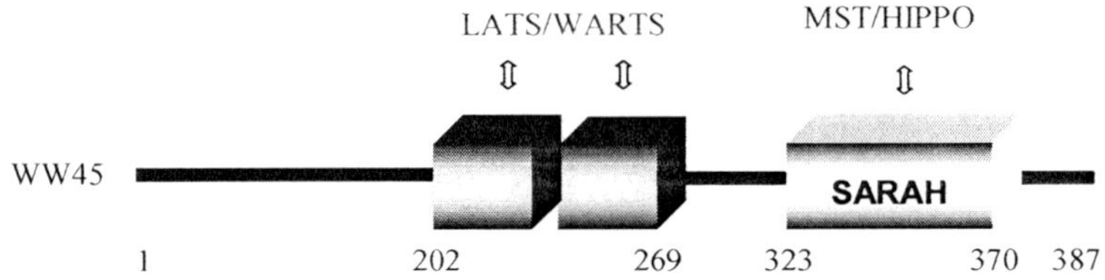

B

```
RAT     MLSRKKTKNEVSKPAEVQGKYVKKETSPLLRNLMPSFIRHGPTIPRRTDLCLPESSASAF 60
MOUSE   MLSRKKTKNEVSKPAEVQGKYVKKETSPLLRNLMPSFIRHGPTIPRRTDLCLPDSSATAF 60
HUMAN   MLSRKKTKNEVSKPAEVQGKYVKKETSPLLRNLMPSFIRHGPTIPRRTDICLPDSSPNAF 60
        *************************************************:***:**..**

RAT     SASGDGVVSRNQSFLRTPVQRTPHEVMRRESNRLSAPSSYLVRSLADVPREYGSS-QSFL 119
MOUSE   SASGDGVVSRNQSFLRTAIQRTPHEVMRRESHRLSAP-SYLVRSLADVPRECGSS-QSFL 118
HUMAN   STSGD-VVSRNQSFLRTPIQRTPHEIMRRESNRLSAP-SYLARSLADVPREYGSS-QSFV 117
        *:*** **********.:**.***:*****:***** ***.********  *** ***:

RAT     TEVNFAVENGDSGSRYFYSDNFFDGQRRRPLGDRAQEDYRYYEYNHDLFQRMPQNQGRHT 179
MOUSE   TEVNFAVENGDSGSRYFFSDNFFDGQRRRPLGDRAQEDYRYYEYNHDLFQRMPQSQGRHT 178
HUMAN   TEVSFAVENGDSGSRYYYSDNFFDGQRKRPLGDRAHEDYRYYEYNHDLFQRMPQNQGRHA 177
        ***.************::*********:*******:*****************.*****:

RAT     SGIGRVTATSLGNLTNHGSEDLPLPPGWSVDWTMRGRKYYIDHNTNTTHWSHPLEREGLP 239
MOUSE   SGIGRVTATSLGNLTNHGSEDLPLPPGWSVDWTMRGRKYYIDHNTNTTHWSHPLEREGLP 238
HUMAN   SGIGRVAATSLGNLTNHGSEDLPLPPGWSVDWTMRGRKYYIDHNTNTTHWSHPLEREGLP 237
        ******:*****************************************************

RAT     PGWERVESSEFGTYYVDHTNKRAQYRHPCAPSVPRYDQPPPITYQPQQTERNQSLLVPAN 299
MOUSE   PGWERVESSEFGTYYVDHTNKRAQYRHPCAPSVPRYDQPPPITYQPQQTERNQSLLVPAN 298
HUMAN   PGWERVESSEFGTYYVDHTNKKAQYRHPCAPSVPRYDQPPPVTYQPQQTERNQSLLVPAN 297
        *********************:*******************:******************

RAT     PYHTAEIPDWLQVYARAPVKYDHILKWELFQLADLDTYQGMLKLLFMKELEQIVKLYEAY 359
MOUSE   PYHTAEIPDWLQVYARAPVKYDHILKWELFQLADLDTYQGMLKLLFMKELEQIVKLYEAY 358
HUMAN   PYHTAEIPDWLQVYARAPVKYDHILKWELFQLADLDTYQGMLKLLFMKELEQIVKMYEAY 357
        *******************************************************:****

RAT     RQALVTELENRKQRQQWYAQHHGKKFLS 387
MOUSE   RQALLTELENRKQRQQWYAQQHGKTFLS 386
HUMAN   RQALLTELENRKQRQQWYAQQHGKNF-- 383
        ****:***************:***.*
```

C

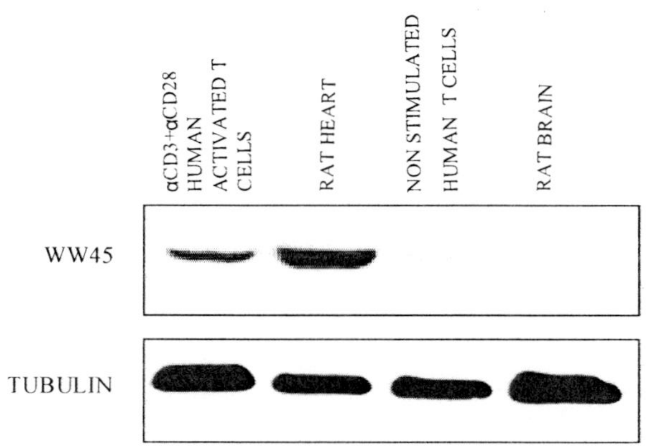

Figure 5. Domain distribution of rat WW45, alignments of rat, mouse and human WW45 aminoacid sequences and western blot analysis of endogenous human and rat WW45. (A). Domain distribution of rat WW45 as predicted by SMART. Rat WW45 sequence contains two WW domains and one SARAH domain for its putative interaction with Lats (named Warts in Drosophila) and Mst (named Hippo in Drosophila) respectively. (B). Sequence alignment of rat, mouse and human WW45 amino acid sequences was performed with ClustalW. The consensus line obtained by ClustalW analysis means :"*" = identical or conserved residues in all sequences in the alignment, ":" = indicates conserved substitutions and "." = indicates semi-conserved substitutions. Location of WW and SARAH domains are shown in black and grey shadow areas respectively. Location of the peptides used in the preparation of antibodies for WW45 protein is shown underlined. (C). Western blot analysis of with anti-WW45 antibody mix. Lysates of naive $CD3^+$ T cells unstimulated or activated in the presence of αCD3 and αCD28 antibodies for 48 h, adult rat heart or adult rat brain were subjected to western blot analysis with a mix of two anti-WW45 antibodies or antitubulin antibodies.

Rat WW45 expression is upregulated in antigen-specific stimulated memory Th1 clone cells that express RANKL and induce bone resorption in vivo: In the next set of experiments, stimulation of rat G23 Th1 clone memory cells with its specific antigen *Actinobacillus actinomycetemcomitans* Omp29 and APC was performed *in vitro* to determine whether the expression of WW45 could be upregulated in an antigen-specific fashion as previously reported for RANKL [8, 14]. As shown in Figure 6A, rat G23 Th1 clone cells exhibited more than 3-fold higher WW45 or RANKL mRNA expression when activated in the presence of APC and their specific antigen than when cultured with APC alone for 48 h. Similar results were found after 24 h or 72 h of stimulation (data not shown). However, when G23 Th1 clone cells were activated in the presence of kaliotoxin or CTLAIg, that have been shown to decrease T cell-mediated inflammatory bone resorption *in vivo* [8, 13, 14] by either attenuating the calcium signaling or the B7/CD28 costimulatory signaling pathways respectively [89, 95], the antigen-specific upregulation of WW45 and RANKL expression was significantly decreased (Figure 6A). Western blot analysis with antibodies for either WW45 or RANKL (Figure 6B) also demonstrated that their levels of expression were increased 3-fold or more upon antigen-specific stimulation at the protein level. In agreement with the results obtained by RT-PCR analysis, upregulation in the protein expression of

WW45 or RANKL could be significantly decreased by treatments with either kaliotoxin or CTLAIg (Figure 6B). Since blockade of calcium signaling during early stages of T cell activation by kaliotoxin or blockade of the B7/CD28 co-stimulatory pathway with CTLAIg, affected RANKL and WW45 upregulation in a similar fashion, WW45 and RANKL may be sharing common upstream regulators during their antigen-specific upregulation of the memory T cell clone used in these studies. These results were also in agreement with the ability of cyclosporine to downregulate the upregulation of WW45 or RANKL expression upon T cell activation of human naïve T cells (Figure 1D).

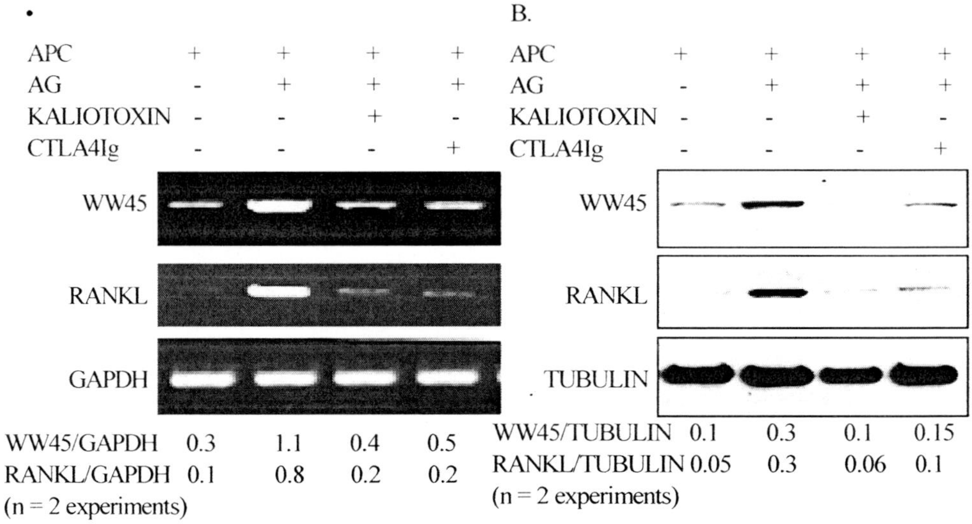

Figure 6. Antigen-specific upregulation of WW45 and RANKL in the presence and absence of kaliotoxin or CTLA4Ig. Rat memory Th1 clone cells were stimulated by APC and specific antigen in the presence and absence of kaliotoxin or CTLA4Ig for 72 h. (A) Semiquantitative RT-PCR analysis. RT-PCR analyses were performed to monitor changes in mRNA expression for WW45 and RANKL. GAPDH amplification was performed for normalization of mRNA quantity. The intensity of the amplified bands was normalized with that of GAPDH and the means of two independent experiments are shown at the bottom of the figure of a representative experiment (B) Western blot analysis. Total protein lysates were subjected to SDS-PAGE and western blotting with antibodies for WW45 and RANKL. Detection of tubulin was used for normalization purposes. Results of the normalized WW45 and RANKL levels obtained from two independent experiments are shown at the bottom of the figure of a representative one.

Downregulation of WW45 expression in rat Th1 clone cells decreases their antigen-specific upregulation of RANKL expression: In the next set of experiments, the targeted downregulation of WW45 expression was performed by using an antisense approach. Treatment of Th1 clone T cells with liposome-delivered phosphorothioate antisense oligonucleotides for WW45 ($WW45^{ANT}$) at concentrations ranging 0.25 to 1 μM, decreased the antigen-specific upregulation of WW45 at the mRNA (Figure 7A) and protein levels (Figure 7B). Interestingly, expression of RANKL was also downregulated at a similar extent as that of WW45 (Figure 7A and 7B). In contrast, delivery of a control scrambled oligonucleotide ($WW45^{SC}$) did not cause any significant effects in the expression of either

WW45, RANKL or the loading controls (either GAPDH or tubulin) (Figure 7A and 7B). These results suggested that changes in the expression of WW45 might influence the expression of the cytokine RANKL by activated/memory T cells directly by either acting as an upstream regulator of its expression or indirectly by compensatory mechanisms that require further investigation.

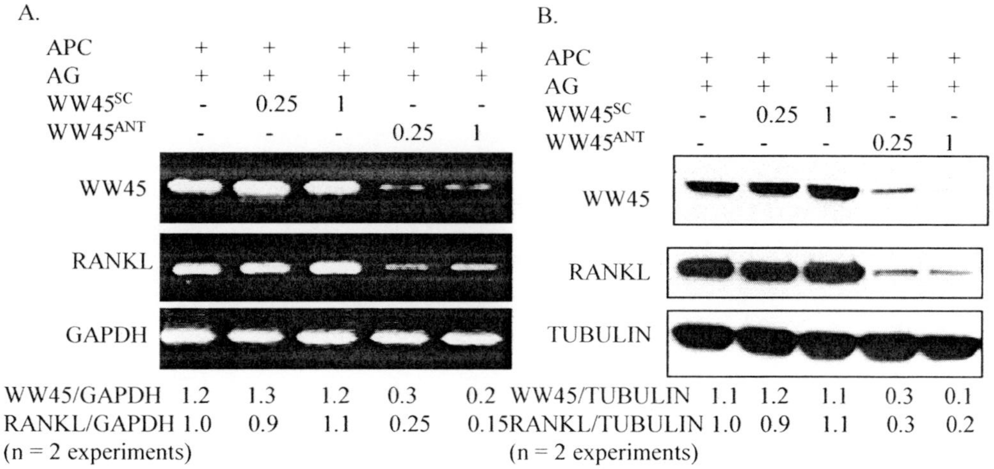

Figure 7. Downregulation of WW45 expression in activated memory rat Th1 clone cells by $WW45^{ANT}$ decreases WW45 and RANKL expression levels *in vitro*. Rat memory Th1 clone cells were pretreated with PBS or oligofectamine-delivered $WW45^{ANT}$ or $WW45^{SC}$ (at the concentrations of 0.25 or 1 μM) for 2 h and then stimulated by APC and specific antigen for 72 h. (A) Semiquantitative RT-PCR analysis. RT-PCR analyses were performed to monitor changes in mRNA expression for WW45 and RANKL. GAPDH amplification was performed for normalization of mRNA quantity. The intensity of the amplified bands was normalized with that of GAPDH and the mean of two independent experiments are shown at the bottom of the figure of a representative one. (B) Western blot analysis. Total protein lysates were subjected to SDS-PAGE and western blotting with antibodies for WW45 and RANKL. Detection of tubulin was used for normalization purposes. Results of the normalized WW45 and RANKL levels obtained from two independent experiments are shown at the bottom of the figure of a representative one.

Downregulation of WW45 expression in activated memory Th1 clone cells decreases their ability to induce osteoclast differentiation and formation of bone resorption pits in vitro: The results obtained in the previous section suggested that downregulation of WW45 expression could lead to the downregulation of RANKL expression in antigen-specific stimulated G23 T clone cells. Since RANKL plays a key role in mediating the induction of osteoclastogenesis and bone resorption by activated/memory T cells [8, 14, 77-80], downregulation of WW45 might be expected to decrease the ability of G23 T cells to induce osteoclast differentiation and the formation of bone resorption pits *in vitro*. To test this hypothesis, a co-culture system was established with fixed G23 T cells (which were activated in the presence of $WW45^{ANT}$ or $WW45^{SC}$) with the committed osteoclast precursor cell line MOCP-5. The influence of T cell-derived RANKL in inducing osteoclast differentiation and bone resorption *in vitro* was tested by TRACP staining and by monitoring the formation of bone resorption pits by the MOCP-5 osteoclast precursor cell line. G23 T cells cultured in the absence of the specific antigen did not induce the formation of multinucleated TRACP^{+} cells (data not shown) or form resorption areas in hydroxyapatite discs (data not shown) as previously described [14]. However, fixed G23 T cells activated in the presence of their specific antigen and irradiated APC, were able to induce the formation of large TRACP^{+} cells with 3 or more nuclei (Figure 8) and bone resorption pits by MOCP-5 cells *in vitro* (Figure 9). In support to the initial hypothesis, co-cultures of MOCP-5 cells with G23 T cells which were activated in the presence of WW45^{ANT}, produced significantly fewer and smaller multinucleated TRACP^{+} cells (Figure 8). In addition, resorption pit area was lower for those co-cultures (Figure 9) than for those established with G23 T cells activated in the presense of saline or WW45^{SC}.

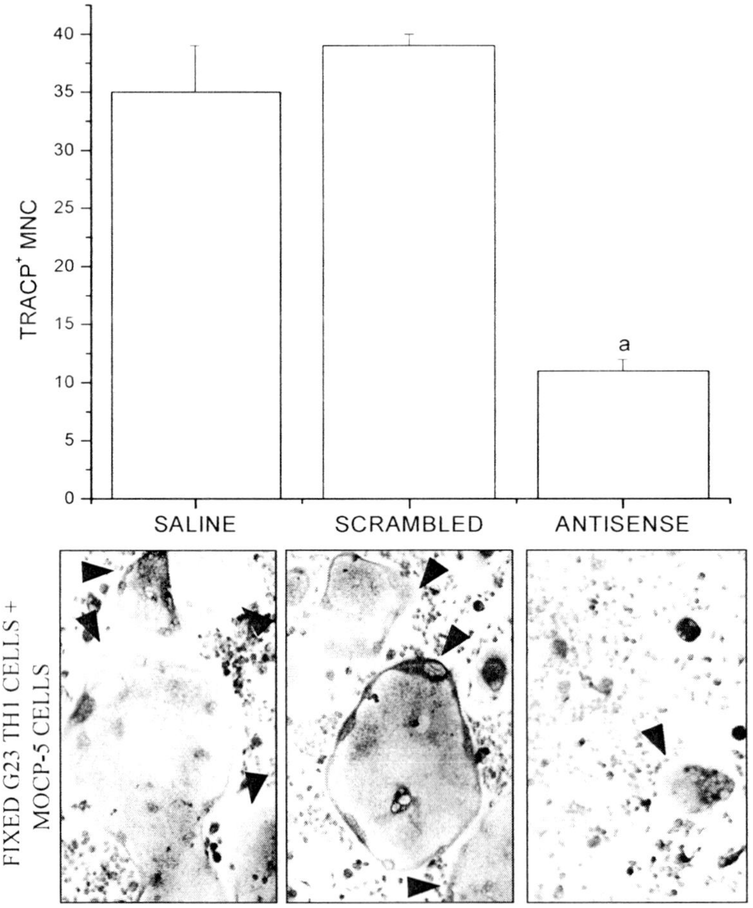

Figure 8. The ability of activated memory rat Th1 clone cells to induce osteoclastogenesis *in vitro* is decreased upon WW45 downregulation by WW45^{ANT} treatment. 5×10^5 G23 T cells were treated with WW45^{ANT}, WW45^{SC} or saline and stimulated with APC and antigen for 72 h. Then, 5×10^2 MOCP-5 cells/well were cultured on 96 well-plates with 10^5 fixed G23 T cells/well in α-MEM with 15% FBS and 10 ng/ml M-CSF for 7 days. Differentiated MOCP-5 cells were then identified by cytochemical staining and quantified as TRACP^{+} MNC. Each point in the graph was expressed as the mean ± SE of three independent experiments. [a]Significantly different with respect to control co-cultures of MOCP-5 and activated G23 T cells treated with saline or WW45^{SC} ($^{a}p<0.05$). Representative experiment is shown at the bottom of the graph with black arrows pointing at multinucleated osteoclasts

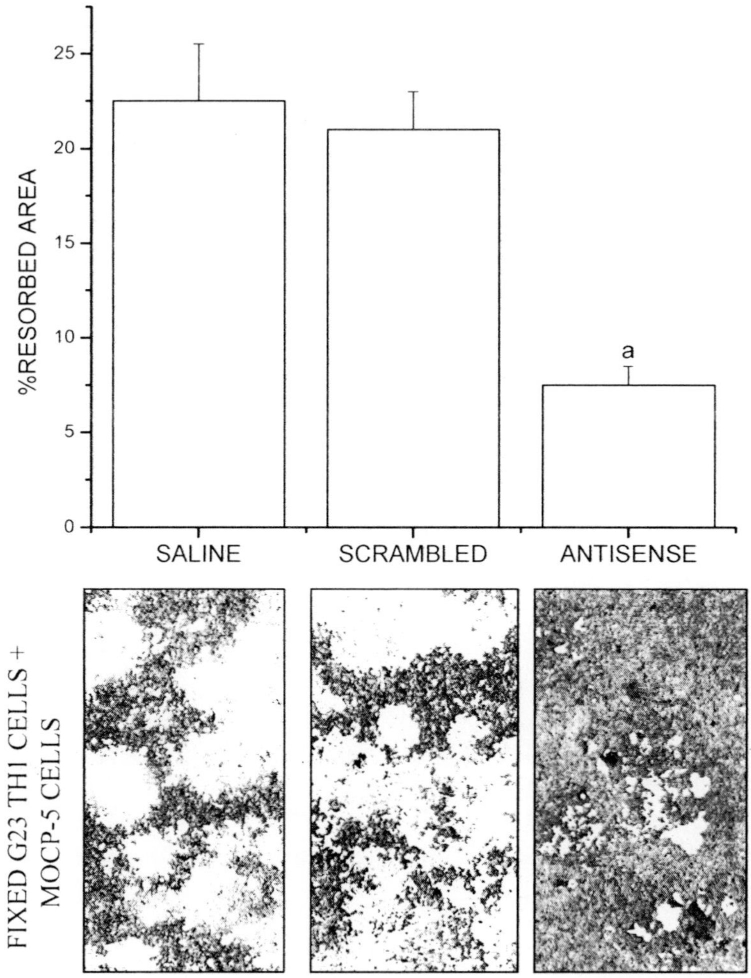

Figure 9. The ability of activated memory rat Th1 clone cells to induce the formation of bone resorptive pits *in vitro* is decreased upon WW45 downregulation. 5 x 10^5 G23 T cells were treated with $WW45^{ANT}$, $WW45^{SC}$ or saline and activated with APC and antigen for 72 h. Then, 5 x 10^2 MOCP-5 cells/well were cultured together with 10^5 fixed G23 T cells/well in α-MEM with 15% FBS and 10 ng/ml M-CSF on BD BioCoat Osteologic discs for 7 days. The resorption areas on each disc were quantified and expressed as % resorbed area = Resorbed area per disc/ total disc area x 100. Each point in the graph was expressed as mean ± SE of three independent experiments [a]Significantly different with respect to control co-cultures of MOCP-5 and activated G23 T cells treated with saline or $WW45^{SC}$ ([a]$p<0.05$). Representative images of resorption pits (white areas) in osteologic discs.

Immunological influence of downregulating WW45 in activated memory Th1 clone cells in a rat model of experimental periodontal disease: Since WW45 downregulation decreased RANKL expression by activated G23 T cells *in vitro* (Fig 7) as well as their ability to induce osteoclast differentiation in co-culture experiments with osteoclast precursor cells (Fig 8 and Fig 9), the immunological influence of decreasing WW45 expression on adoptively transferred G23 clone T cells was tested in a rat model for experimental periodontal disease previously described [13, 14] (Figure 10). In this model, local gingival stimulation of G23 clone T cells by MHC class II^+/$B7^+$ APC leads to increased migration of G23 clone T cells to gingival tissue and cervical lymph nodes. High levels of memory T cells can be detected by RT-PCR in the gingival tissue that receives gingival challenge but not in the control side that is only injected with saline (Figure 10). Some G23 T cells migrate to the spleen and maintain responsiveness to Omp29 plus APC *in vitro* up to 10 days after adoptive T cell transfer [13, 14]. Furthermore, both the increase in the serum IgG2a antibody levels to the specific antigen *A. actinomycetemcomitans* Omp29 and the antigen-specific periodontal bone resorption can be easily detected 10 days after T cell transfer [13, 14]. Since WW45 exhibits a ubiquitous distribution in adult tissues [3], systemic treatment with antisense oligonucleotides would be expected to affect the normal function of many different cell types, leading to a variety of effects *in vivo* that would be difficult to interpret. To specifically analyze the involvement of the upregulation of WW45 in activated/memory T cells in antigen-specific T cell-induced bone resorption, downregulation of WW45 by antisense treatment was only performed in the G23 clone T cells that were going to be used in the adoptive transfer (Figure 10). To that end, G23 clone T cells were activated with APC and antigen in the presence of saline, $WW45^{ANT}$ or $WW45^{SC}$ for 72 h and later used in the adoptive transfer of three different groups of animals (Figure 10).

Once the adoptive transfer was performed on day 0 (Figure 10), the downregulation of WW45 expression in the activated/memory T cells that had migrated to the gingival tissues was tested on days 0 or 2 days after the adoptive transfer. Based upon the results obtained by RT-PCR analysis, whole gingival tissue isolated from the experimental sides on day 0 (Figure 11) or from the control sides on days 0 or 2 (Figure 11) did not express detectable levels of RANKL or WW45 mRNA. However, $CD3^+$ T cells or mononuclear cells isolated from the experimental sides of $WW45^{SC}$-treated group (Figure 11) or saline-treated animals (data not shown), expressed significant levels of RANKL and WW45 mRNA on day 2 (Figure 11). Consistent with the ability of WW45 to influence the antigen-specific upregulation of RANKL expression, $CD3^+$ T cells isolated from the experimental sides of the $WW45^{ANT}$-treated group showed lower RANKL mRNA levels that those isolated from the control groups on day 2 (Figure 11). Since gingival $CD3^+$ T cells from the experimental sides of the three groups of rats, expressed comparable levels of GAPDH mRNA at 2 days after the adoptive transfer, a similar number of $CD3^+$ T cells appeared to have migrated and to have been retained within the gingival tissues in the three different groups of animals. Therefore, expression levels of WW45 does not appear to play a role in regulating the migration or adhesion of pathogenic memory T cells in the rat model for experimental disease described in this chapter.

Ten days after adoptive transfer, T cells were also recovered from the spleens of saline-treated, $WW45^{ANT}$-treated and $WW45^{SC}$-treated groups of rats and tested for their ability to proliferate in the presence of specific antigen and irradiated APC *in vitro*. Results indicated that the antigen-specific-stimulation index of splenocytes from the $WW45^{ANT}$-treated group of animals was about 50% lower than that of the splenocytes from the saline or $WW45^{SC}$-treated groups (Figure 12A).

To further examine the *in vivo* effects of downregulating WW45 expression on the immunological influence of G23 T cells in recipient animals, 10 days after transfer of the activated Th1-clone cells, the serum IgG2a antibody levels to Omp29 were determined by ELISA. Results showed that the sera from the $WW45^{ANT}$-treated group of rats exhibited about 25 % lower levels of IgG2a antibody to Omp29 than the saline or $WW45^{SC}$-treated groups (Figure 12B).

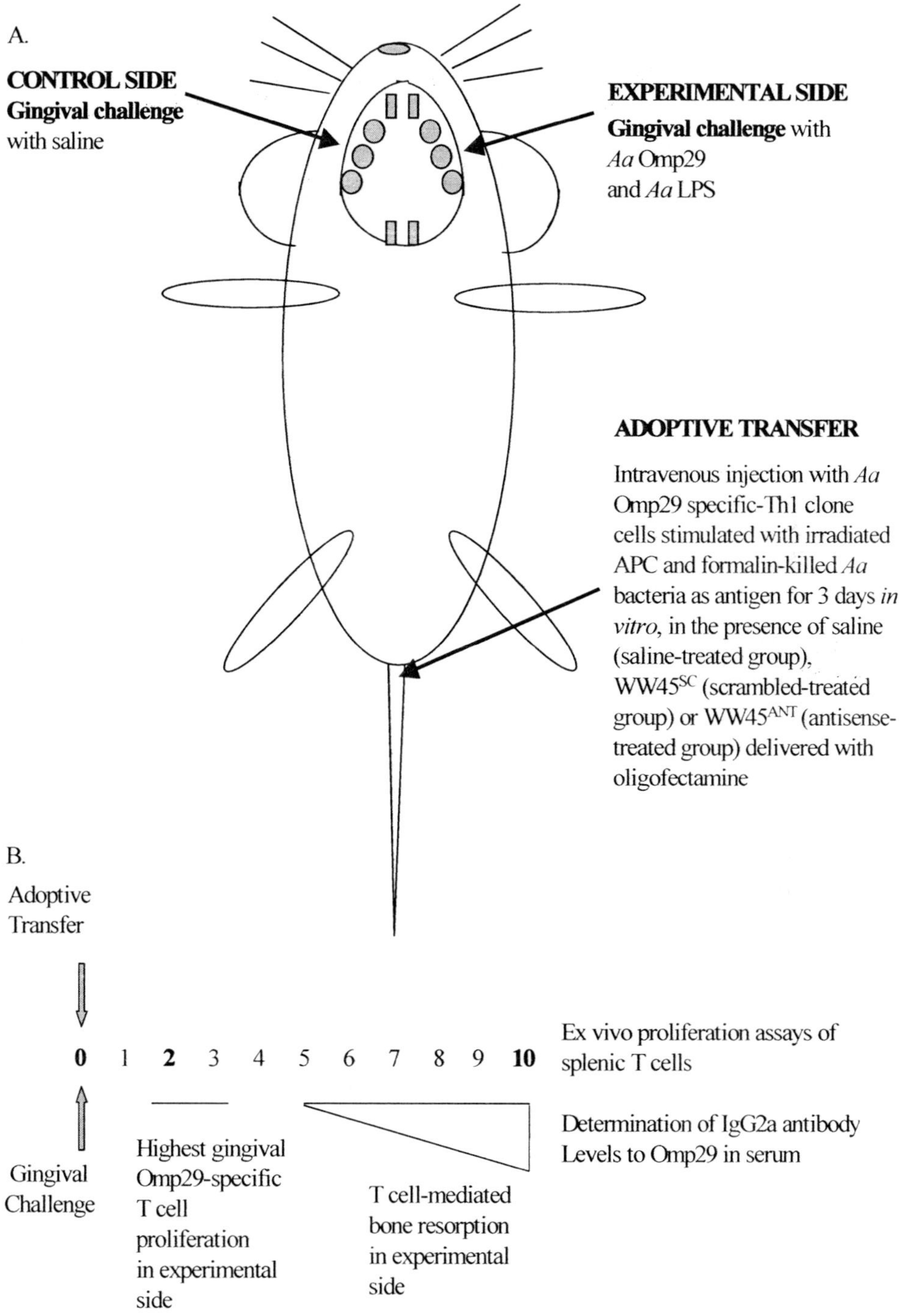

Figure 10. Schematic diagram of the experimental model for periodontal disease used in this study. A. Gingival challenge and adoptive transfer of three different groups of rats, the saline-treated, the $WW45^{ANT}$-treated and the $WW45^{SC}$-treated groups. B. Collection of tissues or serum on days 0, 2 or 10 to evaluate the immunological influence of decreasing WW45 expression on adoptively transferred G23 clone T cells activated in the presence of saline, $WW45^{ANT}$ or $WW45^{SC}$ *in vitro*.

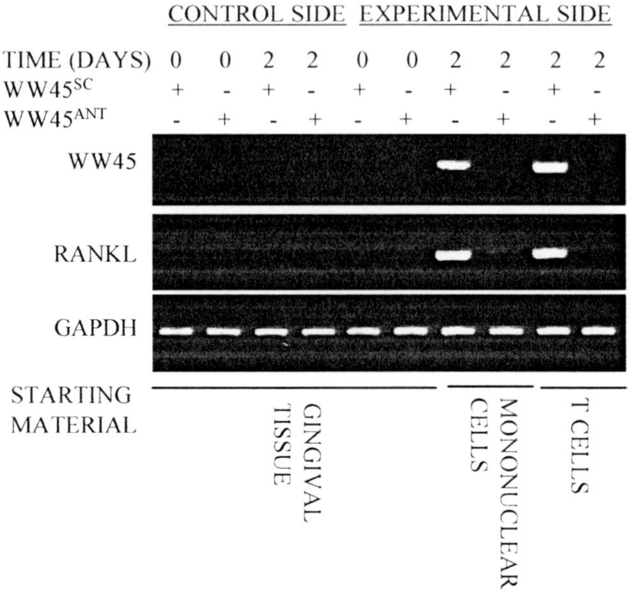

Figure 11. Gingival infiltrates express less WW45 and RANKL in the $WW45^{ANT}$-treated group than in $WW45^{SC}$-treated group two days after the adoptive transfer. RT-PCR with 150-200 pg of RNA per reaction was used to monitor the expression levels of WW45 and RANKL by $CD3^{+}$ T lymphocytes, mononuclear infiltrates or total gingival tissue from the control or experimental sides on days 0 and 2 after the adoptive transfer (n=4 rats per time point). The mRNA expression levels of WW45, RANKL and GAPDH were monitored in samples obtained from rats adoptively transferred with either $WW45^{ANT}$- or $WW45^{SC}$-treated G23 activated clone cells.

In the last set of experiments, it was tested whether the inflammatory alveolar bone resorption induced by adoptive transfer of activated G23 Th1 clone cells that that expressed lower levels of WW45 due to the antisense treatment, could have also been decreased as compared to the saline or $WW45^{SC}$-treated groups. Measurements of experimental alveolar bone resorption indicated that the $WW45^{ANT}$-treated group exhibited 90% lower values than the saline or $WW45^{SC}$-treated groups (Figure 12C).

In conclusion, these results support the notion that downregulation of WW45 expression in activated memory T cells inhibits T cell-mediated inflammatory bone resorption in experimental periodontal disease by decreasing RANKL expression and the pool of activated T cells. Since WW45 pattern of expression is different upon naïve or memory T cell activation, the role of WW45 in controlling proliferation and/or apoptosis is expected to be different during primary and secondary responses of the immune system and during normal and pathological T cell-mediated immunoregulation. Additional experimental research needs to be done to determine the mechanisms by which activation of WW45/Lats/Mst signaling complex may be involved in regulating normal T cell and bone homeostasis. This line of investigation may also help us develop new therapeutic approaches to control aberrant T cell-mediated immunoregulation and excessive bone resorption. However, considering the broad variety of diseases characterized by the aberrant coupling of proliferation, differentiation and apoptosis, additional studies on WW45/Lats/Mst signaling pathway in adult tissues are expected to lead to exciting new advances in other pathologies such as cancer, neurological disorders and cardiovascular disease.

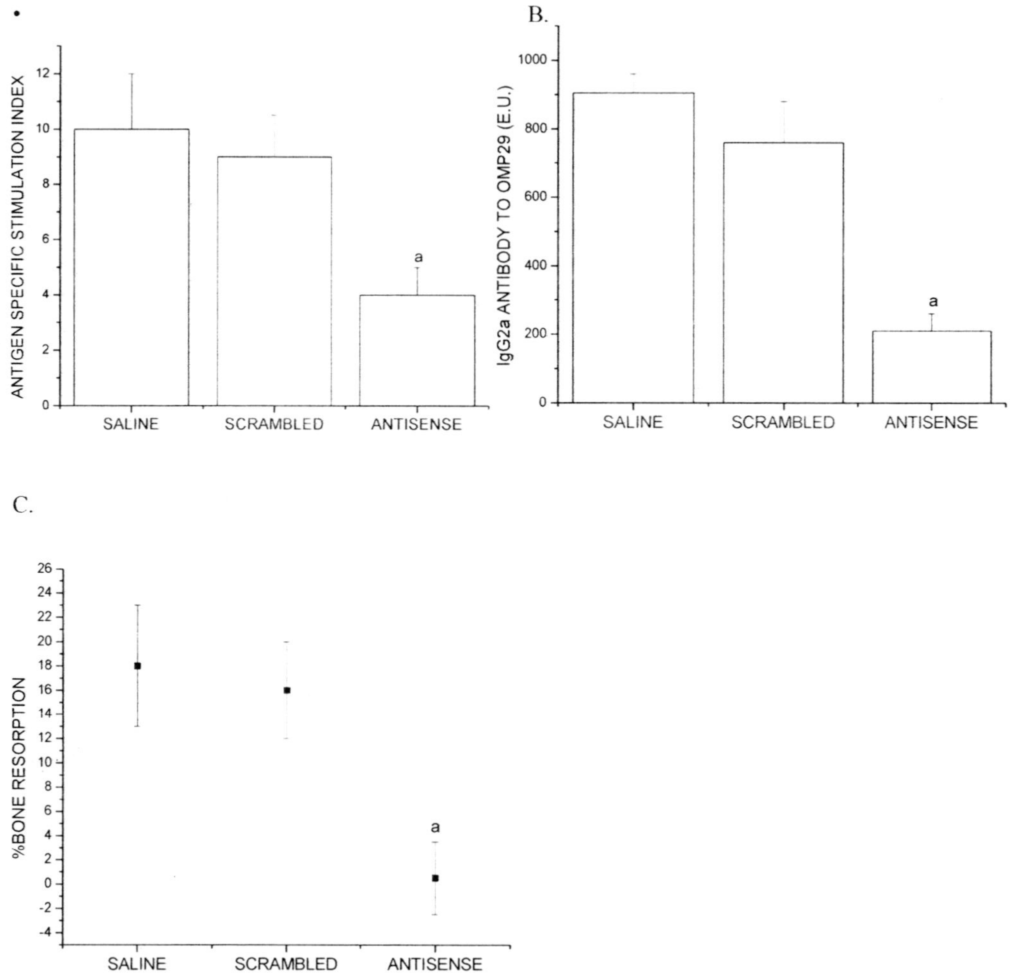

Figure 12. The WW45^{ANT}-treated group of rats exhibited lower antigen-specific stimulation index, lower serum IgG2a antibody levels to Omp29 and lower alveolar bone resorption than the control groups. Ten days after gingival injection of Omp29/LPS and adoptive-transfer of the three different groups of rats (n=6 per group) with G23 T cells activated in the presence of saline, WW45^{ANT} or WW45^{SC}, animals were sacrificed. Then, spleens, serum samples and heads were collected for further analysis. (A) Splenic cells were co-cultured with irradiated APC with or without Omp29 as an antigen for 3 days. The proliferation of antigen specific T cells was measured by [^{3}H]Thymidine incorporation (cpm). The antigen-specific-stimulation index was expressed as a ratio of the cpm obtained with cells cultured in the presence and absence of antigen (mean ± S.E.M.). [a]significantly different respect to the saline-treated or the WW45^{SC}-treated groups (One-way Anova, $p<0.001$). (B) Serum samples were used to measure specific IgG2a response to the purified *A. actinomycetemcomitans* Omp29 antigen by ELISA. ELISA Units (EU) were calculated based on rat hyperimmune serum to the Omp29. [a]significantly different from the saline-treated group (One-way Anova, $P<0.001$). (C) Rat heads were processed to measure experimental periodontal bone resorption as previously described [14]. Bone resorption was expressed as % bone resorption = [(total CEJ-AL measurements of left maxilla - total CEJ-AL measurements of right maxilla) / total CEJ-AL measurements of right maxilla] x 100. [a]significantly reduced as compared to the saline-treated or the WW45^{SC}-treated groups (One-way Anova, $p<0.001$).

Acknowledgments

The experimental work described in this chapter was partly supported by the J. W. Hein Fellowship from the Forsyth Institute (Boston, MA).

References

[1] Tapon, N; Harvey, KF; Bell, DW; Wahrer, DC; Schiripo, TA; Haber, DA; Hariharan, IK. (2002) Salvador promotes both cell cycle exit and apoptosis in Drosophila and is mutated in human cancer cell lines. *Cell*, 110, 467-478.

[2] Kango-Singh, M; Nolo, R; Tao, C; Verstreken, P; Hiesinger, PR; Bellen, HJ; Halder, G. (2002) Shar-pei mediates cell proliferation arrest during imaginal disc growth in *Drosophila. Development*, 129, 5719-5730.

[3] Valverde, P. (2000) Cloning, expression, and mapping of hWW45, a novel human WW domain-containing gene. *Biochem Biophys Res Commun.*, 276, 990-998.

[4] Taubman, MA; Kawai, T. (2001) Involvement of T-lymphocytes in periodontal disease and in direct and indirect induction of bone resoption. *Cri Rev Oral Biol Med*, 12, 125-135.

[5] Taubman, MA; Valverde, P; Han, X; Kawai, T. (2005) Immune response: the key to bone resorption in periodontal disease. *J Periodontol.*, 76(11 Suppl), 2033-2041.

[6] Wong, BR; Rho, J; Arron, J; Robinson, E; Orlinick, J; Chao, M; Kalachikov, S; Cayani, E; Bartlett, FS III; Frankel, WN; Lee, SY; Choi, Y. (1997) TRANCE is a novel ligand of the tumor necrosis factor receptor family that activates c-Jun N-terminal kinase in T cells. *J Biol Chem*, 272, 25190-25194.

[7] Wong, BR; Josien, R; Lee, SY; Sauter, B; Li, H-L; Steinman, RM; Choi, Y. (1997) TRANCE (Tumor Necrosis Factor [TNF]-related activation-induced cytokine), a new TNF family member predominantly expressed in T cells, is a dendritic cell-specific survival factor. *J Exp Med*, 186, 2075-2080.

[8] Valverde, P; Kawai, T; Taubman, MA. (2005) Potassium channel-blockers as therapeutic agents to interfere with bone resorption of periodontal disease. *J Dent Res.*, 84, 488-499.

[9] Mogi, M; Otogoto, J; Ota, N; Togari, A. (2004) Differential expression of RANKL and osteoprotegerin in gingival crevicular fluid of patients with periodontitis. *J Dent Res*, 83, 166-169.

[10] Yasuda, H; Shima, N; Nakagawa, N; Yamaguchi, K; Kinosaki, M; Mochizuki, S; Tomoyasu, A; Yano, K; Goto, M; Murakami, A; Tsuda, E; Morinaga, T; Higashio, K; Udagawa, N; Takahashi, N; Suda, T. (1998) Osteoclast differentiation factor is a ligand for osteoprotegerin/osteoclastogenesis-inhibitory factor and is identical to TRANCE/RANKL. *Proc Natl Acad Sci* USA, 95, 3597-3602.

[11] Lacey, DL; Timms, E; Tan, HL; Kelley, MJ; Dunstan, CR; Burgess, T; Elliott, R; Colombero, A; Elliott, G; Scully, S; Hsu, H; Sullivan, J; Hawkins, N; Davy, E; Capparelli, C; Eli, A; Oian, YX; Kaufman, S; Sarosi, I; Shalhoub, V; Senaldy, G; Guo, J; Delaney, J; Boyle, WJ. (1998) Osteoprotegerin ligand is a cytokine that regulates osteoclast differentiation and activation. *Cell*, 93, 165-176.

[12] Kong, YY; Yoshida, H; Sarosi, I; Tan, HL; Timms, E; Capparelli, C; Morony, S; Oliveira-dos-Santos, AJ; Van, G; Itie, A; Khoo, W; Wakeham, A; Dunstan, CR; Lacey, DL; Mak, TW; Boyle, WJ; Penninger, JM. (1999) OPGL is a key regulator of osteoclastogenesis, lymphocyte development and lymph-node organogenesis. *Nature,* 397, 315-323.

[13] Kawai, T; Eisen-Lev, R; Seki, M; Eastcott, JW; Wilson, ME; Taubman, MA. (2000) Requirement of B7 costimulation for Th1-mediated inflammatory bone resorption in experimental periodontal disease. *J Immunol,* 164, 2102-2109.

[14] Valverde, P; Kawai, T; Taubman, MA. (2004) Selective blockade of voltage-gated potassium channels reduces inflammatory bone resorption in experimental periodontal disease. *J Bone Miner Res.,* 19,155-164.

[15] Sudol, M; Chen, HI; Bougeret, C; Einbond, A; Bork, P. (1995) Characterization of a novel protein-binding module the WW domain. *FEBS Letters,* 369, 67-71.

[16] Sudol, M. (1996) Structure and function of the WW domain. *Prog. Biophys. Mol. Biol.,* 65, 113-132.

[17] Sudol, M; Bork, P; Einbond, A; Kastury, K; Druck, T; Negrini, M; Huebner, K; Lehman, D. (1995) Characterization of the mammalian YAP (Yes-associated protein) gene and its role in defining a novel protein module, the WW domain. *J Biol Chem.,* 270, 14733-14741.

[18] Ermekova, KS; Zambrano, N; Linn, H; Minopoli, G; Gertler, F; Russo, T; Sudol, M. (1997) The WW domain of neural protein FE65 interacts with proline-rich motifs in Mena, the mammalian homolog of Drosophila enabled. *J Biol Chem.,* 272, 32869-32877.

[19] Bedford, MT; Reed, R; Leder, P. (1998) WW domain-mediated interactions reveal a spliceosome-associated protein that binds a third class of proline-rich motif: the proline glycine and methionine-rich motif. *Proc. Natl. Acad. Sci.* U.S.A, 95, 10602-10697.

[20] Lu PJ; Zhou, XZ; Shen, M; Lu, KP (1999) Function of WW domains as phosphoserine- or phosphothreonine-binding modules. *Science*, 283, 1325-1328.

[21] Scheel, H; Hofmann, K. (2003) A novel interaction motif, SARAH, connects three classes of tumor suppressor. *Curr. Biol.,* 13, R899-R900.

[22] Harvey, KF; Pfleger, CM; Hariharan, IK. (2003) The Drosophila Mst ortholog, hippo, restricts growth and cell proliferation and promotes apoptosis. Cell, 114, 457-467.

[23] Wu, S; Huang, J; Dong, J; Pan, D. (2003) Hippo encodes a Ste-20 family protein kinase that restricts cell proliferation and promotes apoptosis in conjunction with salvador and warts. *Cell,* 114, 445-456.

[24] Udan RS; Kango-Singh, M; Nolo, R; Tao, C; Halder, G. (2003) Hippo promotes proliferation arrest and apoptosis in the Salvador/Warts pathway. Nat Cell Biol., 5, 914-920.

[25] Pantalacci, S; Tapon, N; Leopold, P. (2003) The Salvador partner Hippo promotes apoptosis and cell-cycle exit in Drosophila. *Nat Cell Biol.,* 5, 921-927.

[26] Justice, RW; Zilian, O; Woods, DF; Noll, M; Bryant, PJ. (1995) The Drosophila tumor suppressor gene warts encodes a homolog of human myotonic dystrophy kinase and is required for the control of cell shape and proliferation. *Genes Dev.,* 9, 534-546.

[27] Xu, T; Wang, W; Zhang, S; Stewart, RA; Yu, W. (1995) Identifying tumor suppressors in genetic mosaics: the Drosophila lats gene encodes a putative protein kinase. *Development*, 121, 1053-1063.

[28] Jia, J; Zhang, W; Wang, B; Trinko, R; Jiang, J. (2003) The Drosophila Ste20 family kinase dMST functions as a tumor suppressor by restricting cell proliferation and promoting apoptosis. *Genes Dev.*, 17, 2514-2519.

[29] Nishiyama, Y; Hirota, T; Morisaki, T; Hara, T; Marumoto, T; Iida, S; Makino, K; Yamamoto, H; Hiraoka, T; Kitamura, N; Saya, H. (1999) A human homolog of Drosophila warts tumor suppressor, h-warts, localized to mitotic apparatus and specifically phosphorylated during mitosis. *FEBS Lett.*, 459, 159-165.

[30] Tao, W; Zhang, S; Turenchalk, GS; Stewart, RA; St John, MA; Chen, W; Xu, T. (1999) Human homologue of the Drosophila melanogaster lats tumour suppressor modulates CDC2 activity. *Nat Genet.*, 21, 177-181.

[31] Hori, T; Takaori-Kondo, A; Kamikubo, Y; Uchiyama, T. (2000) Molecular cloning of a novel human protein kinase, kpm, that is homologous to warts/lats, a Drosophila tumor suppressor. *Oncogene*, 19, 3101-3109.

[32] Yabuta, N; Fujii, T; Copeland, NG; Gilbert, DJ; Jenkins, NA; Nishiguchi, H; Endo, Y; Toji, S; Tanaka, H; Nishimune, Y; Nojima, H. (2000) Structure, expression, and chromosome mapping of LATS2, a mammalian homologue of the Drosophila tumor suppressor gene lats/warts. *Genomics*, 63, 263-270.

[33] Taylor, LK; Wang, HR; Erikson, RL. (1996) Newly identified stress-responsive protein kinases, Krs-1 and Krs-2. Proc. *Natl. Acad. Sci.* USA., 93, 10099-10104.

[34] Graves, JD; Gotoh, Y; Draves, KE; Ambrose, D; Han, DK; Wright, M; Chernoff, J; Clark, EA; Krebs, EG. (1998) Caspase-mediated activation and induction of apoptosis by the mammalian Ste20-like kinase Mst1. *EMBO J.*,17, 2224-2234.

[35] Yoo, NY; Soung, YH; Lee, JW; Park, WS; Kim, SY; Nam, SW; Han, JHH; Kim, SH; Lee, JY; Lee, SH. (2003) Mutational analysis of Salvador gene in human carcinomas. *APMIS,* 111, 595-598.

[36] Li, Y; Pei, J; Xia, H; Ke, H; Wang, H; Tao, W. (2003) Lats2, a putative tumor suppressor, inhibits G1/S transition. *Oncogene*, 22, 4398-4405.

[37] Yang, X; Yu, K; Hao, Y; Li, DM; Stewart, R; Insogna, KL; Xu, T. (2004) LATS1 tumor suppressor affects cytokinesis by inhibiting LIMK1. Nat Cell Biol., 6, 609-617.

[38] St John, MA; Tao, W; Fei, X; Fukumoto, R; Carcangiu, ML; Brownstein, DG; Parlow, AF; McGrath, J; Xu, T. (1999) Mice deficient of Lats1 develop soft-tissue sarcomas, ovarian tumours and pituitary dysfunction. *Nat. Genet.*, 21, 182-186.

[39] Hisaoka, M; Tanaka, A; Hashimoto, H. (2002) Molecular alterations of h-warts/LATS1 tumor suppressor in human soft tissue sarcoma. *Lab Invest.*, 82, 1427-1435.

[40] Reszka, AA; Halasy-Nagy, JM; Masarachia, PJ; Rodan, GA. (1999) Bisphosphonates act directly on the osteoclast to induce caspase cleavage of mst1 kinase during apoptosis. A link between inhibition of the mevalonate pathway and regulation of an apoptosis-promoting kinase. *J Biol Chem,* 274, 34967-34973.

[41] Watabe, M; Kakeya, H; Onose, R; Osada, H. (2000) Activation of MST/Krs and c-Jun N-terminal kinases by different signaling pathways during cytotrienin A-induced apoptosis. *J Biol Chem,* 275, 8766-8771.

[42] Lee, KK; Ohyama, T; Yajima, N; Tsubuki, S; Yonehara, S. (2001) MST, a physiological caspase substrate, highly sensitizes apoptosis both upstream and downstream of caspase activation. *J Biol Chem.*, 276, 19276-19285.

[43] Lee, KK; Yonehara, S. (2002) Phosphorylation and dimerization regulate nucleocytoplasmic shuttling of mammalian STE20-like kinase (MST). *J Biol Chem.*, 277, 12351-12358.

[44] Cheung, WL; Ajiro, K; Samejima, K; Kloc, M; Cheung, P; Mizzen, CA; Beeser, A; Etkin, LD; Chernoff, J; Earnshaw, WC; Allis, CD. (2003) Apoptotic phosphorylation of histone H2B is mediated by mammalian sterile twenty kinase. *Cell*, 113, 507-517.

[45] Ishizaki, Y; Jacobson, MD; Raff, MC. (1998) A role for caspases in lens fiber differentiation. *J Cell Biol.*, 140, 153-158.

[46] Weil, M; Raff, MC; Braga, VM. (1999) Caspase activation in the terminal differentiation of human epidermal keratinocytes. *Curr Biol.*, 9, 361-364.

[47] Sun, R; Ravid, K. (1999) Role of a serine/threonine kinase, Mst1, in megakaryocyte differentiation. *J Cell Biochem*, 76, 44-60.

[48] Fernando, P; Kelly, JF; Balazsi, K; Slack, RS; Megeney, LA. (2002) Caspase 3 activity is required for skeletal muscle differentiation. *Proc Natl Acad Sci* USA, 99:11025-11030.

[49] Weber, GF; Menko, AS. (2005) The canonical intrinsic mitochondrial death pathway has a non-apoptotic role in signaling lens cell differentiation; *J of Biol Chem*, 280, 22135-22145.

[50] Alam, A; Cohen, LY; Aouad, S; Sekaly, RP. (1999) Early activation of caspases during T lymphocyte stimulation results in selective substrate cleavage in nonapoptotic cells. *J Exp Med.*, 190, 1879-1890.

[51] Chan, EH; Nousiainen, M; Chalamasetty, RB; Schafer, A; Nigg, EA; Sillje, HH. (2005) The Ste20-like kinase Mst2 activates the human large tumor suppressor kinase Lats1. *Oncogene*, 24, 2076-2086.

[52] Yamamoto, S; Yang, G; Zablocki, D; Liu, J; Hong, C; Kim, S; Soler, S; Odashima, M; Thaisz, J; Yehia, G; Molina, CA; Yatani, A; Vatner, DE; Vatner, SF; Sadoshima, J. (2003) Activation of Mst1 causes dilated cardiomyopathy by stimulating apoptosis without compensatory ventricular myocyte hypertrophy. *Journal of Clinical Investigation*, 111, 1463-1474.

[53] Ullman, KS; Northrop, JP; Verweij, CL; Crabtree GR. (1990) Transmission of signals from the T lymphocyte antigen receptor to the genes responsible for cell proliferation and immune function: the missing link. *Annu Rev Immunol.*, 8, 421-452.

[54] Krogsgaard, M; Huppa, JB; Purbhoo, MA; Davis, MM. (2003) Linking molecular and cellular events in T-cell activation and synapse formation. *Semin Immunol.*, 15, 307-315.

[55] Randriamampita, C; Trautmann, A. (2004) Ca^{2+} signals and T lymphocytes; "new mechanisms and functions in Ca^{2+} *signalling"*. *Biol Cell.*, 96, 69-78.

[56] Crabtree, GR. (1999) Generic signals and specific outcomes: signaling through Ca^{2+}, calcineurin, and NFAT. Cell, 96, 611-614.

[57] Baier, G. (2003) The PKC gene module: molecular biosystematics to resolve its T cell functions. *Immunol Rev.*, 192, 64-79.

[58] Cantrell, D. (1996) T cell antigen receptor signal transduction pathways. *Annu Rev Immunol.*, 14, 259-274.

[59] Im, SH; Rao, A. (2004) Activation and deactivation of gene expression by Ca^{2+}/calcineurin-NFAT-mediated signaling. *Mol. Cells.*, 18, 1-9.

[60] Macian, F; Garcia-Cozar, F; Im, SH; Horton, HF; Byrne, MC; Rao, A. (2002) Transcriptional mechanisms underlying lymphocyte tolerance. *Cell*, 109, 719-731.

[61] Roberts, AI; Devadas, S; Zhang, X; Zhang, L; Keegan, A; Greeneltch, K; Solomon, J; Wei, L; Das, J; Sun, E; Liu, C; Yuan, Z; Zhou, JN; Shi, Y. (2003) The role of activation-induced cell death in the differentiation of T-helper-cell subsets. *Immunol Res.*, 28, 285-293.

[62] Janssen, O; Sanzenbacher, R; Kabelitz, D. (2000) Regulation of activation-induced cell death of mature T-lymphocyte populations. *Cell Tissue Res.*, 301, 85-99.

[63] Baumann, S; Krueger, A; Kirchhoff, S; Krammer, PH. (2002) Regulation of T cell apoptosis during the immune response. *Curr Mol Med.*, 2, 257-72.

[64] Sprent, J; Surh, CD. (2002) T cell memory. *Annu Rev Immunol.*, 20, 551-579.

[65] Lenardo, MJ. (1991) Interleukin-2 programs mouse alpha beta T lymphocytes for apoptosis. *Nature*, 353, 858-861.

[66] Wang, R; Rogers, AM; Rush, BJ; Russell, JH. (1996) Induction of sensitivity to activation-induced death in primary $CD4^+$ cells: a role for interleukin-2 in the negative regulation of responses by mature $CD4^+$ T cells. *Eur J Immunol.*, 26, 2263-2270.

[67] Van Parijs, L; Refaeli, Y; Lord JD; Nelson, BH; Abbas, AK; Baltimore, D. (1999) Uncoupling IL-2 signals that regulate T cell proliferation, survival and Fas-mediated activation-induced cell death. *Immunity*, 11, 281-288.

[68] Boehme, SA; Lenardo, MJ. (1993) Propriocidal apoptosis of mature T lymphocytes occurs at S phase of the cell cycle. *Eur J Immunol.*, 23, 1552-1560.

[69] Refaeli, Y; Van Parijs, L; London, CA; Tschopp, J; Abbas, AK. (1998) Biochemical mechanisms of IL-2-regulated Fas-mediated T cell apoptosis. *Immunity*, 8, 615-623.

[70] Sadlack, B; Merz, H; Schorle, H; Schimpl, A; Feller, AC; Horak, I. (1993) Ulcerative colitis-like disease in mice with a disrupted interleukin-2 gene. *Cell*, 75, 253-261.

[71] Willerford, DM; Chen J; Ferry, JA, Davidson, L; Ma, A; Alt, FW. (1995) Interleukin-2 receptor alpha chain regulates the size and content of the peripheral lymphoid compartment. *Immunity*, 3, 521-530.

[72] Teitelbaum, SL. (2000) Bone resorption by osteoclasts. *Science*, 289, 1504-1508.

[73] Ducy, P; Schinke, T; Karsenty, G. (2000) The osteoblast: a sophisticated fibroblast under central surveillance. *Science*, 289,1501-1504.

[74] Akatsu, T; Murakami, T; Nishakawa, N; Ono, K; Shinomiya, N; Tsuda, E; Mochizuki, S; Yamaguchi, K; Kinosaki, M; Higashio, K; Yamamoto, M; Motovoshi, K; Nagata, N. (1998) Osteoclastogenesis-inhibitory factor suppresses osteoclast survival by interfering in the interaction of stromal cells with osteoclasts. *Biochem Biophys Res Commun.*, 250, 229-234.

[75] Bharti, AC; Takada, Y; Shishodia, S; Aggarwal, BB. (2004) Evidence that receptor activator of nuclear factor (NF)-kappaB ligand can suppress cell proliferation and induce apoptosis through activation of a NF-kappaB-independent and TRAF6-dependent mechanism. *J Biol Chem*, 13, 279, 6065-6076.

[76] Wu, X; Pan, G; McKenna, MA; Zayzafoon, M; Xiong, WC; McDonald, JM. (2005) RANKL regulates Fas expression and Fas-mediated apoptosis in osteoclasts. *J Bone Miner Res.*, 20, 107-116.

[77] Theill, LE; Boyle, WJ; Penninger, JM. (2002) RANKL and RANK: T cells, Bone loss and mammalian evolution. *Annu Rev Immunol*, 20, 795-823.

[78] Walsh, MC; Kim, N; Kadono, Y; Rho, J; Lee, SY; Lorenzo, J; Choi, Y. (2006) Osteoimmunology: Interplay between the immune system and bone metabolism. *Annu. Rev. Immunol.,* 24, 2.1-2.31.

[79] Weitzmann, MN; Cenci, S; Rifas, L; Haug, J; Dipersio, J; Pacifici, R. (2001) T cell activation induces human osteoclast formation via receptor activator of nuclear factor κB ligand-dependent and independent mechanisms. *J Bone Miner Res,* 16, 328-337.

[80] Wang, R; Zhang, L; Shang, X; Moreno, J; Celluzzi, C; Tondravi, M; Shi, Y. (2002) Regulation of activation-induced receptor activator of NF-kB ligand (RANKL) expression in T cells. *Eur J Immunol,* 32, 1090-1098.

[81] Teng, YA; Nguyen, H; Gao, X; Kong, Y; Gorczynski, RM; Singh, B; Ellen, RP; Penninger, JM. (2000) Functional human T-cell immunity and osteoprotegerin ligand control alveolar bone destruction in periodontal infection. J Clin Invest, 106, 59-67.

[82] Teng, YA. (2003) The role of acquired immunity and periodontal disease progression. *Crit Rev Oral Biol Med,* 14, 237-252.

[83] Kong, YY; Feige, U; Sarosi, I; Bolon, B; Tafuri, A; Moroni, S; Capparelli, C; Li, J; Elliott, R; McCabe, S; Wong, T; Campagnuolo, G; Moran, E; Bogoch, ER; Van, G; Nguyen, LT; Ohashi, PS; Lacey, DL; Fish, E; Boyle, WJ; Penninger, JM. (1999) Activated T cells regulate bone loss and joint destruction in arthritis via OGPL. *Nature,* 402, 304-309.

[84] Kawai, T; Seki, M; Hiromatsu, K; Eastcott, JW; Watts, GF; Sugai, M; Smith, DJ; Porcelli, SA; Taubman, MA. (1999) Selective diapedesis of Th1 cells induced by endothelial cell RANTES. *J Immunol.,* 163, 3269-3278.

[85] Takeichi, O; Haber, J; Kawai, T; Smith, DJ; Moro, I; Taubman, MA. (2000) Cytokine profiles of T-lymphocytes from gingival tissues with pathological pocketing. *J Dent Res.,* 79, 1548-55.

[86] Komatsuzawa, H; Kawai, T; Wilson, M; Taubman, MA; Sugai, M; Siginaka, H. (1999) Cloning of the gene encoding the *Actinobacillus actinomycetemcomitans* serotype b OmpA-like outer membrane protein. *Infect Immun,* 67, 942-945.

[87] Perry, MB; MacLean, MM; Gmur, R; Wilson, ME. (1996) Characterization of the O-polysaccharide structure of lipopolysaccharide from *Actinobacillus actinomycetemcomitans* serotype b. *Infect Immun.,* 64, 1215-1219.

[88] Kawai, T; Shimauchi, H; Eastcott, JW; Smith, DJ; Taubman, MA. (1998) Antigen direction of specific T-cell clones into gingival tissues. *Immunology,* 93, 11-19.

[89] Cahalan, MD; Chandy, KG. (1998) Ion channels in the immune system as targets for immunosuppression. *Current Opinion in Biotechnology.* 8, 749-756.

[90] Cahalan, MD; Wulff, H; Chandy, KG. (2001) Molecular properties and physiological roles of ion channels in the immune system. *J Clin Immunol.,* 21, 235-252.

[91] Chandy, KG; Wulff, H; Beeton, C; Pennington, M; Gutman, GA; Cahalan, MD. (2004) K^+ channels as targets for specific immunomodulation. *Trends Pharmacolog Sci.,* 25, 280-289.

[92] Chandy, KG; Wulff, H; Beeton, C; Calabresi, PA; Gutman, GA; Pennington, M. (2006) Kv1.3 Potassium channels: Physiology, Pharmacology and Therapeutic indications. In *Voltage-Gated ion channels as drug targets.* Edited by D. Triggle. WILEY-VCH Verlag Gmbh & Co. KGaA, Weinheim.

[93] Beeton, C; Barbaria, J; Giraud, P; Devaux, J; Benoliel, A; Gola, M; Sabatier, JM; Bernard, D; Crest, M; Beraud, E. (2001) Selective blocking of voltage-gated K+

channels improves experimental autoimmune encephalomyelitis and inhibits T cell activation. *J Immunol.,* 166, 936-944.

[94] Beeton, C; Wulff, H; Barbaria, J; Clot-Faybesse, O; Pennington, M; Bernard, D; Cahalan, MD; Chandy, KG; Beraud, E. (2001) Selective blockade of T lymphocyte K^+ channels ameliorates experimental autoimmune encephalomyelitis, a model for multiple sclerosis. *Proc Natl Acad Sci* USA, 98, 13942-13947.

[95] Judge, TA; Tang, A; Turka, LA. (1996) Immunosuppression through blockade of CD28:B7-mediated costimulatory signals. *Immunol Res.,* 15, 38-49.

[96] Neurath, MF; Pettersson, S; Meyer zum Buschenfelde, KH; Strober, W. (1996) Local administration of antisense phosphorothioate oligonucleotides to the p65 subunit of NF-kappa B abrogates established experimental colitis in mice. *Nat Med.,* 2, 998-1004.

[97] Roshak, AK; Capper, EA; Stevenson, C; Eichman, C; Marshall, LA. (2000) Human calcium-independent phospholipase A2 mediates lymphocyte proliferation. *J Biol Chem.,* 275, 35692-35698.

[98] Valverde, P; Obin, MS; Taylor, A. (2004) Role of Gas6/Axl signaling in lens epithelial cell proliferation and survival. *Exp Eye Res.,* 78, 27-37.

[99] Valverde, P. (2005) Effects of Gas6 and hydrogen peroxide in Axl ubiquitination and downregulation. *Biochem Biophys Res Commun.,* 333, 180-185.

[100] Valverde, P; Tu, Q; Chen, J. (2005) BSP and RANKL induce osteoclastogenesis and bone resorption synergistically. *J Bone Miner Res.,* 20, 1669-1679.

[101] Chen, W; Li, YP. (1998) Generation of mouse osteoclastogenic cell lines immortalized with SV40 large T antigen. *J Bone Miner Res.,* 13, 1112-1123.

[102] Shahinian, A; Pfeffer, K; Lee, KP; Kundig, TM; Kishihara, K; Wakeham, A; Kawai, K; Ohashi, PS; Thompson, CB; Mak, TW. (1993) Differential T cell costimulatory requirements in CD28-deficient mice. *Science,* 261, 609-612.

[103] Chambers, CA; Allison, JP. (1999) Costimulatory regulation of T cell function. *Curr Opin Cell Biol.,* 11, 203-210.

[104] Schwartz, RH. (1990) A cell culture model for T lymphocyte clonal anergy. *Science.* 15, 248, 1349-1356.

[105] Baroja, ML; Lorre, K; Van Vaeck, F; Ceuppens, JL. (1989) The anti-T cell monoclonal antibody 9.3 (anti-CD28) provides a helper signal and bypasses the need for accessory cells in T cell activation with immobilized anti-CD3 and mitogens. *Cell Immunol.,* 15, 120, 205-217.

[106] Clipstone, NA; Crabtree, GR. (1993) Calcineurin is a key signaling enzyme in T lymphocyte activation and the target of the immunosuppressive drugs cyclosporine A and FK506. *Ann N Y Acad Sci.,* 696, 20-30.

In: Focus on Signal Transduction Research
Editor: Gavin McAlpine, pp. 103-122

ISBN 1-60021-376-6

Chapter 4

NMDA RECEPTOR HYPOFUNCTION IN SCHIZOPHRENIA: FROM MOUSE TO MAN

Stephen I. Deutsch*,[1,2] John Mastropaolo,[1] Barbara L. Schwartz,[1,2] Richard B. Rosse,[1,2] and Brooke Gaskins[1]

[1]Mental Health Service Line, Department of Veterans Affairs Medical Center NW, Washington, DC

[2]Department of Psychiatry, Georgetown University School of Medicine, Washington, DC

ABSTRACT

Glutamate is one of the most abundant neurotransmitters in the brain; it may be responsible for mediating synaptic neurotransmission at 40 percent or more of the brain's synapses. The glutamatergic signal is recognized by both metabotropic and ion channel receptors. The N-methyl-D-aspartic acid (NMDA) receptor is an example of a glutamate-gated ion channel receptor; the binding of glutamate to this receptor increases the likelihood that its channel will transiently assume an open-configuration, allowing the influx of calcium ions. Importantly, there are allosteric modulatory sites, including an obligatory co-agonist site that recognizes glycine, on the NMDA receptor complex that influence glutamate's ability to promote calcium ion conductance. Abnormalities of NMDA receptor-mediated neurotransmission are implicated in a variety of major neuropsychiatric disorders, especially schizophrenia. Specifically, NMDA receptor hypofunction (NRH) is implicated in the pathophysiology of schizophrenia because of the ability of phencyclidine (PCP) to precipitate a schizophreniform psychosis. PCP is a noncompetitive NMDA receptor antagonist that binds to a hydrophobic site within the channel; it is referred to as an open-channel blocker. Descriptively, the PCP-model of schizophrenia is the best pharmacological model of this disorder with symptoms manifest in all of the relevant domains of psychopathology, including positive (e.g.,

* Corresponding Author: Stephen I. Deutsch, M.D., Ph.D. Associate Chief of Staff for Mental Health Department of Veterans Affairs Medical Center, 50 Irving Street, NW, Washington, DC 20422; 202 745 8156 (Office); 202 745 8169 (Fax); E-Mail Address: Stephen.Deutsch@med.va.gov

hallucinations), negative (e.g., affective flattening), cognitive (e.g., abnormalities of attention), mood (e.g., dysphoria) and motor symptoms (e.g., posturing).

Because of PCP's pharmacological action as a noncompetitive NMDA receptor antagonist and the PCP-model of schizophrenia, our laboratory has been engaged in characterizing and quantifying behaviors in mice, including genetically-inbred strains, elicited by MK-801 (dizocilpine), a high-affinity analogue of PCP that binds to the same hydrophobic channel domain. Presumably, blockade of the NMDA receptor by MK-801 is a pharmacological strategy for creating NRH in the intact animal. We characterized two behavioral outcome measures: the dose-dependent ability of MK-801 to raise the threshold voltage required for the precipitation of tonic hindlimb extension and elicit irregular episodes of intense jumping behavior (referred to as "popping") in mice. Moreover, 24 hours after mice are forced to swim for up to 10 minutes in cold water, the ability of MK-801 to antagonize electrically precipitated seizures is reduced. Thus, stress affects the endogenous tone of NMDA receptor-mediated neurotransmission. Genetically inbred strains of mice differ in their behavioral sensitivity to MK-801 on these two measures; interestingly, the BALB/c inbred mouse strain is more sensitive than other inbred strains and the NIH Swiss outbred mouse strain. Stress and genetic strain differences interact to affect behavioral sensitivity to MK-801. Importantly, these animal models serve to test the ability of glycinergic interventions to modulate NMDA receptor-mediated neurotransmission in the intact animal. Thus, they are used as screening procedures for NMDA receptor agonist interventional strategies that may be developed as medications for the treatment of schizophrenia. A series of glycinergic interventions have been tested in these models. Also, these paradigms are useful in demonstrating that drugs, whose direct actions are on other neurotransmitter systems, affect NMDA receptor-mediated neurotransmission in the intact animal. Preliminary clinical trials of a few NMDA receptor agonist interventions have been conducted in patients with schizophrenia. In general, the medications are well-tolerated and the results suggest that the adjuvant therapeutic efficacy of specific glycinergic interventions may be influenced by state of illness and the specific antipsychotic medication chosen for maintenance pharmacotherapy (e.g., clozapine). The chapter will concentrate on the results of our laboratory and clinical investigations, whose primary aim was modulation of NMDA receptor-mediated neurotransmission in mice and man.

Key Words: Glutamate, Glycine, NMDA Receptor, Stress, BALB/c Mice, Schizophrenia

Anatomy and Biochemical Pharmacology of the NMDA Receptor Complex

The ion channel that is a part of the N-methyl-D-aspartate (NMDA) receptor complex, a type of glutamate-gated receptor ion channel, has properties of both ligand- and voltage-gating. The channel is named for NMDA, the glutamate analogue that binds preferentially to this receptor. Importantly, a variety of allosteric modulatory sites, covalent modifications of the complex (e.g., phosphorylation and nitrosylation), gene expression, and environmental and developmental factors influence the likelihood that glutamate will be effective in promoting channel opening. The complex regulation of NMDA receptor-gated channel opening makes it both susceptible to pathologic disruption by a variety of mechanisms and affords opportunities for pharmacologic manipulations (Bonhaus et al., 1987; Fagg, 1987; Foster and Wong, 1987).

The channel is comprised of individual polypeptide subunits with multiple transmembranous domains that align themselves to create a channel. The polypeptide receptor subunits are of two types, designated NR1 and NR2. The "obligatory glycine co-agonist site" is located on the NR1 subunit, whereas the site that binds glutamate is on the NR2 subunit. Although these sites are distinct from each other, they interact so that the binding of glycine affects glutamate's affinity for the receptor and the efficiency of coupling between its binding and channel opening. There are four classes of NR2 receptors, designated A through D, that differ in terms of anatomic distribution and their influence on the pharmacological properties of the receptor, including desensitization kinetics and the "modulatory" properties of glycine. Also, subunit expression is regulated by stage of development; in fact, an NR3 subunit appears transiently during development (Deutsch et al., 1998, 2001; Millan, 2005).

Ordinarily, at the resting membrane potential, magnesium ions occupy multiple sites within the channel, obstructing the movement of calcium ions. This magnesium ion blockade is relieved with the depolarization of the membrane, serving as the basis of the voltage-gating of the receptor. Thus, the glutamate-gated activation of NMDA receptors is dependent on membrane depolarization, which may necessitate coordinating the activation of other neurotransmitter receptors with the NMDA receptor. The hydrophobic channel domain possesses binding sites for compounds like phencyclidine (PCP), ketamine and memantine, which are referred to as "open-channel blockers." Interestingly, the functional consequences of blockade or antagonism of the NMDA receptor complex may differ depending upon whether the blockade is due to a competitive (i.e., a compound that binds to glutamate's agonist recognition site on NR2 subunits) or noncompetitive (i.e., a compound that binds to a site within the hydrophobic channel domain) antagonist (Deutsch et al., 1996). Pretreatment with competitive antagonists can prevent channel opening, which is a necessary condition for the action of noncompetitive antagonists.

The complex regulation of channel opening is consistent with the importance of NMDA receptor-mediated neurotransmission to a number of critical processes linked to regions wherein which this receptor is enriched, such as the hippocampus, thalamus and frontal cortex. Moreover, the astrocyte, which enjoys an intimate association with pre-and post-synaptic elements, participates in regulating availability of glutamate, glycine and ligands acting at the so-called Glycine$_B$ co-agonist site (Millan, 2005). In fact, NMDA receptor-mediated neurotransmission is a cogent example of the active participation of glial elements in synaptic transmission.

Glial cells serve as an important potential source of glycine within the area of the synapse; glycine is derived from L-serine by the action of serine hydroxymethltransferase, a key enzyme involved in the regulation of one-carbon metabolism. Further, "excess" glycine is cleared from the synapse by glial cells via several independent, but related mechanisms: glycine-1 transporters (GlyT-1), which transport glycine into the cell in a sodium ion- and chloride ion-dependent manner; small neutral amino acid transporters (SNAT) located on the surface of astrocytes; and the glycine cleavage enzyme system, which is a mitochondrial enzyme complex that oxidatively decarboxylates or "cleaves" glycine such that its alpha carbon enters the one-carbon pool complexed with tetrahydrofolate (Deutsch et al., 1990, 1998; Millan, 2005). D-Serine has been described as a naturally-occurring high-affinity agonist for the $Glycine_B$ co-agonist site. D-Serine is derived from L-serine by the actions of D-serine racemase, an enzyme found within astrocytes in regions of forebrain enriched in NMDA receptors. The clearance of D-serine is via transporters and D-amino acid oxidase, a glial enzyme that deaminates D-serine. Glial cells participate in the "recycling" of glutamate that is released from presynaptic nerve terminals in a depolarization-dependent and calcium ion-dependent manner; in addition to neurons, glial cells possess sodium ion-dependent excitatory amino acid transporters (EAAT) on their surface. The glutamate taken up by glial cells is converted to glutamine, via the action of glutamine synthase, a compound that is released, taken up by neurons, and converted back into glutamate by the action of glutaminase. Glial cells also possess the enzymatic machinery responsible for an alternative pathway for tryptophan metabolism, referred to as the kynurenine pathway (Rosse et al., 1992; Millan, 2005). The kynurenine pathway leads to the formation of kynurenic acid, a competitive antagonist of glycine at the $Glycine_B$ co-agonist binding site, and quinolinic acid, an excitotoxic compound. Conceivably, the synthesis and release of kynurenic acid is a "local" mechanism for regulating NMDA receptor-mediated neurotransmission at the level of the synapse (Rosse et al., 1992).

The surface of astrocytes also regulate levels and availability of N-acetyl-aspartate-glutamate (NAAG), a neurally-active dipeptide that influences presynaptic release of L-glutamate, serves as a potential "local" source of L-aspartate and L-glutamate, two excitatory amino acid neurotransmitters, and is a weak NMDA receptor antagonist. NAAG is a substrate for glutamate carboxypeptidase II and III, which are located on astrocyte plasma membranes and cleave NAAG into L-glutamate and N-acetyl-aspartate (NAA), the source of L-aspartate. There are provocative postmortem data suggesting that elevated levels of NAAG can occur in selected brain regions of schizophrenia patients (i.e., temporal cortex, hippocampus, and frontal cortex), because of reduced activity of glutamate carboxypeptidase II (GCPII), the hydrolytic enzyme enriched on astrocytes responsible for forming NAA and L-glutamate (Coyle, 1996). Importantly, the existence of these glial pathways and mechanisms for modulating the gating efficiency of L-glutamate and NMDA receptor-associated ion channel properties offers platforms for developing medications that can adjust the gain or "fine tune" NMDA receptor-mediated neurotransmission (Millan, 2005). Ideally, these medications would be indicated for a variety of neuropsychiatric disorders, including schizophrenia

Glutamate is a major neurotransmitter employed in intracortical projection pathways, conveying this integrated cortical information along efferent projections to subcortical structures in the midbrain and basal ganglia, and to the hippocampus. Glutamate is prominent within the hippocampus, where it is implicated in the induction of long-term potentiation, an electrophysiological analogue of learning and memory, and thalamus, where it may play a role in "thalamic filtering." Apart from their enrichment in anatomic areas that are linked to the psychopathology of schizophrenia (e.g., frontal cortex and hippocampus), the precipitation of a drug-induced schizophreniform psychosis by PCP in susceptible individuals has focused intense interest on a pathogenetic role of the NMDA receptor in schizophrenia (Deutsch et al., 1989, 2001; Coyle, 1996; Hirsch et al., 1997; Tamminga, 1998). L-Glutamate mediates both "fast" synaptic transmission (i.e., ligand-gated channels) and "slow" synaptic transmission (i.e., "metabotropic" receptors coupled to the synthesis of small, soluble intracellular messengers). There are three "types" of glutamate-gated ion channels, designated NMDA, kainic acid (KA), and *alpha*-amino-3-hydroxy-5-methylisoxazole-4-propionic acid (AMPA) receptors.

A significantly diminished density of dendritic spines on layer III pyramidal neurons has been reported in the cerebral cortex , especially frontal and temporal cortex, of autopsied schizophrenia brains (Hirsch et al., 1997). Importantly, glutamate receptors are located on the dendritic spines of layer III pyramidal cells in the cerebral cortex. Thus, glutamatergic projections "link" anatomic areas that are implicated pathophysiologically or by histopathology in schizophrenia (Coyle, 1996; Farber et al., 1998; Jentsch and Roth, 1999).

PATHOPHYSIOLOGY OF NMDA RECEPTOR HYPOFUNCTION IN SCHIZOPHRENIA

Descriptively, the psychosis caused by PCP resembles naturally-occurring schizophrenia with symptoms manifest in all of the relevant domains of psychopathology, including positive (e.g., hallucinations), negative (e.g., affective blunting), cognitive (e.g., inability to abstract and impairments of attention and working memory), mood (e.g., dysphoria) and motor (e.g., mannerisms) symptoms. In fact, the "PCP model of schizophrenia" is widely considered to be the best drug-induced model of this disorder (Deutsch et al., 1989). Because PCP is a noncompetitive NMDA receptor antagonist that interferes with glutamate's ability to promote calcium ion conductance across the channel, NMDA receptor hypofunction (NRH) or a glutamatergic deficiency has been proposed as a pathophysiological mechanism of this disorder (Deutsch et al., 1989). Further, because of reciprocal connections and complex circuitry, a disturbance or "imbalance" of glutamatergic transmission would affect neurotransmission mediated by dopamine, acetylcholine and gamma aminobutyric acid (GABA), among other neurotransmitters that have been implicated in the pathophysiology of this disorder (Deutsch and Hitri, 1993; Deutsch et al., 1995, 2001, 2002, 2003, 2005; Mastropaolo et al., 2004). NMDA receptors exist on the surface of GABAergic inhibitory neurons within the hippocampus; GABA is the major inhibitory amino acid neurotransmitter within the brain. Given this location, a hypothesized consequence of NRH is a failure to "stimulate" these GABAergic inhibitory interneurons, leading to disinhibition of projections originating downstream from the GABAergic nerve terminal. A heuristic model that has been evoked to explain progressive neurodegeneration in at least a subgroup of patients with schizophrenia and is guiding medication development suggests that NRH leads to disinhibition of glutamatergic projections, whose consequence includes excessive stimulation of the AMPA/KA class of glutamate-gated receptor complexes (Deutsch et al., 2001). Excessive stimulation of AMPA/KA receptors could lead to excitotoxic cell death, a process resulting from disruption of unequal ionic gradients across membranes, depletion of energy reserves as ATP-dependent ion pumps attempt to restore these gradients, generation of reactive oxygen species, and cell death from apoptotic and other mechanisms. The proposed existence of disinhibited glutamatergic neurotransmission and the cascade of excitotoxic events offer many opportunities and suggestions for therapeutic interventions, including facilitation of NMDA receptor-mediated neurotransmission, potentiation of GABAergic neurotransmission, antagonism of AMPA/KA receptors, and "quenching" of locally generated reactive oxygen species (Deutsch et al., 2001).

An anatomical basis exists for defective GABAergic inhibitory mechanisms in schizophrenia that is independent of NRH, but could also contribute to progressive excitotoxicity. Specifically, a significant reduction of GABAergic interneurons in layers II, III, V, and VI of the anterior cingulate has been reported in postmortem brain obtained from schizophrenia patients (Benes et al., 1992; Coyle, 1996; Jentsch and Roth, 1999). Biochemical data also support defective GABAergic inhibitory mechanisms in schizophrenia, including reduced cortical density of reuptake sites for ^{3}H-GABA and a reduction in the binding of ^{3}H-nipecotic acid, a biochemical marker of the sodium-dependent GABA-reuptake site, in medial temporal lobe structures (i.e., left polar temporal cortex, and amygdala and hippocampus bilaterally) and basal ganglia. Reduced activity and levels of mRNA for glutamate decarboxylase (GAD), the synthetic enzyme responsible for GABA production, have been reported that could contribute to diminished GABAergic activity (Benes et al., 1992).

Molecular Genetic and Pharmacological Investigations of NRH and Possible Mechanisms of Progressive Excitotoxicity in Schizophrenia

When levels of mRNA of the alternatively spliced short and long forms of the $gamma_2$ subunit of the $GABA_A$ receptor complex were studied in prefrontal cortex from a postmortem cohort of five schizophrenia patients and five controls (Huntsman et al., 1998), the levels of the short form of this subunit were markedly decreased (i.e., 51.7% of control levels) in the schizophrenia patients. The relative overexpression of the long form of the $gamma_2$ subunit may result in defective "transduction" of the GABA signal in at least some patients with schizophrenia.

Neurons in the cingulate cortex (posterior cingulate/retrosplenial cortex) and other corticolimbic regions (including piriform cortex, entorhinal cortex, hippocampus and amygdala) are sensitive to toxic intraneuronal vacuolization caused by noncompetitive NMDA receptor antagonists, such as PCP and MK-801 (dizocilpine) (Olney et al., 1989). Benzodiazepine agonists (i.e., drugs related to diazepam), which potentiate GABAergic inhibitory mechanisms, may protect against this PCP/MK-801-induced toxic intraneuronal vacuolization. Females may be more sensitive to PCP/MK-801-induced damage than males, and dendritic spines may be a specific target. Destruction of dendritic spines would lead to a loss of synaptic complexes and disruption of "connectivity". Thus, NRH leads to diminished GABAergic inhibitory tone, whose downstream consequences include toxic intraneuronal vacuolization and destruction of dendritic spines (Coyle, 1996; Farber et al., 1998).

Discrete subgroups of schizophrenia patients may exist with NRH (due to diminished synthesis or altered "ratios" of specific NMDA receptor subunits resulting in fewer heteromeric NMDA receptors or "combination[s]" with diminished sensitivity to L-glutamate) (Jentsch and Roth, 1999; Moghaddam et al., 1997). Reduced genomic expression of NR1, a specific polypeptide subunit of the NMDA receptor complex, was reported in the superior temporal cortex in an autopsy series of schizophrenia brains (Humphries et al., 1996). Moreover, significant associations were found between levels of NR1 mRNA and measures of cognition, including the Global Deterioration Scale, the Mini-Mental State Examination, and the National Adult Reading Test, which provides information about premorbid IQ scores.

In a postmortem study, regional differences in the relative expression of NMDA receptor subunits were found in postmortem brains from schizophrenia patients with "a chronic, nonremitting pattern of illness (n=15)", controls with no evidence of neurological or psychiatric disorders (n=15), and "neuroleptic-treated controls (n=8)" (Akbarian et al., 1996). The schizophrenia cohort showed a relatively selective increase (greater than 50%) in levels of mRNA for the NR2D subunit in the prefrontal cortex, which may be a compensatory response, making these receptors more excitable.

In another postmortem study, data were consistent with an increased density of both the total population of NMDA receptors and NMDA receptors composed of assemblies of NR1 and NR2B subunits in the superior temporal cortex (Grimwood et al., 1999), which were not observed in the premotor cortex. Although these binding data conflict with the molecular data showing reduced levels of NR1 mRNA in a cognitively impaired subgroup of schizophrenia patients, they are consistent with NMDA receptor abnormalities of a neurodevelopmental or compensatory nature in schizophrenia. Thus, differential gene expression and alternative splicing of individual NMDA receptor subunits may represent a genetic mechanism of NRH. There is also a report consistent with the possible bilateral upregulation of NMDA receptors in the orbital frontal cortex of schizophrenia patients (Simpson et al., 1992). Another postmortem study reported an increased density of the strychnine-insensitive glycine binding site in several cortical regions of schizophrenia brain, consistent with upregulation of the NMDA receptor complex (Ishimaru et al., 1994). Inverse relations were suggested in several postmortem brain regions of schizophrenia patients between (elevated) ^{3}H-KA binding and (diminished) ^{3}H-MK-801 binding (for review, see Deutsch et al., 2001).

A variety of postmortem studies of schizophrenia patients report diminished binding (e.g., ^{3}H-KA) and expression of nonNMDA receptor subunits (e.g., GluR1, GluR2, KA2, and GluR6), especially in hippocampus (Jentsch and Roth, 1999), reflecting, perhaps, anomalous development of medial temporal structures in schizophrenia. AMPA receptor binding may be increased in the caudate nucleus of schizophrenia patients, suggesting that pathological changes in binding may be regionally selective (Noga et al., 1997).

BIOCHEMICAL AND BEHAVIORAL STUDIES OF NRH

Under certain circumstances, PCP is proconvulsant, which may relate to the ability of acute administration of relatively low, subanesthetic doses of ketamine (i.e., 10 to 30 mg/kg, intraperitoneally), a noncompetitive NMDA receptor antagonist, to cause efflux of L-glutamate in the prefrontal cortex of the rat resulting in behavioral excitation and seizures (Moghaddam et al., 1997). Acute antagonism of NMDA receptor-mediated neurotransmission by administration of ketamine (30 mg/kg, intraperitoneally) to rats increased extracellular levels of dopamine in the prefrontal cortex that was blocked by intracortical infusion of CNQX (6-cyano-7-nitroquinoxaline-2,3-dione; 50 μM), an AMPA/KA receptor antagonist. These data support the role of "nonNMDA" glutamatergic mechanisms in cortical release of dopamine. Also, local blockade of AMPA/KA receptors in the prefrontal cortex of the rat by CNQX attenuated the ability of stress to stimulate the release of dopamine. Behavioral correlates of acutely elevated extracellular levels of L-glutamate include impaired working memory and behavioral activation. Moreover, systemically or locally infused AMPA/KA anatagonists are capable of attenuating the acutely disruptive effects of "low dose" noncompetitive NMDA receptor antagonists such as ketamine (for review, see Deutsch et al., 2001).

MK-801-Elicited Behaviors: Strategies for Studying NRH and its Pharmacological Modulation

The ability of PCP to precipitate a schizophreniform psychosis and its pharmacological action as a noncompetitive NMDA receptor antagonist stimulated interest in the characterization of behaviors elicited in animals by noncompetitive NMDA receptor antagonists, such as PCP and MK-801 (dizocilpine). Ideally, these animal models would serve as useful screening paradigms for the identification of candidate compounds that could attenuate the severity of behavioral consequences linked to NRH and, thus, could serve as potential medications for the treatment of schizophrenia. Also, these animal models could contribute to a fuller understanding of the pathophysiological consequences of NRH, serving, perhaps, as animal models of at least some aspects of schizophrenia itself.

A particularly informative animal behavior that we observed and characterized is the ability of MK-801 (dizocilpine) to elicit irregular episodes of intense jumping behavior in mice that has been referred to as "popping" (Deutsch and Hitri, 1993; Rosse et al., 1995). As noted, MK-801 is an open-channel blocker, whose actions depend on accessibility to a hydrophobic domain within the open channel. Thus, administration of MK-801 is a pharmacological mechanism for creating a condition of NRH; also, a dose response relation was demonstrated for its effectiveness in eliciting popping. A series of experiments showed that the elicitation of popping is unique to noncompetitive NMDA receptor antagonists; thus, popping is a behavioral outcome measure that distinguishes between noncompetitive (e.g., MK-801) and competitive NMDA receptor antagonists (i.e., compounds that bind to glutamate's agonist recognition site). Whereas MK-801 was able to elicit popping in a dose-dependent fashion, a derivative of CGP (2-amino-4-methyl-5-phosphono-3-pentenoic acid), a competitive glutamate antagonist, was unable to do so (Deutsch et al., 1996). Moreover, when mice were pretreated with the CGP derivative, MK-801's ability to elicit popping was blocked. Thus, when glutamate-gated opening of the channel is antagonized competitively by the CGP derivative and fewer channels are able to assume the open configuration, MK-801 cannot gain access to its binding site within the hydrophobic channel domain to elicit popping (Deutsch et al., 1996). Environmental factors influence popping behavior as shown by the fact that mice tested in cages with crushed corn-cob bedding on the floor showed a three-fold greater intensity of MK-801-elicited popping behavior than that of mice tested without this bedding (Deutsch and Hitri, 1993). The intensity of MK-801-elicited popping behavior could be attenuated in dose-dependent fashion by both a conventional (i.e., haloperidol) and atypical (i.e., clozapine) antipsychotic medication (Deutsch and Hitri, 1993; Rosse et al., 1995).

In addition to environmental factors and the ability of pharmacological agents to influence MK-801-elicited popping behavior, the laboratory showed that genetic factors contribute significantly to sensitivity to its elicitation (Deutsch et al., 1997). Specifically, the genetically inbred BALB/c strain of mouse was more sensitive to MK-801-elicited popping than three other inbred strains (i.e., C57BL/6, AKR, and DBA/2) and one outbred strain (NIH Swiss). These data on heightened behavioral sensitivity of the BALB/c strain to MK-801-elicited popping is consistent with work showing that this inbred strain is more sensitive to locomotor stimulating effects of PCP and anticonvulsant effects of MK-801 (Freed et al., 1984). The inheritance of sensitivity to the behavioral effects of MK-801/PCP is complex and likely will be shown to be regulated by multiple genes, each of which contributes a small amount to this sensitivity. The deciphering of the mouse genome and the availability of closely spaced positional markers encourages quantitative trait loci (QTL) analyses to identify loci contributing to "sensitivity" and "resistance." Based on our pharmacological studies, we anticipate that at least some of the genetic strain differences will reflect NMDA receptor subunit combinations that confer differences in binding of glutamate, glycine, and MK-801 as well as differences in the efficiency of coupling the binding of glutamate and glycine with channel opening (Deutsch et al., 1997). However, it is also likely that some of the differences between strains reflect their differences in metabolism, distribution, brain uptake or elimination of MK-801. In any event, the statistical strategies for determining relations between genetic loci and behavioral traits in mice might lead to the identification of homologous loci in the human genome conferring sensitivity to PCP-psychosis in particular, and psychosis and schizophrenia in general.

Using MK-801-elicited mouse popping behavior as a behavioral outcome measure, the laboratory explored the ability of a variety of compounds to attenuate this behavioral effect of MK-801, which is a behavioral reflection of NRH in mice. Because NRH is a hypothesized pathophysiological mechanism of schizophrenia and PCP, a noncompetitive NMDA receptor antagonist, precipitates a drug-induced schizophreniform psychosis with much descriptive similarity to schizophrenia, compounds that attenuate MK-801-elicited mouse popping behavior are potential candidates for development as antipsychotic medications. Also, the ability of a compound that does not act at the NMDA receptor complex directly to attenuate an MK-801-elicited behavior may reveal important functional relationships between neurotransmission mediated by the NMDA receptor and other specific receptors. Thus, blockade of dopamine attenuates MK-801-elicited popping, consistent with the intimate relationship that these two neurotransmitter systems enjoy (Deutsch and Hitri, 1993). To date, based on an evolving preclinical literature, our laboratory has shown that the following compounds, which have very different pharmacological actions, attenuate MK-801-elicited popping behavior: inhibitors of neuronal nitric oxide synthase (i.e., 7-nitroindazole), topiramate, galantamine, and anabasine (Deutsch et al., 1996, 2002, 2003; Mastropaolo et al., 2004). These findings stimulated interest in development and preliminary clinical trials of these, related compounds or related medication strategies as sole or adjuvant approaches for the treatment of schizophrenia (Deutsch et al., 1997, 2003; Drapalski et al., 2001; Rosse et al., 2002). For example, topiramate is of interest because it addresses what may be "downstream" consequences of NRH, such as a dampening of GABAergic inhibitory neurotransmission and excessive stimulation of the AMPA/kainate classes of glutamate receptor. Topiramate is approved for use as an anticonvulsant in the pediatric population; its primary mechanisms of action include antagonism of excitatory amino acid receptors (e.g., kainate receptors) and potentiation of GABAergic neurotransmission in a manner that differs from that of the benzodiazepines. Galantamine is of interest because of its action as a positive allosteric modulator of nicotinic acetylcholine receptors; thus, galantamine would be expected to potentiate cholinergic neurotransmission. In addition to its action as a positive allosteric modulator of nicotinic acetylcholine receptors, galantamine is an inhibitor of acetylcholinesterase and, thus, would be expected to increase and prolong the lifetime of acetylcholine within the synapse. There are very exciting and provocative data suggesting that deficient expression of the alpha7 subtype of nicotinic acetylcholine receptor occurs in schizophrenia. Conceivably, galantamine could target deficits of cholinergic neurotransmission, such as deficient expression of the alpha7 nicotinic acetylcholine receptor. Because of the pathophysiological role attributed to the diminished expression of alpha7 nicotinic acetylcholine receptors in schizophrenia (e.g., impaired sensory inhibition and deficits of voluntary smooth pursuit eye movements), our laboratory examined the ability of anabasine, a direct acting alpha7 nicotinic acetylcholine receptor agonist, to influence MK-801-elicited mouse popping behavior. Anabasine was able to attenuate popping behavior. Unfortunately, the alpha7 nicotinic acetylcholine receptor desensitizes rapidly upon exposure to agonist; thus, anabasine is not likely to be effective when administered chronically, which would be necessary in an illness like schizophrenia (Deutsch et al., 2005). However, our data support consideration of alpha7 nicotinic acetylcholine receptor agonist strategies that would not be limited by the rapid desensitization of this receptor. From a practical clinical perspective, our group has been conducting a pilot investigation of a strategy that combines galantamine, taking advantage of its property as a positive allosteric modulator of nicotinic acetylcholine receptors in general, with a selective alpha7 nicotinic acetylcholine receptor agonist in an effort to overcome the problem with receptor desensitization. Because of agonist-induced receptor desensitization, an alpha7 nicotinic acetylcholine receptor agonist could become a "functional" antagonist when administered for a chronic neuropsychiatric disorder. From a theoretical perspective, it is hoped that galantamine will preserve the receptor in a sensitive, as opposed to refractory, state and enhance the efficiency of coupling between the binding of the selective alpha7 nicotinic acetylcholine receptor agonist and channel opening. Importantly, our animal paradigm of MK-801-elicited popping behavior provided a compelling preclinical rationale for pursuing alpha7 nicotinic acetylcholine receptor agonist strategies for the treatment of schizophrenia. In fact, the MK-801-elicited mouse popping paradigm also provided the preclinical rational for clinical examination of topiramate and galantamine as sole adjuvant medication interventions (Deutsch et al., 2005).

Our laboratory has also investigated MK-801's ability to raise the threshold voltage for the elicitation of tonic hindlimb extension in mice (an antiseizure effect of the drug). This paradigm reflects the endogenous tone of NMDA receptor-mediated neurotransmission. Moreover, we have used changes in the dose-response relation for antagonizing seizure elicitation with MK-801 to demonstrate the effects of environmental manipulations, pharmacological interventions and genetic factors on NMDA receptor-mediated neurotransmission. These results have immediate relevance and implications for our understanding of the pathophysiology of schizophrenia and development of novel pharmacological interventions that are not dependent directly on the selective antagonism of the dopamine type 2 (D2) receptor. An unexpected early finding that has been replicated several times is the fact that 24 hours after mice are forced to swim for up to 10 minutes in cold water (6^0C), the ability of MK-801 to antagonize electrically precipitated seizures is significantly reduced. These data suggest that 24 hours after exposure of mice to a profound stressor, fewer NMDA receptor-associated channels are in the open configuration and/or channel properties are changed such that MK-801 no longer binds to its site within the hydrophobic domain of the open channel. The latter changes could result from genetic expression, resulting in altered combinations of receptor subunits conferring changed pharmacological properties, covalent modifications of the receptor (such as changes in the state of phosphorylation), or changes in levels of endogenous modulators of the NMDA receptor (e.g., polyamines and neurosteroids), among other mechanisms. Our data support both possibilities: fewer channels in the open configuration and changes in channel properties.

The stress-induced reduction in the ability of MK-801 to antagonize the elicitation of tonic hindlimb extension in mice could be a behavioral manipulation that results in NRH. In order to examine whether or not stress affects "functional" properties of the channel, we tested the ability of a series of open-channel blockers to raise the threshold voltage for the elicitation of tonic hindlimb extension in mice, in addition to MK-801 serving as the comparator control (i.e., PCP, ketamine and memantine), in nonhandled control and stressed mice. Interestingly, a single session of cold water swim stress reduced the antiseizure efficacies of MK-801 and memantine without affecting PCP and ketamine when tested 24 hours later. Thus, there is the possibility that stress affects the charge, size and/or shape characteristics of the channel such that entry to the channel is "denied" to molecules sharing shape and charge distribution characteristics of MK-801 and memantine, whereas entry is permitted to molecules whose features resemble those of PCP and ketamine.

Similar to our findings using the MK-801-elicited mouse popping paradigm, genetic strain differences were observed between four inbred (i.e., BALB/c, C57BL/6, AKR, and DBA/2) and the outbred NIH Swiss mouse strains in terms of the ability of MK-801 to antagonize the electrical precipitation of tonic hindlimb extension. The BALB/c strain was most sensitive to the antiseizure effects of MK-801. Again, the heightened sensitivity of the BALB/c strain could reflect a higher percentage of its NMDA-associated ionophores in the "active" or open configuration, compared with the other strains, or the fact that its channels are constructed from combinations of constituent receptor subunit polypeptides that result in channel domains with relatively greater affinity for MK-801. However, these differences may also be due, in part, to differences between strains in the peritoneal absorption (MK-801 is given intraperitoneally in these experiments), blood-brain barrier penetrability, biodistribution, metabolism or clearance of MK-801. In any event, our data suggest that genetic factors contribute significantly to both seizure proneness and the ability of MK-801 to antagonize seizure elicitation (Deutsch et al., 1998).

Glycinergic agonist interventions have been proposed as a treatment for the presumptive NRH occurring in schizophrenia; it is hoped that agonists for the strychnine-insensitive glycine binding site on the NMDA receptor would "fine tune" or adjust the "gain" of deficient NMDA receptor-mediated neurotransmission. From a theoretical perspective, glycinergic agonist interventions would also be devoid of the potential for excitotxicity that directly-acting glutamatergic interventions would possess. Our laboratory has shown that the effectiveness of glycinergic agonist interventions for enhancing NMDA receptor-mediated neurotransmission is dependent on an interaction of genetic and environmental factors. Specifically, D-serine, a directly-acting glycine agonist, was not able to influence the dose-response relation for MK-801's ability to raise the threshold voltage for the elicitation of electrical seizures in nonhandled and stressed NIH Swiss outbred mice and nonhandled BALB/c genetically-inbred mice. However, D-serine and sarcosine, a glycine reuptake inhibitor, reduced MK-801's antiseizure efficacy in stressed BALB/c mice, the inbred strain with heightened behavioral sensitivity to MK-801. Thus, in the context of acute stress (i.e., subjecting animals to a single session of cold water swim stress 24 hours prior to testing MK-801's antiseizure efficacy), the ability of glycinergic interventions to "antagonize" MK-801's pharmacological-induction of NRH may be dependent on special properties of the genetic substrate. Even in the context of stress, D-serine was ineffective when administered to the NIH Swiss outbred strain, whereas glycinergic interventions were effective in stressed BALB/c genetically inbred mice. With respect to the heuristic value of MK-801's ability to antagonize electrically-precipitated seizures, these data have two practical implications: firstly, the acutely stressed BALB/c mouse strain may be especially valuable for the identification of potentially effective glycinergic interventions and, secondly, the "NMDA receptor pharmacology" of the acutely stressed BALB/c mouse may inform biochemical and pharmacological studies of acutely exacerbated patients with schizophrenia.

Clinical Trials of NRH Modulation

Our group has been involved in the clinical and preclinical investigation of NMDA receptor hypofunction for more than a decade. This research has included investigation of NMDA receptor agonist interventions for schizophrenia, Alzheimer's disease and cognitive function in normal subjects, including healthy elderly adults. The following briefly summarizes some of our studies, including purely cognitive investigations. The latter are relevant because they have led to the inclusion of unique cognitive tasks that may be sensitive to NMDA receptor agonist interventions in general.

We conducted early trials of pharmacologic interventions designed to facilitate NMDA receptor-mediated neural transmission at the level of the strychnine-insensitive glycine binding site in patients with schizophrenia. We examined the use of high dose glycine, milacemide (an acylated "prodrug"of glycine that readily crosses the blood brain barrier and is converted to glycine in the brain), and d-cycloserine (a partial glycine agonist) as adjuvant medications combined with antipsychotic medications (Rosse et al., 1989, 1990, 1991, 1996). None of the findings from our above mentioned studies suggested significant pharmacotherapeutic utility for these agents at the doses and treatment intervals studied. However, later studies, using higher doses of glycine and d-cycloserine than those employed in our studies did show some promising results (e.g., Javitt et al., 1999; Goff et al., 1999a,b). d-Cycloserine has been shown to improve negative symptoms of schizophrenia when combined with conventional neuroleptics, but it worsens negative symptoms when combined with the atypical antipsychotic clozapine (Goff et al., 1999a,b). It should be noted that in our study, d-cycloserine was given as an adjuvant to the antipsychotic medication molindone, which has been shown to have atypical-like antipsychotic features (e.g., Robertson et al., 1994). In general, the literature on glycinergic interventions is mixed with some studies reporting favorable therapeutic effects. Clearly, dosing issues and the specific antipsychotic medication prescribed in combination with the NMDA receptor agonist intervention influence outcome. The efficacy of these interventions may also be influenced by state (e.g., exacerbated versus remitted state) and chronicity of illness. Finally, the domains of psychopathology or specific symptoms that may be sensitive to NMDA receptor agonist-induced changes are still unclear. Nonetheless, the hypothesis is a compelling one and there are many additional medications that have yet to be explored.

Additionally, we examined the effect of a tryptophan-depletion diet in patients with schizophrenia (Rosse et al., 1992). Tryptophan is a precursor of kynurenic acid, a naturally occurring antagonist that binds to the strychnine-insensitive glycine binding site on the NMDA receptor complex. It is also a precursor to quinolinic acid, a naturally occurring excitotoxin that binds to the NMDA agonist recognition site and has been implicated in the pathogenesis of various neurologic diseases. Thus, depletion of L-tryptophan, by virtue of its ability to diminish centrally available kynurenic and quinolinic acid, might have therapeutic benefit in patients with schizophrenia. After approximately a week on the diet, patients with schizophrenia revealed improved Stroop test performance, but no change in the severity of their symptoms. Other data confirmed the lack of clinical efficacy of tryptophan depletion diets in schizophrenia patients (Sharma et al., 1997).

Cognition and Medication Trials

Our group has demonstrated the effectiveness of drugs that interact with the NMDA-type of glutamate receptor on memory and language measures. We demonstrated in healthy young and older adults that administration of a glycine prodrug (milacemide) facilitated the number of words retrieved and decreased the latency to retrieve those words in a semantic memory task called word retrieval (Schwartz et al., 1991). Milacemide also improved healthy young and older adults' ability to remember the source or context of previously presented information (Schwartz et al., 1992). Remembering source or contextual information is essential to declarative forms of memory such as recall and recognition. In a subsequent study, we observed that the partial glycine agonist, d-cycloserine, improved word retrieval abilities in a priming paradigm in patients with probable Alzheimer's disease (Schwartz et al., 1996).

Our group conducted an open-label investigation to evaluate the therapeutic efficacy, safety, and tolerability of adjuvant topiramate in 12 patients with schizophrenia and schizoaffective disorder (Deutsch et al., 2003; see also case report by Drapalski et al., 2001). An optimal dose of topiramate was determined for each of the 12 patients during a slow four-week titration procedure. Patients were maintained on topiramate and their stable antipsychotic medications for eight weeks, after which topiramate was tapered and discontinued. Patients were followed for an additional four weeks on their stable antipsychotic medications. Clinical measures of efficacy (e.g., Positive and Negative Syndrome Scale), cognitive measures (e.g., recall and recognition, verbal fluency, word retrieval, Trails A and B), and safety measures (e.g., postural sway, weight) were assessed during four phases of the study: baseline, four weeks on adjuvant topiramate (week 4), eight weeks on adjuvant topiramate (week 8), and at follow-up (four weeks after topiramate was discontinued). The results showed that topiramate administration (average dose = 110 mg/day) significantly decreased scores on the Positive and Negative Syndrome Scale. Specifically, total scores at baseline were significantly higher than total scores at week 4 and week 8 on topiramate. After topiramate was discontinued at follow-up, total PANSS scores increased compared with week 8. Scores on the negative subscale of the PANSS showed a similar pattern. Negative symptoms were higher at baseline relative to week 4 and marginally so at week 8 ($P = .06$). After topiramate was discontinued, negative scores at follow-up showed a marginally significant increase compared with those at week 8 ($P = .052$). For the general psychopathology scale, there was a significant difference between baseline and week 8 scores, and between week 8 and follow-up scores. Topiramate did not have a significant effect on scores from the positive symptom subscale of the PANSS, the Quality of Life Scale, or the CGI scale. Adjuvant topiramate did not have global deleterious effects on cognitive function. Consistent with previous reports, topiramate decreased the number of words produced in the verbal fluency test; however, this effect was selective and reversible after the drug was discontinued. There was also a trend for topiramate to decrease working memory measured by the Letter-Number Sequencing Test in the WAIS-III (Wechsler, 1997). It is noteworthy, however, that topiramate did not have a detrimental effect on recall or recognition performance. In fact, an exploratory look at recognition d' scores (measure of recognition discrimination) showed that patients' recognition after eight weeks on topiramate was significantly higher than at baseline. Although topiramate was associated with decreased word finding, these recognition data point to the possibility that adjuvant topiramate might also have salutary effects on cognition in schizophrenia. Because topiramate administration can cause dizziness and unsteadiness, we used a measure of *postural stability* to assess the safety of topiramate. Data collected with the postural stability test revealed that topiramate did not increase instability or balance problems in schizophrenia patients.

In summary, the data are provocative, supporting continued exploration of NMDA receptor agonist strategies for the treatment of schizophrenia. As discussed, the therapeutic efficacy of at least some of these interventions may be influenced by state of illness (e.g., exacerbated versus remitted state) and the specific antipsychotic medication that is co-prescribed. Also, there may be discrete domains of psychopathology or specific symptoms that are sensitive to the therapeutic actions of NMDA receptor agonist interventions.

Conclusions

The descriptive studies of the psychosis associated with PCP and the demonstration of stereoselective, saturable binding sites for PCP in the brain led to the unfounded suggestion that a competitive "PCP antagonist" would be a desirable medication strategy for the treatment of schizophrenia. However, in the late 1980's, when it was shown that the PCP binding site was itself located within a hydrophobic channel domain of a glutamate receptor-associated ionophore, it became apparent that a competitive "PCP antagonist" would also be an open-channel blocker; thus, a competitive "PCP antagonist" might also cause NRH, a proposed mechanism of psychosis. The NMDA receptor complex possesses several allosteric modulatory sites that are distinct from the agonist recognition site for glutamate, in addition to a binding site for glycine, the obligatory co-agonist. The existence of these sites represents an opportunity for the development of ligands that are capable of increasing the "potency" of L-glutamate for the promotion of channel opening and calcium ion conductance. Conceivably, positive allosteric modulatory ligands of the NMDA receptor will be developed as medications for treating the presumed NRH in patients with schizophrenia in a manner analogous to the development of benzodiazepine agonists for the treatment of anxiety, seizures, sleep disturbance and muscle spasticity. Benzodiazepine agonists increase the likelihood that GABA will be effective in promoting chloride ion conductance through its associated ionophore; they improve the efficiency of coupling between the binding of GABA and opening of the chloride ion channel.

In the mouse, we have characterized two behaviors that result from the administration of MK-801 (dizocilpine), a noncompetitive NMDA receptor antagonist that causes NRH acutely: MK-801's antagonism of electrically precipitated tonic hindlimb extension and elicitation of mouse popping behavior. Using these animal paradigms, we have shown that environmental stress, pharmacological interventions and genetic factors influence NMDA receptor-mediated neurotransmission. Importantly, these MK-801-elicited behaviors were used successfully to test the ability of drugs to address NRH in the intact animal. These interventions work to enhance NMDA receptor-mediated neurotransmission, potentiate GABAergic neurotransmission, and antagonize nonNMDA excitatory amino acid receptors. Additionally, these procedures support the clinical examination of selective alpha7 nicotinic acetylcholine receptor agonist interventions and inhibitors of neuronal nitric oxide synthase for the treatment of schizophrenia.

It is expected that future medications for the treatment of schizophrenia will target the negative symptom and cognitive domains of psychopathology; they will be less dependent on the antagonism of dopamine for their primary pharmacological mechanism of action. The enhancement of NMDA receptor-mediated neurotransmission and correcting the downstream consequences of NRH, such as dampened GABAergic neurotransmission and excessive and dysregulated stimulation of nonNMDA excitatory amino acid receptors, will guide and serve as the rationale for many of the medication strategies that will be tested in the future. Our work over the past decade represents a useful interchange of experiments and information derived from the mouse and man.

REFERENCES

Akbarian S, Sucher NJ, Bradley D, Tafazzoli A, Trinh D, Hetrick WP, Potkin SG, Sandman CA, Bunney WE Jr., Jones EG. Selective alterations in gene expression for NMDA receptor subunits in prefrontal cortex of schizophrenics. *Journal of Neuroscience* 1996; 16: 19-30.

Benes FM, Vincent SL, Alsterberg G, Bird ED, SanGiovanni JP. Increased GABA receptor binding in superficial layers of cingulated cortex in schizophrenics. *Journal of Neuroscience*. 1992; 12: 924-929.

Billingslea EN, Mastropaolo J, Rosse RB, Bellack AS, Deutsch SI. Interaction of stress and strain on glutamatergic neurotransmission: Relevance to schizophrenia. *Pharmacology Biochemistry and Behavior* 2003; 74(2): 351-356.

Bonhaus SW, Burge BC, McNamara JO. Biochemical evidence that glycine allosterically regulates an NMDA receptor-coupled ion channel. *Eur J Pharmacol* 1987; 142: 489-490.

Coyle JT. The glutamatergic dysfunction hypothesis for schizophrenia. *Harvard Review of Psychiatry* 1996; 3: 241-253.

Deutsch SI, Mastropaolo J, Schwartz BL, Rosse RB, Morihisa JM. A "glutamatergic hypothesis" of schizophrenia: rationale for pharmacotherapy with glycine. *Clinical Neuropharmacology* 1989; 12: 1-13.

Deutsch SI, Deutsch LH, Weizman R. Nonketotic hyperglycinemia: a paradigm for the application of neuroscience to the understanding and treatment of a developmental disorder. In: Deutsch SI, Weizman A, Weizman R, eds. *Application of basic neuroscience to child psychiatry*. New York: Plenum Publishing Corporation, 1990, pp. 165-180.

Deutsch SI, Hitri A: Measurement of an explosive behavior in the mouse induced by MK-801, A PCP analogue. *Clinical Neuropharmacology* 1993; 16(3): 251-257.

Deutsch SI, Park CH, Lukacs LG, Morn C, Koetzner L, Mastropaolo J. MK-801 alters the $GABA_A$ receptor complex and potentiates flurazepam's antiseizure efficacy. *Pharmacology Biochemistry and Behavior* 1995; 51(4): 909-915.

Deutsch SI, Rosse RB, Paul SM, Tomasino V, Koetzner BA, Morn CB, Mastropaolo J. 7-Nitroindazole and metylene blue, inhibitors of neuronal nitric oxide synthase and no-stimulated guanylate cyclase, block MK-801 elicited behaviors in mice. *Neuropsychopharmacology* 1996; 15: 37-43.

Deutsch SI, Rosse RB, Riggs RL, Koetzner L, Mastropaolo J. The competitive NMDA antagonist CPP blocks MK-801-elicited popping behavior in mice. *Neuropsychopharmacology* 1996; 15:329-331.

Deutsch SI, Rosse RB, Paul SM, Riggs RL, Mastropaolo J. Inbred mouse strains differ in sensitivity to popping elicited by MK-801. *Pharmacology Biochemistry and Behavior* ; 57(1/2): 315-317.

Deutsch SI, Rosse RB, Schwartz BL, Fay-McCarthy M, Rosenberg P, and Fearing K. Methylene blue adjuvant therapy of schizophrenia. *Clinical Neuropharmacology* 1997; 20(4): 357-363.

Deutsch SI, Mastropaolo J, Riggs RL, and Rosse RB. The antiseizure efficacies of MK-801, phencyclidine, ketamine and memantine are altered selectively by stress. *Pharmacology Biochemistry and Behavior* 1997; 58(3): 709-712.

Deutsch SI, Rosse RB, and Mastropaolo J. Behavioral approaches to the functional assessment of NMDA-mediated neural transmission in intact mice. *Clinical Neuropharmacology* 1997; 20(5):375-384.

Deutsch SI, Rosse RB, Mastropaolo J. Current status of NMDA antagonist interventions in the treatment of nonketotic hyperglycinemia. *Clinical Neuropharmacology* 1998; 21(2): 71-79.

Deutsch SI, Mastropaolo J, Powell DG, Rosse RB, Bachus SE. Inbred mouse strains differ in their sensitivity to an antiseizure effect of MK-801. *Clinical Neuropharmacology* 1998; 21(4): 255-257.

Deutsch SI, Mastropaolo J, Rosse RB. Neurodevelopmental consequences of early exposure to phencyclidine and related drugs. *Clinical Neuropharmacology* 1998; 21(6): 320-332.

Deutsch SI, Rosse RB, Schwartz BL, Powell DG, and Mastropaolo J. Stress and a glycinergic intervention interact in the modulation of MK-801-elicited mouse popping behavior. *Pharmacology Biochemistry and Behavior* 1999; 62(2): 395-398.

Deutsch SI, Rosse RB, Schwartz BL, Mastropaolo J. A revised excitotoxic hypothesis of schizophrenia: therapeutic implications. *Clinical Neuropharmacology* 2001; 24: 43-49.

Deutsch SI, Rosse RB, Billingslea EB, Bellack AS, Mastropaolo J. Topiramate antagonizes MK-801 in an animal model of schizophrenia. *European Journal of Pharmacology* 2002; 449(1): 121-125.

Deutsch SI, Schwartz BL, Rosse RB, Mastropaolo J, Marvel CL, Drapalski AL. Adjuvant topiramate administration: A pharmacological strategy for addressing NMDA receptor or hypofunction in schizophrenia. *Clinical Neuropharmacology* 2003; 26(4): 199-206.

Deutsch SI, Rosse RB, Billingslea EN, Bellack AS, Mastropaolo J. Modulation of MK-801–elicited mouse popping behavior by galantamine is complex and dose-dependent. *Life Sciences* 2003; 73(18): 2355-2361.

Deutsch SI, Rosse RB, Schwartz BL, Weizman A, Chilton M, Arnold DS, Mastropaolo J. Therapeutic implications of selective alpha7 nicotinic receptor abnormality in schizophrenia. *Israel Journal of Psychiatry* 2005; 42(1): 33-44.

Drapalski AL, Rosse RB, Peebles RP, Schwartz BL, Marvel CL, Deutsch SI. Topiramate improves deficit symptoms in a patient with schizophrenia when added to a stable regimen of antipsychotic medication. *Clinical Neuropsychopharmacology* 2001; 24: 290-294.

Fagg GE. Phencyclidine and related drugs bind to the activated N-methyl-D-aspartate receptor-channel complex in rat brain membranes. *Neurosci Lett* 1987; 76: 221-227.

Farber NB, Newcomer JW, Olney JW. The glutamate synapse in neuropsychiatric disorders. *Progress in Brain Research* 1998; 116:421-437.

Foster AC, Wong EHF. The novel anticonvulsant MK-801 binds to the activated state of the N-methyl-D-aspartate receptor in rat brain. *Br J Pharmacol* 1987; 91: 403-409.

Freed WJ, Crump S, Jeste DV. Genetic effects on PCP-induced stimulation in recombinant inbred strains of mice. *Pharmacol Biochem Behav* 1984; 21(1): 159-162.

Gibbs JW, Sombati S, DeLorenzo RJ, Coulter DA. Cellular actions of topiramate: Blockade of kainite-evoked inward currents in cultured hippocampal neurons. *Epilepsia*. 2000; 41 (Suppl. 1): S10-16.

Goff DC, Henderson DC, Evins AE, Amico E. A placebo-controlled crossover trial of D-cycloserine added to clozapine in patients with schizophrenia. *Biological Psychiatry* 1999; 45(4): 512-514.

Goff DC, Tsai G, Levitt J, Amico E, Manoach D, Schoenfeld DA, Hayden DL, McCarley R, Coyle JT. A placebo-controlled trial of D-cycloserine added to conventional neuroleptics in patients with schizophrenia. *Archives of General Psychiatry* 1999; 56: 21-27.

Grimwood S, Slater P, Deakin JFW, et al. NR2B-containing NMDA receptors are up-regulated in temporal cortex in schizophrenia. *NeuroReport* 1999; 10: 461-465.

Hirsch SR, Das I, Garey LJ, deBelleroche J. A pivotal role for glutamate in the pathogenesis of schizophrenia, and its cognitive dysfunction. *Pharmacology Biochemistry and Behavior* 1997; 56: 797-802.

Humphries C, Mortimer A, Hirsch S, de Belleroche J. NMDA receptor mRNA correlation with antemortem cognitive impairment in schizophrenia. *Neuroreport* 1996; 7(12): 2051-2055.

Huntsman, MM, Tran BV, Potkin SG, Bunney WE Jr, Jones EG. Altered ratios of alternatively spliced long and short $gamma_2$, subunit mRNAs of the gamma-amino butyrate type A receptor in prefrontal cortex of schizophrenics. *Proceedings of the National Academy of Sciences USA* 1998; 95: 15066-71.

Ishimaru M, Kurumaji A, Toru M. Increases in strychnine insensitive glycine binding sites in cerebral cortex of chronic schizophrenics: evidence for glutamate hypothesis. *Biological Psychiatry* 1994; 35: 84-95.

Javitt DC, Balla A, Sershen H, Lajtha A. Reversal of phencyclidine-induced effects by glycine and glycine transport inhibitors. *Biological Psychiatry* 1999: 45(6): 668-679.

Jentsch JD, Roth RH. The neuropsychopharmacology of phencyclidine: from NMDA receptor hypofunction to the dopamine hypothesis of schizophrenia. *Neuropsychopharmacology* 1999; 20(3): 201-225.

Jentsch JD, Taylor JR, Roth RH. Subchronic phencyclidine administration increases mesolimbic dopaminergic system responsivity and augments stress- and psychostimulant-induced hyperlocomotion. *Neuropsychopharmacology* 1998: 19: 105-113.

Long KD, Mastropaolo J, Rosse RB, Deutsch SI. Exogenously administered D-serine failed to potentiate the ability of MK-801 to antagonize electrically precipitated seizures in nonhandled control and stressed mice. *European Neuropsychopharmacology*, in the press.

Long KD, Mastropaolo J, Rosse RB, Manaye KF, Deutsch SI. Modulatory effects of D-serine and sarcosine on NMDA receptor-mediated neurotransmission are apparent after stress in the genetically inbred BALB/c mouse strain. *Brain Research Bulletin* 2006; 69: 626-630.

Martin R, Kuznjiecky R, Ho S, Hetherington H, Pan J, Sinclair K, Gilliam F, Faught E. Cognitive effects of topiramate, gabapentin, and lamotrigine in healthy young adults. *Neurology* 1999; 52: 321-327.

Mastropaolo J, Rosse RB, Deutsch SI. Anabasine, a selective nicotinic acetylcholine receptor agonist, antagonizes MK-801 –elicited mouse popping behavior, an animal model of schizophrenia. *Behavioural Brain Research* 2004; 153(2): 419-422.

Millan MJ. N-methyl-D-aspartate receptors as a target for improved antipsychotic agents: novel insights and clinical perspectives. *Psychopharmacology* 2005; 179: 30-53.

Moghaddam B, Adams B, Verma A, Daly D. Activation of glutamatergic neurotransmission by ketamine: a novel step in the pathway from NMDA receptor blockade to dopaminergic and cognitive disruptions associated with the prefrontal cortex. *Journal of Neuroscience* 1997; 17(8): 2921-2927.

Noga JT, Hyde TM, Herman MM, Spurney CF, Bigelow LB, Weinberger DR, Kleinman JE. Glutamate receptors in the postmortem striatum of schizophrenic, suicide, and control brains. *Synapse* 1997; 27: 168-176.

Norris DO, Mastropaolo J, O'Connor DA, Novitzki MR, Deutsch SI. Glycinergic interventions potentiate the ability of MK-801 to raise the threshold voltage for tonic hindlimb extension in mice. *Pharmacology Biochemistry and Behavior* 1992; 43(1): 609-612.

Norris DO, Mastropaolo J, O'Connor DA, Cohen JM, and Deutsch SI. A glycinergic intervention potentiates the antiseizure efficacies of MK-801, flurazepam, and carbamazepine. *Neurochemical Research* 1994; 19(2): 161-165.

Olney JW, Farber NB. Glutamate receptor dysfunction and schizophrenia. *Archives of General* Psychiatry 1995; 52: 998-1007.

Olney JW, Labruyere J, Price MT. Pathological changes induced in cerebrocortical neurons by phencyclidine and related drugs. *Science* 1989; 244: 1360-1362.

Robertson GS, Matsumura H, Fibiger HC.Induction patterns of Fos-like immunoreactivity in the forebrain as predictors of atypical antipsychotic activity. *Journal of Pharmacology and Experimental Therapeutics* 1994; 271(2): 1058-1066.

Rosse RB, Theut SK, Banay-Schwartz M, Leighton M, Scarcella E, Cohen CG, Deutsch SI. Glycine adjuvant therapy to conventional neuroleptic treatment in schizophrenia: an open-label, pilot. *Clinical Neuropharmacology* 1989; 12(5): 416-424.

Rosse RB, Schwartz BL, Leighton MP, Davis RE, Deutsch SI. An open-label trial of milacemide in schizophrenia: An NMDA intervention strategy. *Clinical Neuropharmacology* 1990: 13(4): 348-354.

Rosse RB, Schwartz BL, Davis RE, Deutsch SI. An NMDA intervention strategy in schizophrenia with "low dose" milacemide. *Clinical Neuropharmacology* 1991; 14(3): 268-272.

Rosse RB, Schwartz BL, Zlotolow S, Banay-Schwartz M, Trinidad AC, Peace TD, Deutsch SI. Effects of a low-tryptophan diet as an adjuvant to conventional neuroleptic therapy in schizophrenia. *Clinical Neuropharmacology* 1992; 15(2): 129-141.

Rosse RB, Mastropaolo J, Koetzner L, Morn CB, Sussman DM, Deutsch SI. Computerized measurement of MK-801 – elicited hyperactivity and popping in mice. *Clinical Neuropharmacology* 1995; 18: 448-457.

Rosse RB, McCarthy M, Kendrick K, Deutsch SI. D-Cycloserine adjuvant therapy to molindone in the treatment of schizophrenia. *Clinical Neuropharmacology* 1996; 19(5): 444-450.

Rosse RB, Deutsch SI. Adjuvant galantamine administration improves negative symptoms in a patient with treatment-refractory schizophrenia. *Clinical Neuropharmacology* 2002; 25(5): 272-275.

Schwartz BL, Hashtroudi S, Herting RL, Handerson H, Deutsch SI. Glycine prodrug facilitates memory retrieval in humans. *Neurology* 1991; 41: 1341-1343.

Schwartz BL, Hashtroudi S, Herting RL, Deutsch SI. The effects of milacemide on item and source memory. *Clinical Neuropharmacology* 1992; 15(2): 114-119.

Schwartz BL, Hashtroudi S, Herting RL, Schwartz P, Deutsch SI. D-Cycloserine enhances implicit memory in Alzheimer patients. *Neurology* 1996; 46: 420-424.

Schwartz BL, Hashtroudi S, Herting RL, Fakouhi TD, Deutsch SI. The NMDA receptor complex: enhancement of memory in aging and dementia. In: Herrmann D, McEvoy C, Hertzog C, Hertel P, Johnson M, eds. *Basic and Applied Memory Research: Volume 2, Practical Applications*. Mahwah, New Jersey: Lawrence Erlbaum Associates, Inc., 1996, pp. 423-437.

Schwartz BL, Parker ES, Hovaguimian A, Williams A, Rosse RB, Deutsch SI. Context-dependent memory impairment in schizophrenia: Analysis of the frontal dysfunction hypothesis. International Congress on Schizophrenia Research. March 29- April 2, submitted abstract.

Schwartz BL, Howard DV, Howard JH Jr, Hovaguimian A, Deutsch SI. Implicit learning of visuospatial sequences in schizophrenia. *Neuropsychology* 2003; 17(3): 517-533.

Shank RP, Gardocki JF, Streeter AJ, Maryanoff BE. An overview of the preclinical aspects of topiramate: Pharmacology, pharmacokinetics, and mechanism of action. *Epilepsia* 2000; 41 (Suppl. 1): S3-9.

Sharma RP, Shapiro LE, Kamath SK, Soll EA, Watanabe MD, Davis JM. Acute dietary tryptophan depletion: effects on schizophrenic positive and negative symptoms. *Neuropsychobiology* 1997; 35(1): 5-10.

Simpson MD, Slater P, Royston MC, Deakin JF. Alterations in phencyclidine and sigma binding sites in schizophrenic brains. Effects of disease process and neuroleptic medication. *Schizophrenia Research* 1992; 6: 41-48.

Tamminga CA. Schizophrenia and glutamatergic transmission. *Critical Reviews in Neurobiology* 1998; 12: 21-36.

Weinberger DR, Lipska BK. Cortical maldevelopment, anti-psychotic drugs, and schizophrenia: a search for common ground. *Schizophrenia Research* 1995; 16: 87-110.

In: Focus on Signal Transduction Research
Editor: Gavin McAlpine, pp. 123-144
ISBN 1-60021-376-6

Chapter 5

Envelope-Mediated Stress Signalling in Bacteria

Joanne Cummins*[1]*, Laurent Michel*[1]*,
***Fergal O'Gara*[1] *and Christine Baysse*[1*]**
[1]BIOMERIT Research Centre, Microbiology Department,
National University of Ireland, Cork, Ireland

Abstract

The ubiquitous occurrence of bacteria relies on their ability to persist under various stressful conditions and changing environments. These bacteria initiate an adaptive response to the stresses they face by modulating their gene transcription programme. The rapidity of the physiological adaptation is critical for their survival in hostile ecological niches. Therefore bacteria have evolved early envelope-initiated signalling processes for responding to the imposition of environmental stresses and to directly monitor changes in the environment. The bacterial cell envelope comes into direct contact with the environment and consequently it is ideally suited for sensing environmental changes and triggering adaptive responses.

Envelope stress responses have been described in a variety of Gram-negative bacteria and serve important roles in colonization and virulence. The two-component systems CpxAR and BaeSR, and the extracytoplasmic sigma factor σ^E are the best understood regulatory components involved in the maintenance of the bacterial envelope in response to envelope perturbations such as adhesion to abiotic surface or alteration in membrane protein folding and biogenesis (Raivio TL, Mol Microbiol. 2005; 56(5):1119-28).

An emerging theme is the activation of a regulatory cascade as an outcome of changes in the cytoplasmic membrane fluidity. Many environmental stressors, such as xenobiotics, variations in temperature, pressure or pH, instantly alter the membrane fluidity by acting on the biochemical properties of the acyl components of the

[*] Corresponding author: c.baysse@ucc.ie

phospholipids. Certain regulatory systems are though to be triggered by a modification of membrane fluidity as exemplified by the multiple stress sensor kinase HiK33 of the cyanobacterium *Synechocystis* sp. PCC 6803 and the two-component system DesK/DesR of the Gram-positive bacterium *Bacillus subtilis* (Los DA and Murata N, Biochim Biophys Acta. 2004; 1666(1-2):142-57). New data suggest that membrane-based signal transduction systems are widely distributed in microorganisms. We have recently demonstrated that the Gram-negative bacterium *Pseudomonas aeruginosa* also recognizes alterations in its membrane fluidity as a trigger for adaptive cellular responses (Baysse C *et al.*, Microbiology. 2005; 151(Pt 8):2529-42).

This chapter will review the current knowledge of early stress detection mechanisms in bacteria and discuss novel fundamental issues in membrane-initiated signalling processes.

INTRODUCTION

Of all living organisms, bacteria and Archaea occupy the greatest variety of niches, including the extremes of environment that cannot be tolerated by other life forms. This ubiquitous occurrence of bacteria in nature relies on their ability to persist under changing environments and various stressful conditions including extremes in temperature, osmolarity, pH and many physical and chemical stresses. Microorganisms have evolved highly sophisticated mechanisms to respond to changes in environmental parameters, including many different stress signalling systems which initiate an adaptive response to the adverse conditions they face by modulating the gene transcription programme. When stresses damage cellular constituents, cells induce the expression of genes whose products will counteract the damage. The rapidity of the physiological adaptation is critical for their survival and numerous studies performed under physiologically stressful conditions have elucidated how cellular damages are sensed and how transcription factors are activated to direct stress responses. Depending upon whether the cytoplasm or the bacterial envelope is subjected to damages, these stress responses can be broadly classified into cytoplasmic and *extracytoplasmic or envelope stress responses (ESRs).* The cytoplasmic stress response has been extensively investigated and includes the heat shock response as well as the response to generation of toxic radicals that damage most cellular macromolecules including proteins, lipids and DNA [Zheng and Storz, 2000]. The ESR is only recently been studied in detail and it is evident that it is much more diverse and comprehensive than the cytoplasmic stress response system. The roles of these physiological responses were first related to maintenance of envelope integrity and biogenesis. Recent studies have linked envelope stress responses with pathogenesis. The bacterial cell envelope is the direct interface between the cell and the environment and hence is essential for sensing changing conditions and triggering rapid adaptive responses. Cells must be able to transmit the information from the cell surface to the cytoplasm. The two envelope-mediated signalling mechanisms most common in bacteria are two-component systems (TCSs) and extracytoplasmic function (ECF) σ factors. Both systems consist of two proteins, a transmembrane sensor protein (a histidine kinase and an anti-σ factor, respectively) that detects a specific stimulus, and a cognate cytoplasmic effector protein (a response regulator and a σ factor respectively) that mediates the cellular response.

The role of the ESRs has been the most intensively investigated in Gram-negative bacteria but analogous systems have also been described in Gram-positives species. The best characterized of these responses are the two-component systems CpxAR and BaeSR, and the extracytoplasmic sigma factor σ^E of *Escherichia coli*, which are induced by envelope perturbations such as adhesion to abiotic surface or alteration in membrane protein folding and biogenesis [Rowley *et al.*, 2006].

An emergent theme is the activation of these signalling pathways as an outcome of changes in the cytoplasmic membrane fluidity. Alteration of membrane fluidity has been reported as a possible signal to initiate a differential gene regulation in some bacteria [Los and Murata, 2004]. Bacterial membrane fluidity is instantly affected by many environmental stresses that directly affect the structural characteristics and fluidity of membranes by altering the biochemical properties of the acyl components of membrane lipids. Phospholipids can form different phases, e.g. a liquid crystalline phase and a gel phase. Interconversions (or "phase transition") between these phases can occur depending on the prevailing conditions, including hyperosmotic stress, extremes of temperature, presence of pollutants such as pentachlorophenol or phenolic compounds [Mikami and Murata, 2003]. Additionally, alteration in the state phase of lipids can be caused by the presence of ions (Ca^{2+}, Mg^{2+} and Sr^{2+}) [Rilfors *et al.*, 1984] or by an increase in the external pressure that causes the membrane lipid to pack more tightly [Hazel and Williams, 1990]. Finally, starvation and growth phase can also affect the status of the membrane [Denich *et al.*, 2003]. In *E. coli* cells at stationary phase of growth, membranes were found to have less lipid motility and an increased lipid-protein ratio which has a stabilizing effect upon the membrane [Souzu, 1986]. It seems likely that some regulatory systems perceive these alterations *via* sensor proteins located in the cytoplasmic membrane. Interestingly, such stress sensor proteins have been discovered in Archaea, as well as Gram-positive bacteria, suggesting that membrane-based signal transduction systems may be widely distributed in microorganisms

I. Type of Environmental Stresses Challenging Bacterial Viability

Microbes are exposed to a succession of variable physico-chemical parameters such as temperature, osmolarity, pH, reactive oxygen species, nutrient availability and inorganic ions concentration which may affect the physiological status and properties of the cells. The major effect induced by high temperature is protein denaturation, but heat stress also affects the transmembrane proton gradient, leading to a decrease of the intracellular pH [Piper, 1993; Piper *et al.*, 1997; Weitzel *et al.*, 1987]. In the opposite context, diminution in temperature instigates severe physiological disturbances in microorganisms such as decrease in membrane fluidity, changes in the level of DNA supercoiling, and formation of stable secondary structures in DNA and RNA that impair replication, transcription and translation. Acidic conditions may affect several cell functions such as capacity for nutrient acquisition and energy generation, cytoplasmic pH homeostasis and protection of proteins and DNA [Booth *et al.*, 2002]. Exposure to variation in osmolarity can be harmful for bacterial cells since a sudden increase in the extracellular solute concentration results in movement of water from the cell to the outside, ultimately leading to dramatic changes in cell volume.

Oxidative stress refers to a condition in which reactive by-products of oxygen, such as superoxide anion radical (O_2^-), hydrogen peroxide (H_2O_2) and hydroxyl radicals (·OH) damage specific molecules or the whole cell fitness [McCormick *et al.*, 1998]. Most of such derivatives result from the reduction of molecular oxygen catalyzed by several membrane-associated respiratory chain enzymes. Many of these factors change markedly during the transition from the external environment to the host, between different locations in the host and between different hosts. The first signal encountered by pathogenic bacteria upon entry into a mammalian host is an increase in temperature from that of the environment to the physiological temperature of the body. Passage from natural to host associated environments also implies other cellular stresses such as changes in osmolarity and indeed, osmolarity increases during passage in the intestinal tract and the bloodstream. In the same manner, microorganisms which establish themselves via the intestine as well as all gastrointestinal organisms have to pass through the stomach and therefore must be able to survive extreme shifts in pH. Invasive pathogens not only have to adapt to the changing environments within the host but also have to counteract the defence mechanisms of the host during the course of infection. In fact microbes are also exposed to toxic reactive oxygen species and nitrogen intermediates released by the host phagocytic cells: reactive oxygen species (ROS) are used by immune cells as an oxidative stress-generating weapon during phagocytosis. Plant also react to bacterial infection by an oxidative burst generating H_2O_2 and O_2^- [Buchmeier *et al.*, 1995].

The mammalian host also specifically limits iron availability to defend itself against invading pathogens. Thus pathogenic bacteria must adapt to a wide variety of changing environmental conditions during the passage through a host organism [Kazmierczak *et al.*, 2005].

To establish a food-borne infection in a human host, a microbe must also adapt to a wide variety of stress conditions such as stresses generated by food processing. Although preservation techniques are becoming milder due to consumers demands for higher quality and less heavily processed food, pathogens still have to face many stresses generated to preserve food products. These include osmotic stress, pH, temperature stress, and high pressure stress.

Free-living bacteria are also frequently exposed to challenges in their natural environment. Indeed several types of stresses may occur in soil, particularly in the rhizosphere. They include temperature stress, water stress, pH, oxidative stress, presence of pollutants and stress caused by competition with other organisms.

II. ENVELOPE SIGNALLING: THE EARLY DETECTION OF STRESSES

The cell envelope is a protective barrier, allowing bacteria to growth in hostile environments. It is also the first exposed to environmental stresses and therefore the first element to signal the emergence of environmental changes. The bacterial membrane has emerged as an initial target for dealing with stresses [Russel, 1989]. Microbial envelopes are responsible for a plethora of physiological processes such as regulating the movement of substances in and out the cells, stabilising protein structure for a correct functioning of membrane embedded enzymes, and providing a matrix for many biological reactions.

Therefore, the maintenance of envelope structure in the face of a constantly changing environment is an important challenge for all bacteria. Most of ESR pathways are involved in the protection and repair of the cell envelope during stressful conditions. However, emerging data suggest that these ESRs impact on a variety of biological processes, important for stress adaptation, virulence and persistence in hostile environments.

E. coli, a bacterium capable of persisting in diverse environments, has provided a valuable model to study stress adaptation in single cell organisms. The following chapters described mainly the ESRs described in *E. coli*, i.e. σ^E (σ^{24}, CpxAR and BaeSR pathways, as well as recent data on membrane fluidity sensing in prokaryotes via newly described transmembrane sensors (Hik33, DesK).

1. The σ^E -signalling Pathway

1.1. The σ^E -Pathway in E. Coli

σ^E in *E. coli* is the best-characterised member of the ESR. It was first discovered because it was found to be required for high temperature-dependent expression. σ^E is a member of the ECF subfamily of alternative sigma factors. The σ^E -pathway is activated by several stimuli most notably misfolded outer membrane proteins (OMP) such as OmpC and OmpF, and by mutations in genes encoding protein folding factors for the envelope (e.g. *surA* or *dsbC*) (Figure 1) [Missiakas *et al.*, 1996]. It has recently been shown that reduced acylation of lipid A due to mutations in *htrB* or *msbB* which encode two acyltransferases also results in increased activity of the σ^E regulon [Tam and Missiakas, 2005].

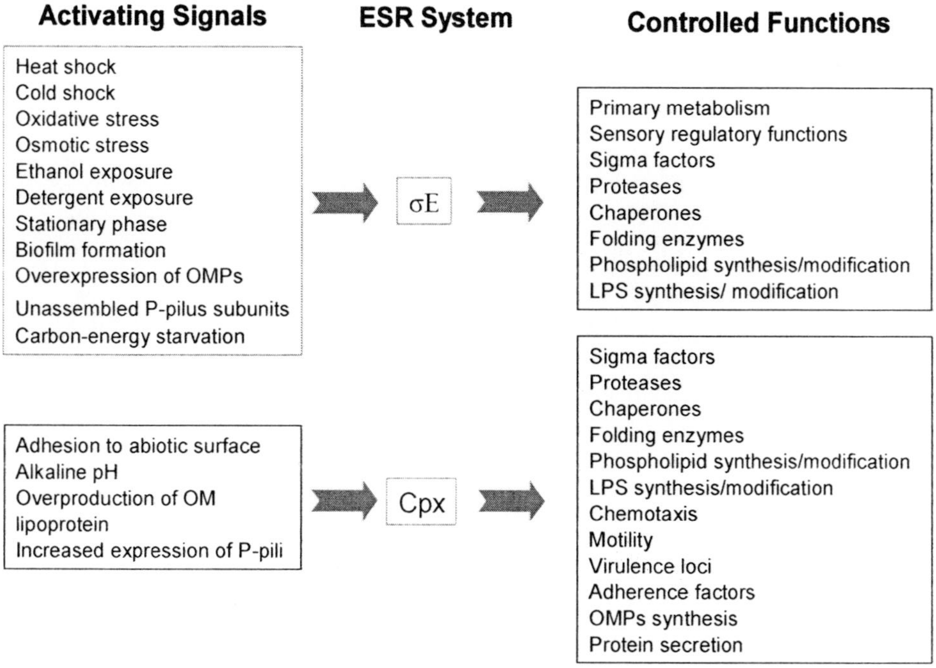

Figure 1. Examples of conditions known to activate σ^E and Cpx -mediated ESRs in bacteria, and known function of gene product of the σ^E and Cpx regulons. From Rowley et al., 2006.

σ^E is encoded by *rpoE* contained in an operon with three other genes, *rseABC*. These genes code for proteins that are involved in post-translationally regulating σ^E. RseA and RseB are the two major negative regulators of σ^E. RseA is a membrane spanning protein which contains a cytoplasmic N-terminal domain that functions as an anti-sigma factor of σ^E [Alba and Gross, 2004]. RseA functions by blocking the major binding determinants on σ^E for RNA polymerase [Campbell *et al.*, 2003]. RseB is a soluble periplasmic protein that binds to the periplasmic domain of RseA and it is thought that RseB may increase stability or activity of RseA and *rseB* mutants showed a 2-fold increase in basal expression of σ^E [De Las Penas *et al.*, 1997]. RseC is an inner membrane protein that plays a minor role in positively regulating σ^E. Rsec is also involved in the reduction of the iron-sulfur cluster of the transcriptional regulator SoxR that governs the superoxide response regulon, therefore participating in the cellular response to oxidative stress [Koo *et al.*, 2003].

RseA activity has been demonstrated to be regulated by proteolysis. Upon heat shock and/or overproduction of OmpC, RseA half life decreases while *rpoE* expression increases [Ades *et al.*, 1999]. Two inner membrane proteins are involved in the sequential cleavage of RseA, they are DegS and RseP (also known as YaeL). DegS is a member of the HtrA/DegP family of proteases and is essential for viability in *E. coli* [Alba *et al.*, 2001; Clausen *et al.*, 2002]. RseP is a Zn^{2+} metalloprotease that is also essential for viability [Dartigalongue *et al.*, 2001b; Kanehara *et al.*, 2001]. Data suggests that RseA is first cleaved within its periplasmic domain by DegS and then the intermediate state is cleaved by RseP [Alba *et al.*, 2001; Kanehara *et al.*, 2001].

DegS can sense and activate the σ^E dependent stress response by the means of a PDZ protein interaction domain. PDZ domains are found in diverse signalling proteins in bacteria, yeast, plants, insects and vertebrates. The term PDZ is derived from the first three eukaryotic proteins in which these domains were found: PSD-95 (a 95 kDa protein involved in signalling in the post-synaptic density), Dlg (the Drosophila discs large protein), and ZO1 (the zonula occludens 1 protein involved in maintaining epithelial cell polarity). PDZ domains consist of 80 to 90 amino acids comprising six beta-strands and two alpha-helices compactly arranged in a globular structure. *In vitro* studies have shown that the PDZ domain of DegS recognizes a conserved sequence found at the carboxy termini of outer membrane porins [Walsh *et al.*, 2003]. Accumulation of C-termini of unfolded OMPs antagonize PDZ-dependent autoinhibition, therefore once the PDZ domain of DegS is activated it initiates the proteolytic cascade that degrades RseA [Walsh *et al.*, 2003]. It is known that RseP also contains a PDZ domain and may be negatively regulated in the same manner as DegS. Kanerhara et al. (2003) has shown that deletion of the PDZ domain of RseP allowed it to cleave RseA in the absence of cleavage of DegS [Kanehara *et al.*, 2003]. In the periplasmic domain of RseA is a glutamine-rich portion which is involved in inhibiting RseP dependent degradation of RseA. Data suggest that cleavage of RseA by DegS may remove these inhibitory glutamine-rich regions therefore allowing RseP to further degrade RseA in the cytoplasm [Kanehara *et al.*, 2003]. Further degradation of the RseA cytoplasmic part is performed by the ClpXP protease [Flynn *et al.*, 2004].

Regulated intramembrane proteolysis (RIP) refers to a conserved eukaryotic and prokaryotic regulatory mechanism by which membrane-bound proteins are cleaved to release factors into the bacterial cytoplasm or into the eukaryotic cytosol [Urban and Freeman, 2002]. These released factors regulate different processes by activating gene transcription. The release of σ^E from the membrane-associated protein RseA by the successive activity of DegS and RseP can be considered as a RIP process [Douchin *et al.*, 2006] (Figure 2).

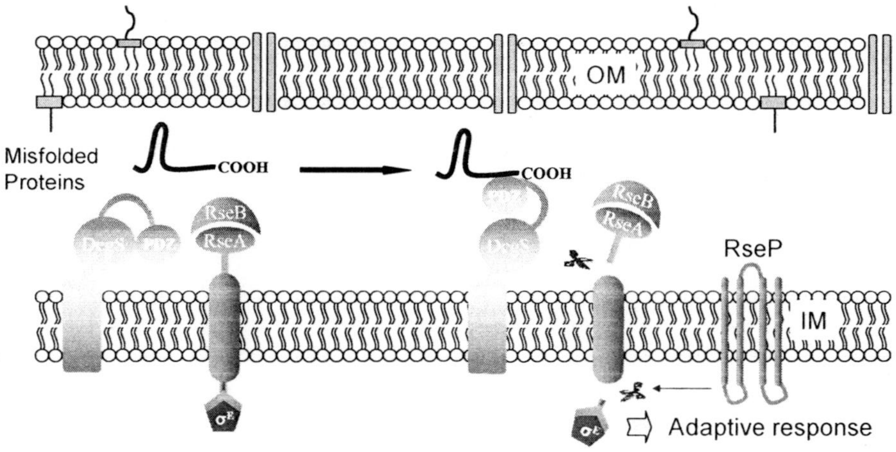

Figure 2. Model for regulation for σ^E in *E. coli* by misfolded proteins (adapted from Rowley *et al.*, 2006). The mechanism of σ^E activation is due to the sequential degradation of RseA by DegS and RseP (YaeL). In the absence of misfolded OMP proteins the PDZ domain negatively regulates DegS expression, therefore preventing RseA degradation and allowing RseA to sequester σ^E. In the presence of misfolded proteins the PDZ domain of DegS interacts with the C-termini of the OMP. This activated DegS protease activity, RseA cleavage and hence allows the release of σ^E from RseA and the subsequent expression of σ^E-dependent genes. OM, Outer Membrane; IM, Inner Membrane.

1.2. The σ^E-signalling Pathway in other Bacteria

The σ^E operon is highly conserved in other bacteria. The invasive intestinal pathogen *Salmonella typhimurium* contains highly conserved σ^E, RseA and RseB proteins. In *Salmonell*a the *rpoE rseABC* operon is auto-regulated as it is in *E. coli* [Miticka *et al.*, 2003]. The difference between these homologues is that, in *Salmonella, rpoE* is not necessary for survival [Johnson *et al.*, 1991]. The *rpoE* mutant of *S. enterica* serovar Typhimurium has reduced virulence and reduced ability to colonise host organs and survive within macrophages. The *rpoE* mutant is also more sensitive to ROS and antimicrobial peptides, components of the host defence mechanisms [Humphreys *et al.*, 1999; Testerman *et al.*, 2002].

σ^E also appears to contribute to oxidative stress resistance in other Gram-negative pathogens as exemplified by *Haemophilus influenzae*. *rpoE* expression in *H. influenzae* was shown to increase inside macrophages. Moreover, the survival of a *rpoE* mutant of *H. influenzae* was reduced relative to that of the wild type in macrophages [Craig *et al.*, 2002].

The opportunistic pathogen *Pseudomonas aeruginosa* possesses an orthologue of σ^E known as AlgU or AlgT. The organisation of the *algU* operon is identical to what is found in *E. coli*. It contains three downstream genes *mucABC* that are involved in regulating *algU* expression [Mathee *et al.*, 1997]. AlgU is a key regulator of alginate synthesis, and therefore controls the chronic infection of lungs by alginate-producing (mucoid) *P. aeruginosa* strains, which is the leading cause of mortality among cystic fibrosis patients [Hershberger *et al.*,

1995]. σ^E is found in a wide variety of organisms including human pathogens such as *Vibrio cholerae* and animal pathogens such as *Actinobacillus pleuropnemoniae* underlining the importance of σ^E for survival and pathogenesis [Rowley *et al.*, 2006].

1.3. The σ^E Regulon

The σ^E regulon controls the heat-shock response, one of the first described stress response in bacteria. The heat-shock response is characterized by the induction of a large set of proteins (heat-shock proteins - HSPs) upon exposure to conditions in which proteins are denatured (i.e., high temperature, alcohols, and heavy metals). The heat-shock response is universal, and has been studied in many systems such as bacteria, yeast, insects [Michaud *et al.*, 1997], worm [Rose and Rankin, 2001] and mammals [Christians *et al.*, 2002]. Heat-shock proteins are highly conserved among species and have been studied extensively in Gram-positive (*Bacillus subtilis*) and Gram-negative bacteria (i.e., *E. coli*, *Agrobacterium tumefaciens*).

HSPs comprise molecular chaperones (e.g., GroEL, GroES, DnaK, and DnaJ) and ATP-dependent proteases (e.g., ClpP, Lon, DegP, FtsH and HslVU) that play a critical role in the repair of protein folding and in protein degradation under normal and stress conditions [Dougan *et al.*, 2002].

GroEL and DnaK with their cofactors constitute the major chaperone systems promoting protein folding in the *E. coli* cytosol. The DnaK/GroEL chaperone network is proposed to assists refolding of proteins that are unfolded by stress treatment and proteins whose size is too large for the GroEL ring underneath GroES [Buchberger *et al.*, 1996].

σ^E control the levels of HSPs (i.e the cytoplasmic branch of the heat-shock response) via the control of *rpoH* expression, encoding the alternative sigma factor σ^H (σ^{32}). σ^H is regulated at several levels. Secondary structures of *rpoH* mRNA are relieved at high temperatures, promoting translation, and hence resulting in more σ^H activity in the cell [Morita *et al.*, 1999]. σ^H, which is normally unstable, being degraded by FtsH [Tomoyasu *et al.*, 1995], is transiently stabilized upon heat stress [Straus *et al.*, 1987]. The transcriptional regulation of *rpoH* is very complex since the gene can be transcribed from at least four promoters, three of them are recognized by the vegetative σ^D (σ^{70}), and the fourth is recognized by σ^E [Erickson *et al.*, 1987; Nagai *et al.*, 1990].

The activity of σ^H is also negatively regulated by DnaK and its partner chaperone DnaJ which bind to σ^H, and inhibit the formation of RNAP-σ^H [Tomoyasu *et al.*, 1998]. The complex is released via the activity of GrpE, a co-chaperone for DnaK which acts as a nucleotide exchange factor.

In addition, σ^E positively regulates *rpoD* encoding the housekeeping sigma factor σ^D (σ^{70}) [Dartigalongue *et al.*, 2001a] and *rpoN* encoding the sigma factor controlling nitrogen metabolism σ^N (σ^{54}) [Rhodius *et al.*, 2006]. Altogether the σ^E regulon encompasses more than 40 genes, involved in proteolysis and folding of proteins, primary metabolism, phospholipid and LPS biogenesis/modification and regulatory functions [Dartigalongue *et al.*, 2001a; Rhodius *et al.*, 2006] (Figure 1).

2. Two-component Regulatory Systems Involved in Envelope Stress Signalling

2.1. CpxAR pathway

2.1.1. CpxAR Pathway in *E. Coli*

TCS are commonly used by bacteria to sense changes in the environment. In *E. coli* there are two known TCS that respond to extracytoplasmic stresses, the CpxAR system and the BaeSR system.

The Cpx system was originally discovered due to the negative impact of a *cpx* mutation on pilus expression in *E. coli*. This mutant also had pleoitropic phenotypes such as temperature sensitivity, isoleucine and valine auxotrophy and altered inner and outer membrane composition [McEwen and Silverman, 1980a; McEwen and Silverman, 1980b]. The Cpx system consists of a membrane localised sensor kinase (CpxA) and a cytoplasmic response regulator (CpxR). The Cpx system is induced by several stimuli, including overexpression of the lipoprotein NlpE, increased pH, alteration in phospholipid composition, accumulation of enterobacterial common antigen (ECA) intermediate lipid II in the inner membrane and misfolded pili subunits that aggregate in the periplasmic space [Danese *et al.*, 1998; Jones *et al.*, 1997; Mileykovskaya and Dowhan, 1997; Nakayama and Watanabe, 1998; Snyder *et al.*, 1995; Thanassi *et al.*, 1998] (Figure 1). Most of these stimuli results in misfolded envelope proteins. This signal is very similar to what occurs in σ^E but each system appears to respond to a distinct class. The σ^E pathway responds to misfolded OMP and not overexpression of NlpE while the opposite is true for Cpx pathway. Interestingly accumulation of P pili subunit (PapG) activates both the σ^E and Cpx pathway, while another (PapE) only activates the Cpx pathway [Isaac *et al.*, 2005].

Envelope stress response results in autophosphorylation of CpxA on a conserved histidine in the cytoplasmic kinase domain. CpxA this in turn phosphorylates CpxR on a conserved aspartate residue. The phosporylated CpxR recognises a conserved DNA site (consensus sequence 5'-GTAAA-N_5-GTAAA-3') that is located upstream of genes belonging to the Cpx regulon [De Wulf *et al.*, 2002; Raivio and Silhavy, 2001] .

The regulation of the Cpx pathway involves the small periplasmic protein CpxP, encoded by a gene located upstream and in the opposite orientation of the *cpx* operon, [Danese *et al.*, 1998]. In the absence of envelope stress CpxP interacts with the sensing domain of CpxA and maintains the pathway in an off state. Inducing signals such as misfolded proteins (PapE) result in CpxP being titrated away from CpxA. DegP mediates the proteolysis of CpxP and relieves the inhibition of the Cpx response. This allows the free CpxA sensing domain to further induce expression of the *cpx* regulon [Buelow and Raivio, 2005; Isaac *et al.*, 2005]. While *cpxP* overexpression results in downregulation of Cpx mediated gene, it is not required for signal transduction. The Cpx pathway can still be activated by envelope stresses in absence of CpxP or when CpxP is overexpressed [DiGiuseppe and Silhavy, 2003; Raivio *et al.*, 1999]. This indicates that CpxP is more likely involved in fine tuning the Cpx response (Figure 3).

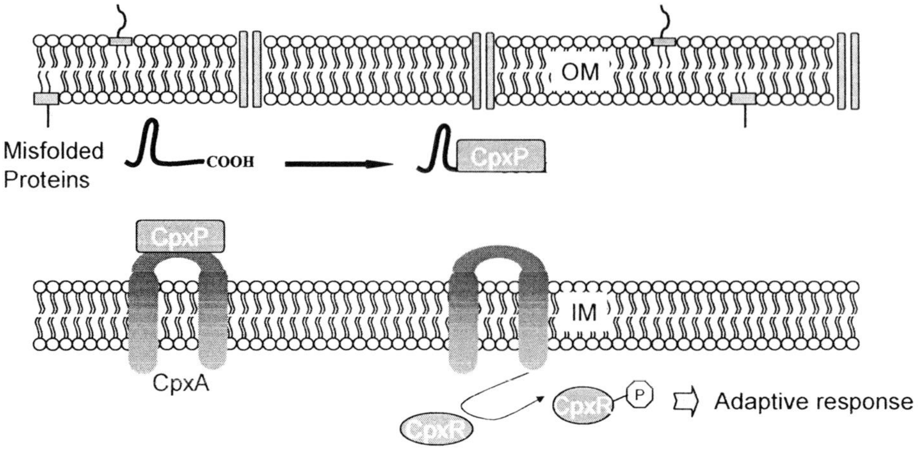

Figure 3. Model for activation of Cpx stress response pathway in *E. coli* by misfolded proteins. Misfolded proteins in the periplasm result in CpxP being titrated away from the sensor CpxA, which in turn undergoes autophosphorylation and transfers the phosphate group to the response regulator CpxR. Phosphorylated CpxR triggers the adaptive response by activating the expression of target genes. OM, Outer Membrane, IM, Inner Membrane.

2.1.2. The Cpx Pathway in other Organisms

The Cpx pathway is not as widely distributed in bacteria than the σ^E regulatory system. It is mainly found in Gram negative pathogens. All three components of the Cpx apparatus are present in *Salmonella typhii* and *Yersinia pestis*. In *S. typhii*, CpxR, CpxA and CpxP share 100%, 100% and 88% similarity respectively with their *E. coli* counterparts, while in *Yersinia*, CpxR, CpxA and CpxP share 88%, 82% and 47% similarity respectively with the corresponding proteins of *E. coli* [De Wulf *et al.*, 2000]. In *Salmonella* the Cpx system affects the host invasiveness in a pH dependent manner [Nakayama *et al.*, 2003]. In *Shigella* spp. the Cpx system regulates the type three secretion system genes (TTSS) in response to pH stress [Tobe *et al.*, 1993]. In *Legionella pneumophilia* the causative agent of Legionnaire's disease, CpxR is involved in regulating the *icm-dot* genes which are necessary for infection of amoebae, macrophages and animals [Gal-Mor and Segal, 2003; Roy, 2002].

2.1.3. The Cpx Regulon

The size of the Cpx regulon and the diversity of the physiological functions linked to the controlled genes suggest that the role of Cpx is more complex than the sole maintenance of the cell envelope integrity. First known members of the Cpx regulon included the disulfide oxidase DsbA, the protease and chaperone DegP, the peptidyl-prolyl isomerase PpiA, and Spy, a periplasmic protein of unknown function. A genome-wide profiling of promoter recognition by the two-component response regulator CpxR in *E. coli* revealed that the Cpx regulon may contain more than 100 members [De Wulf *et al.*, 2002]. Amongst the physiological functions controlled by the Cpx regulon are the folding and degradation of proteins, cell motility (chemotaxis), multidrug resistance, phospholipid biosynthesis, pili biosynthesis and conjugation [Dorel *et al.*, 2006] (Figure 1). The Cpx regulon has been implicated in the virulence of several bacterial pathogens by regulating the expression of surface structures associated with bacterial virulence such as pili and TTSS. The Cpx pathway plays a key role in the regulation of adhesion-induced gene expression, which is an important initial step in biofilm formation [Otto and Silhavy, 2002]. Some authors suggest that the Cpx pathway would serve to the settling of biofilm communities in hostile environments [Dorel *et al.*, 2006].

2.2. The BaeSR System

A second TCS was recently discovered in *E. coli* to also play a role in the ESR. This system is known as BaeSR (*b*acterial *a*daptive r*e*sponse) [Raffa and Raivio, 2002] and was originally found to complement mutants in genes encoding EnvZ (osmosensor), PhoR (phosphate sensor) and CreC (TCS-sensor linked to carbon and energy metabolism). Elimination of BaeR resulted in decreased ability to survive envelope stress in *E. coli* [Raffa and Raivio, 2002]. As with all TCS this system contains a sensor kinase (BaeS) and a response regulator (BaeR). Signals that the Bae system specifically responds to include β-lactams, novobiocin, sodium dodecyl sulphate and bile salts [Baranova and Nikaido, 2002]. Some of the signals that induce BaeSR also overlap with those of the Cpx system; these overlapping signals include overexpression of PapG, spheroplast formation and indole treatment [Raffa and Raivio, 2002]. It has been shown that BaeSR is more critical to indole induction than CpxAR [Raffa and Raivio, 2002]. BaeSR modulates drug resistance in *E. coli* by regulating the expression of genes encoding drug transporters. The response regulator BaeR modulates expression of *mdtABC* and *acrD* by binding to the consensus sequence (5'-TTTTTCTCCATDATTGGC-3') upstream of these multi-drug efflux pump genes [Nishino *et al.*, 2005]. The whole picture of BaeSR regulation was investigated by microarray analysis: the expression of 59 genes of *E. coli*, involved in chemotaxis, motility, maltose and multidrug transport, and two-component signal transduction was activated in a *baeR* overproducing strain [Nishino *et al.*, 2005]. Further work may attempt to determine the physiological role and specificity of the BaeSR system amongst the 3 envelope stress signal pathways of *E. coli.*

2. 3. Membrane Fluidity Sensors

The Gram-positive bacterium *B. subtilis* possesses a TCS capable of sensing and altering membrane fluidity in response to cold stress. This TCS consists of the sensor DesK and the response regulator DesR. This system, known as the Des pathway, introduces a double bond into the acyl chains of membrane phospholipids to restore fluidity to the membrane at levels suitable for growth at low temperature. This double bond is inserted by the activation of a specific membrane bond delta 5 acyl lipid desaturase (delta5-Des) [Diaz *et al.*, 2002].

DesK is a bifunctional enzyme with both kinase and phosphatase activities, which can assume different signal states in response to changes in membrane fluidity [Aguilar *et al.*, 2001; Cybulski *et al.*, 2002]. When lipids are in a disordered state (at physiological or high temperature) DesK remains in a phosphate dominant state, causing the dephosphorylation of DesR. When there is an increase in ordered lipids (at low temperature) DesK is in a kinase dominant state. It is the C-terminal domain of DesK that acts as an autokinase [Albanesi *et al.*, 2004]. It was demonstrated that the transmembrane segment of DesK is essential to sense change in membrane fluidity and to regulate the ratio of kinase to phosphatase activity. Activation of DesK kinase activity occurs upon rigidification of the cytoplasmic membrane and subsequent movement of the transmembrane domains close together. Upon this stimulus DesK undergoes autophosphorylation and transfers the phosphoryl group to DesR. DesR belongs to the NarL group of response regulators. While DesR is a homodimer in solution, when it becomes phophorylated it behaves as a dimmer; but when the concentration of DesR-P increases it behaves as a tetramer [Cybulski *et al.*, 2004]. Phosphorylation of DesR promotes the hierarchical occupation of two adjacent non identical DesR-P DNA-binding sites. This results in a shift towards the tetrameric active form and promotes the recruitment of RNA polymerase to the *des* promoter [Cybulski *et al.*, 2004; Mansilla and de Mendoza, 2005]. The final result is the desaturation of acyl chains of membrane phospholipids. Feedback regulation is provided by the newly synthesized unsaturated fatty acids (UFAs) which act as negative signal for *des* transcription.

The histidine kinase Hik33 in the photosynthetic organisms *Synechocystis* sp PCC6803 acts as both a cold sensor similar to DesK but it can also act as an osmosensor [Mikami *et al.*, 2002; Suzuki *et al.*, 2001]. While the function of DesK is restricted to the homeoviscous adaptation of membrane in response to stress, Hik33 plays a wider role in multi-stress adaptation.

Hik33 contains several conserved regions, a type P linker, a leucine zipper and a PAS domain. The type P linker contains two helical domains assumed to transduce stress signals via intramolecular structural changes [Mikami and Murata, 2003]. It is believed that activation of Hik33 occurs in response to a rigidification of the membrane lipid bilayer followed by a change in the conformation of the transmembrane domains and a subsequent dimerisation of the cytoplasmic domains [Los and Murata, 2004; Murata and Suzuki, 2006] (Figure 4).

Genome-wide pattern of transcription analysis revealed that Hik33 regulates the differential expression of genes inducible by both osmostic stress and cold stress. The set of genes induced by Hik33 in response to osmotic stress and low temperature are distinct but partially overlapping [Mikami *et al.*, 2002].

Osmotic stress induce the expression of 257 genes in wild-type *Synechocystis* sp cells, the expression of about 80 % of these genes being controlled by Hik33. Out of the 45 cold inducible genes, 65% have their expression affected by a mutation in Hik33 [Mikami *et al.*, 2002].

The Hik33-regulated cold inducible genes are involved in gene regulation, regulation of photosynthesis, oxidative stress response and membrane functions. The genes controlled by Hik33 in response to osmotic stress are also involved in gene regulation and regulation of photosynthesis. In addition, this latter class include genes for structural maintenance and function of the cell wall, phosphate transport system, protein folding (chaperones) and protein turnover (proteases) [Mikami *et al.*, 2002].

However, it is not yet established that multi-stress sensors capable of reacting to changes in membrane fluidity are widely distributed in microorganisms, and so far they have only been described in cyanobacteria [Los and Murata, 2004; Paithoonrangsarid *et al.*, 2004]. Recent researches have established a mechanistic link between the alterations in membrane composition/fluidity of *P. aeruginosa* and the resulting activation of global adaptive responses including (p)ppGpp-mediated stringent response and Quorum Sensing [Baysse *et al.*, 2005]. Other emerging data suggest that the sensing of membrane fluidity as an early adaptive signal is widely used in the prokaryotic world [Baysse and O'Gara, 2006]. The presence of membrane fluidity-controlled TCS may explain how ubiquitous bacteria, such as *Pseudomonas,* are able to adapt their physiology and life-cycle to allow growth in diverse environments.

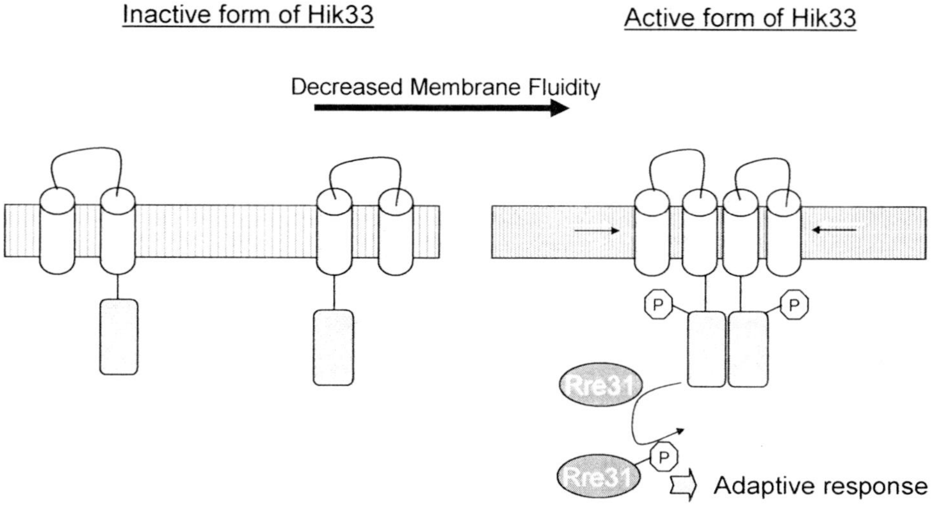

Figure 4. Hypothetical model for the activation of the prokaryotic multistress sensor Hik33 by decreased membrane fluidity and subsequent dimerization of histidine kinase cytoplasmic domains. Adapted from Mikami and Murata (2004). See text for details.

III. The Envelope-stress Regulons and the General Stress Responses Form a Complex Regulatory Network

Envelope-stress signalling are global regulatory systems that control the simultaneous expression of a large number of genes in response to a variety of environmental stresses that affect the membrane integrity. σ^S is the master regulator of the general stress response in most bacteria. σ^S is strongly induced during entry into stationary phase, growth slow-down and/or many stress conditions. It is essential for the expression of multiple stress resistance.

The response of bacteria to different environmental stresses contains considerable overlaps. Examples of these overlaps include co-ordinately regulated products such as alternative sigma factors, heat shock proteins and TCSs. These overlaps in stress responses are the result of a parallel activation of multiple branched systems and/or of crosstalk between pathways [Hellingwerf, 2005]. These phenomena are the result of the interconnection of pathways reacting to different stimuli and of the existence of multi-stress sensors reacting similarly to a variety of signals. Bacteria often have to cope with several environmental stresses simultaneously and survival depends on consequential co-regulations between the respective signalling cascades. The post genomic area and the availability of transcriptome analyse technologies have made the interconnections between distinct ESRs and between ESRs and general stress responses more evident (Figure 5).

1. Cross-talk between General Stress Responses and Envelope Stress Responses

Envelope stress and general stress responses are indeed tightly co-regulated since conditions that affect the membrane integrity often result in a growth-slow down that activate *rpoS*. Similarly, nutrient starvation and stationary phase, as signals for the σ^S pathway, in turn correlate with changes in membrane properties likely to induce an envelope distress. The growth phase and the nutrient status are known to modulate the lipid-protein ratio within the membranes. Stationary phase and nutrient starvation are associated with an increased lipid/protein ratio and in an increase in cardiolipin (CL) to phosphatidylglycerol (PG) ratio. As CL has a higher transition temperature than PG, this alters membrane fluidity. Together these alterations result in less mobile lipids within the phospholipid bilayer and a more rigid structure [DiRusso and Nystrom, 1998].

The transcriptional activation of *cpxRA* by σ^S [De Wulf *et al.*, 1999], as confirmed by transcriptome analysis of *E. coli* wild-type and *rpoS* mutant grown in different stress conditions [Weber *et al.*, 2005], recruits the Cpx system in the general stress response.

In the same way, the starvation response, induced by elevated amount of intracellular (p)ppGpp, links the general stress response to the σ^E/σ^H envelope-stress pathway. The GTP pyrophosphokinase RelA drives the synthesis of (p)ppGpp when activated via the binding of uncharged tRNA to the A site of the ribosome [Block and Haseltine, 1975]. The alarmone (p)ppGpp is considered to be a major signal in response to many kinds of growth perturbations. The role of (p)ppGpp in alternative sigma factor function is linked to the outcome of sigma factor competition. The alternative sigma factors, σ^S, σ^H, and σ^N, compete significantly better against σ^D in the presence of (p)ppGpp [Jishage *et al.*, 2002; Laurie *et al.*, 2003].

(p)ppGpp also controls the Cpx pathway via its action on σ^S since σ^S depends on ppGpp both for expression and function; (p)ppGpp positively affect the translation of *rpoS* via the activity of the partner protein DksA [Brown *et al.*, 2002].

In *P. aeruginosa*, the AlgU (σ^E) system and (p)ppGpp production are tightly co-regulated through a cascade involving the nucleoside diphosphate kinase (Ndk): a two-component regulatory systems AlgR-AlgZ, required for alginate synthesis, is part of the AlgU regulon [Yu *et al.*, 1997]. AlgR positively regulates the nucleoside diphosphate kinase (NdK) synthesis and, as a result, modulates the levels of GTP the metabolic precursor of (p)ppGpp [Kim *et al.*, 1998; Kuroda *et al.*, 1997].

2. Cross-talk between Envelope Stress Responses

2. 1. Cross-Talk between CpxRA and σ^E Pathways

The two envelope-stress systems Cpx and σ^E are distinct in terms of the nature of the activating signals, mechanism of signal transduction, and target gene set. However, significant overlaps exist between the two systems. The Cpx and σ^E systems both respond to protein-envelope stress and are stimulated by unassembled P-pilus subunits; both systems also activate the *degP* operon encoding a periplasmic protease and *skp* expression, encoding a chaperone involved in OMP assembly.

Interestingly, CpxR also negatively controls *rpoE* and *rseABC* genes which indicates that the Cpx response takes precedence over the σ^E response [De Wulf *et al.*, 2002] and thus over the σ^H response since *rpoH* transcription at *rpoHP3* promotory region is σ^E -dependent [Erickson and Gross, 1989]. Moreover, CpxR may also bind an *rpoHP1* promoter fragment that contains a potential CpxR recognition site and hence possibly represses directly the *rpoH* transcription [De Wulf *et al.*, 2002].

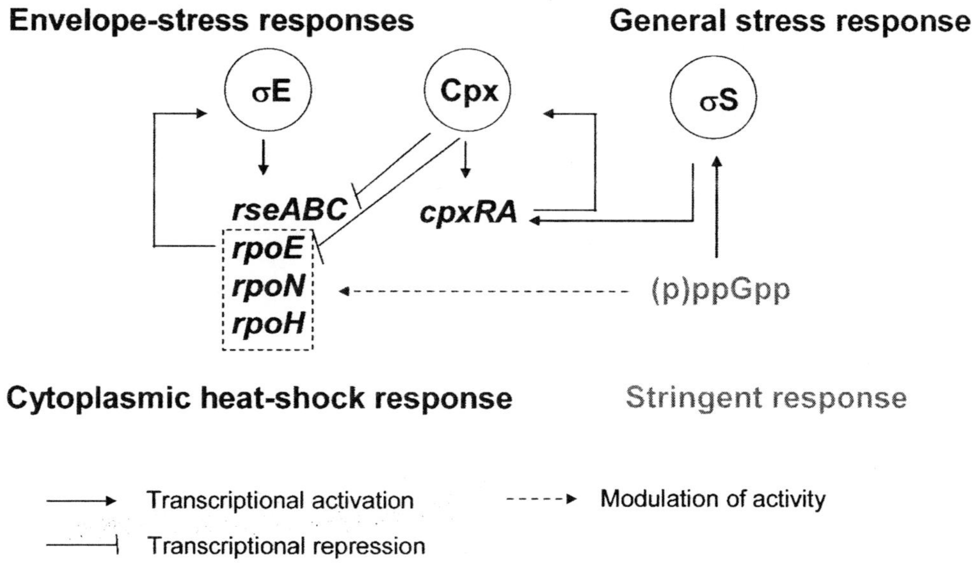

Figure 5. Cross-talks between stress response pathways. This model summarises the co-regulation of ESRs, cytoplasmic stress response, general stress response and stringent response via the activity of alternative sigma factors, TCSs and of the alarmone (p)ppGpp. Details are provided in the text.

2. 2. Hik33 System: One Sensor for Distinct Stress Responses

In *Synechoscystis*, several Histidine kinase (Hik), paired with specific response regulators (Rre), have been shown to react to both salt stress and hyperosmotic stress by inducing two distinct set of genes [Shoumskaya *et al.*, 2005]. Amongst them is the sensor Hik33 (with its cognate regulator Rre31) previously described as a multi-stress sensor capable of sensing changes in the membrane fluidity (see chapter II 2.3.). Interestingly, the current accepted model for two-component regulatory systems can not be applied in these cases since distinct set of genes are regulated by the same TCS couple under different types of stress. This phenomenon suggests that the specificity of a multi-stress sensor, reacting to a universal stress signal such as change of membrane fluidity, is mediated by some unknown factor(s).

The identification of these factors will lead to a better understanding of the genetic flexibility and adaptability of bacteria, and may unravel a yet undiscovered universal prokaryotic system for stress survival.

Conclusion

Many different stress responses have been described and classified according to the cellular localisation of stress targets and the range of stresses to which they respond. This chapter reviews the main extracytoplasmic or envelope stress responses, including newly discovered transmembrane proteins for sensing membrane fluidity.

ESRs have been studied primarily for their role in response to transmembrane protein misfolding and changes in membrane structure. However, it is now clear that the regulatory elements of the ESR networks are associated with each other, and are associated with regulatory element of cytoplasmic stress responses (heat shock, (p)ppGpp) and general stress response (σ^S). Initially ESR roles were linked with envelope maintenance and biogenesis but recent studies have shown their importance for pathogenesis [Raivio, 2005] and stationary phase survival [Costanzo and Ades, 2006].

Another challenge will be to identify the key elements that direct the specificity of stress responses upon unique multi-stress sensor (e.g. HiK33) activation. This will shed further light on how bacteria with small genome and limited number of TCSs, such as *Campylobacter jejuni* [Parkhill *et al.*, 2000], are able to adapt their physiology and life-cycle to allow growth in diverse environments.

References

Ades, S. E., Connolly, L. E., Alba, B. M. and Gross, C. A. (1999). The *Escherichia coli* sigma(E)-dependent extracytoplasmic stress response is controlled by the regulated proteolysis of an anti-sigma factor. *Genes Dev* 13, 2449-2461.

Aguilar, P. S., Hernandez-Arriaga, A. M., Cybulski, L. E., Erazo, A. C. and de Mendoza, D. (2001). Molecular basis of thermosensing: a two-component signal transduction thermometer in *Bacillus subtilis*. *Embo J* 20, 1681-1691.

Alba, B. M. and Gross, C. A. (2004). Regulation of the *Escherichia coli* sigma-dependent envelope stress response. *Mol Microbiol* 52, 613-619.

Alba, B. M., Zhong, H. J., Pelayo, J. C. and Gross, C. A. (2001). *degS* (*hhoB*) is an essential *Escherichia coli* gene whose indispensable function is to provide sigma (E) activity. *Mol Microbiol* 40, 1323-1333.

Albanesi, D., Mansilla, M. C. and de Mendoza, D. (2004). The membrane fluidity sensor DesK of *Bacillus subtilis* controls the signal decay of its cognate response regulator. *J Bacteriol* 186, 2655-2663.

Baranova, N. and Nikaido, H. (2002). The *baeSR* two-component regulatory system activates transcription of the yegMNOB (*mdtABCD*) transporter gene cluster in *Escherichia coli* and increases its resistance to novobiocin and deoxycholate. *J Bacteriol* 184, 4168-4176.

Baysse, C. and O'Gara, F. (2006). Role of membrane structure during stress signalling and adaptation in *Pseudomonas*. In *Pseudomonas: a model system in biology*.In preparation Edited by J. Ramos and A. Filloux. New York: Springer-Verlag.

Baysse, C., Cullinane, M., Denervaud, V., Burrowes, E., Dow, J. M., Morrissey, J. P., Tam, L., Trevors, J. T. and O'Gara, F. (2005). Modulation of quorum sensing in *Pseudomonas aeruginosa* through alteration of membrane properties. *Microbiology* 151, 2529-2542.

Block, R. and Haseltine, A. W. (1975). Purification and properties of stringent factor. *J Biol Chem* 250, 1212-1217.

Booth, I. R., Cash, P. and O'Byrne, C. (2002). Sensing and adapting to acid stress. *Antonie Van Leeuwenhoek* 81, 33-42.

Brown, L., Gentry, D., Elliott, T. and Cashel, M. (2002). DksA affects ppGpp induction of RpoS at a translational level. *J Bacteriol* 184, 4455-4465.

Buchberger, A., Schroder, H., Hesterkamp, T., Schonfeld, H. J. and Bukau, B. (1996). Substrate shuttling between the DnaK and GroEL systems indicates a chaperone network promoting protein folding. *J Mol Biol* 261, 328-333.

Buchmeier, N. A., Libby, S. J., Xu, Y., Loewen, P. C., Switala, J., Guiney, D. G. and Fang, F. C. (1995). DNA repair is more important than catalase for *Salmonella* virulence in mice. *J Clin Invest* 95, 1047-1053.

Buelow, D. R. and Raivio, T. L. (2005). Cpx signal transduction is influenced by a conserved N-terminal domain in the novel inhibitor CpxP and the periplasmic protease DegP. *J Bacteriol* 187, 6622-66230.

Campbell, E. A., Tupy, J. L., Gruber, T. M., Wang, S., Sharp, M. M., Gross, C. A. and Darst, S. A. (2003). Crystal structure of *Escherichia coli* sigmaE with the cytoplasmic domain of its anti-sigma RseA. *Mol Cell* 11, 1067-1078.

Christians, E. S., Yan, L. J. and Benjamin, I. J. (2002). Heat shock factor 1 and heat shock proteins: Critical partners in protection against acute cell injury. *Crit Care Med* 30, S43-S50.

Clausen, T., Southan, C. and Ehrmann, M. (2002). The HtrA family of proteases: implications for protein composition and cell fate. *Mol Cell* 10, 443-455.

Costanzo, A. and Ades, S. E. (2006). Growth Phase-Dependent Regulation of the Extracytoplasmic Stress Factor, {sigma}E, by Guanosine 3',5'-Bispyrophosphate (ppGpp). *J Bacteriol* 188, 4627-4634.

Craig, J. E., Nobbs, A. and High, N. J. (2002). The extracytoplasmic sigma factor, final sigma(E), is required for intracellular survival of nontypeable *Haemophilus influenzae* in J774 macrophages. *Infect Immun* 70, 708-715.

Cybulski, L. E., del Solar, G., Craig, P. O., Espinosa, M. and de Mendoza, D. (2004). *Bacillus subtilis* DesR functions as a phosphorylation-activated switch to control membrane lipid fluidity. *J Biol Chem* 279, 39340-39347.

Cybulski, L. E., Albanesi, D., Mansilla, M. C., Altabe, S., Aguilar, P. S. and de Mendoza, D. (2002). Mechanism of membrane fluidity optimization: isothermal control of the *Bacillus subtilis* acyl-lipid desaturase. *Mol Microbiol* 45, 1379-1388.

Danese, P. N., Oliver, G. R., Barr, K., Bowman, G. D., Rick, P. D. and Silhavy, T. J. (1998). Accumulation of the enterobacterial common antigen lipid II biosynthetic intermediate stimulates *degP* transcription in *Escherichia coli*. *J Bacteriol* 180, 5875-5884.

Dartigalongue, C., Missiakas, D. and Raina, S. (2001a). Characterization of the *Escherichia coli* sigma E regulon. *J Biol Chem* 276, 20866-20875.

Dartigalongue, C., Loferer, H. and Raina, S. (2001b). EcfE, a new essential inner membrane protease: its role in the regulation of heat shock response in *Escherichia coli*. *Embo J* 20, 5908-5918.

De Las Penas, A., Connolly, L. and Gross, C. A. (1997). The sigmaE-mediated response to extracytoplasmic stress in *Escherichia coli* is transduced by RseA and RseB, two negative regulators of sigmaE. *Mol Microbiol* 24, 373-385.

De Wulf, P., Kwon, O. and Lin, E. C. (1999). The CpxRA signal transduction system of *Escherichia coli*: growth-related autoactivation and control of unanticipated target operons. *J Bacteriol* 181, 6772-6778.

De Wulf, P., Akerley, B. J. and Lin, E. C. (2000). Presence of the Cpx system in bacteria. *Microbiology* 146 (Pt 2), 247-248.

De Wulf, P., McGuire, A. M., Liu, X. and Lin, E. C. (2002). Genome-wide profiling of promoter recognition by the two-component response regulator CpxR-P in *Escherichia coli*. *J Biol Chem* 277, 26652-26661.

Denich, T. J., Beaudette, L. A., Lee, H. and Trevors, J. T. (2003). Effect of selected environmental and physico-chemical factors on bacterial cytoplasmic membranes. *J Microbiol Methods* 52, 149-182.

Diaz, A. R., Mansilla, M. C., Vila, A. J. and de Mendoza, D. (2002). Membrane topology of the acyl-lipid desaturase from *Bacillus subtilis*. *J Biol Chem* 277, 48099-48106.

DiGiuseppe, P. A. and Silhavy, T. J. (2003). Signal detection and target gene induction by the CpxRA two-component system. *J Bacteriol* 185, 2432-2440.

DiRusso, C. C. and Nystrom, T. (1998). The fats of *Escherichia coli* during infancy and old age: regulation by global regulators, alarmones and lipid intermediates. *Mol Microbiol* 27, 1-8.

Dorel, C., Lejeune, P. and Rodrigue, A. (2006). The Cpx system of *Escherichia coli*, a strategic signaling pathway for confronting adverse conditions and for settling biofilm communities? *Res Microbiol* 157, 306-314.

Douchin, V., Bohn, C. and Bouloc, P. (2006). Down-regulation of porins by a small RNA bypasses the essentiality of the regulated intramembrane proteolysis protease RseP in *Escherichia coli*. *J Biol Chem* 281, 12253-12259.

Dougan, D. A., Mogk, A. and Bukau, B. (2002). Protein folding and degradation in bacteria: to degrade or not to degrade? That is the question. *Cell Mol Life Sci* 59, 1607-1616.

Erickson, J. W. and Gross, C. A. (1989). Identification of the sigma E subunit of *Escherichia coli* RNA polymerase: a second alternate sigma factor involved in high-temperature gene expression. *Genes Dev* 3, 1462-1471.

Erickson, J. W., Vaughn, V., Walter, W. A., Neidhardt, F. C. and Gross, C. A. (1987). Regulation of the promoters and transcripts of *rpoH*, the *Escherichia coli* heat shock regulatory gene. *Genes Dev* 1, 419-432.

Flynn, J. M., Levchenko, I., Sauer, R. T. and Baker, T. A. (2004). Modulating substrate choice: the SspB adaptor delivers a regulator of the extracytoplasmic-stress response to the AAA+ protease ClpXP for degradation. *Genes Dev* 18, 2292-2301.

Gal-Mor, O. and Segal, G. (2003). Identification of CpxR as a positive regulator of icm and dot virulence genes of *Legionella pneumophila*. *J Bacteriol* 185, 4908-4919.

Hazel, J. R. and Williams, E. E. (1990). The role of alterations in membrane lipid composition in enabling physiological adaptation of organisms to their physical environment. *Prog Lipid Res* 29, 167-227.

Hellingwerf, K. J. (2005). Bacterial observations: a rudimentary form of intelligence? *Trends Microbiol* 13, 152-158.

Hershberger, C. D., Ye, R. W., Parsek, M. R., Xie, Z. D. and Chakrabarty, A. M. (1995). The algT (algU) gene of *Pseudomonas aeruginosa*, a key regulator involved in alginate biosynthesis, encodes an alternative sigma factor (sigma E). *Proc Natl Acad Sci U S A* 92, 7941-7945.

Humphreys, S., Stevenson, A., Bacon, A., Weinhardt, A. B. and Roberts, M. (1999). The alternative sigma factor, sigmaE, is critically important for the virulence of *Salmonella typhimurium*. *Infect Immun* 67, 1560-1568.

Isaac, D. D., Pinkner, J. S., Hultgren, S. J. and Silhavy, T. J. (2005). The extracytoplasmic adaptor protein CpxP is degraded with substrate by DegP. *Proc Natl Acad Sci U S A* 102, 17775-17779.

Jishage, M., Kvint, K., Shingler, V. and Nystrom, T. (2002). Regulation of sigma factor competition by the alarmone ppGpp. *Genes Dev* 16, 1260-1270.

Johnson, K., Charles, I., Dougan, G., Pickard, D., O'Gaora, P., Costa, G., Ali, T., Miller, I. and Hormaeche, C. (1991). The role of a stress-response protein in *Salmonella typhimurium* virulence. *Mol Microbiol* 5, 401-407.

Jones, C. H., Danese, P. N., Pinkner, J. S., Silhavy, T. J. and Hultgren, S. J. (1997). The chaperone-assisted membrane release and folding pathway is sensed by two signal transduction systems. *Embo J* 16, 6394-6406.

Kanehara, K., Akiyama, Y. and Ito, K. (2001). Characterization of the *yaeL* gene product and its S2P-protease motifs in *Escherichia coli*. *Gene* 281, 71-79.

Kanehara, K., Ito, K. and Akiyama, Y. (2003). YaeL proteolysis of RseA is controlled by the PDZ domain of YaeL and a Gln-rich region of RseA. *Embo J* 22, 6389-6398.

Kazmierczak, M. J., Wiedmann, M. and Boor, K. J. (2005). Alternative sigma factors and their roles in bacterial virulence. *Microbiol Mol Biol Rev* 69, 527-543.

Kim, H. Y., Schlictman, D., Shankar, S., Xie, Z., Chakrabarty, A. M. and Kornberg, A. (1998). Alginate, inorganic polyphosphate, GTP and ppGpp synthesis co-regulated in *Pseudomonas aeruginosa:* implications for stationary phase survival and synthesis of RNA/DNA precursors. *Mol Microbiol* 27, 717-725.

Koo, M. S., Lee, J. H., Rah, S. Y., Yeo, W. S., Lee, J. W., Lee, K. L., Koh, Y. S., Kang, S. O. and Roe, J. H. (2003). A reducing system of the superoxide sensor SoxR in *Escherichia coli*. *Embo J* 22, 2614-2622.

Kuroda, A., Murphy, H., Cashel, M. and Kornberg, A. (1997). Guanosine tetra- and pentaphosphate promote accumulation of inorganic polyphosphate in *Escherichia coli*. *J Biol Chem* 272, 21240-21243.

Laurie, A. D., Bernardo, L. M., Sze, C. C., Skarfstad, E., Szalewska-Palasz, A., Nystrom, T. and Shingler, V. (2003). The role of the alarmone (p)ppGpp in sigma N competition for core RNA polymerase. *J Biol Chem* 278, 1494-1503.

Los, D. A. and Murata, N. (2004). Membrane fluidity and its roles in the perception of environmental signals *Biochim Biophy Acta* 1666, 142 - 157.

Mansilla, M. C. and de Mendoza, D. (2005). The *Bacillus subtilis* desaturase: a model to understand phospholipid modification and temperature sensing. *Arch Microbiol* 183, 229-235.

Mathee, K., McPherson, C. J. and Ohman, D. E. (1997). Posttranslational control of the algT (algU)-encoded sigma22 for expression of the alginate regulon in *Pseudomonas aeruginosa* and localization of its antagonist proteins MucA and MucB (AlgN). *J Bacteriol* 179, 3711-3720.

McCormick, M. L., Buettner, G. R. and Britigan, B. E. (1998). Endogenous superoxide dismutase levels regulate iron-dependent hydroxyl radical formation in *Escherichia coli* exposed to hydrogen peroxide. *J Bacteriol* 180, 622-625.

McEwen, J. and Silverman, P. (1980a). Mutations in genes *cpxA* and *cpxB* of *Escherichia coli* K-12 cause a defect in isoleucine and valine syntheses. *J Bacteriol* 144, 68-73.

McEwen, J. and Silverman, P. (1980b). Chromosomal mutations of *Escherichia coli* that alter expression of conjugative plasmid functions. *Proc Natl Acad Sci U S A* 77, 513-517.

Michaud, S., Marin, R., Westwood, J. T. and Tanguay, R. M. (1997). Cell-specific expression and heat-shock induction of Hsps during spermatogenesis in *Drosophila melanogaster*. *J Cell Sci* 110 (Pt 17), 1989-1997.

Mikami, K. and Murata, N. (2003). Membrane fluidity and the perception of environmental signals in cyanobacteria and plants. *Prog Lipid Res* 42, 527-543.

Mikami, K., Kanesaki, Y., Suzuki, I. and Murata, N. (2002). The histidine kinase Hik33 perceives osmotic stress and cold stress in *Synechocystis sp* PCC 6803. *Mol Microbiol* 46, 905-915.

Mileykovskaya, E. and Dowhan, W. (1997). The Cpx two-component signal transduction pathway is activated in *Escherichia coli* mutant strains lacking phosphatidylethanolamine. *J Bacteriol* 179, 1029-1034.

Missiakas, D., Betton, J. M. and Raina, S. (1996). New components of protein folding in extracytoplasmic compartments of *Escherichia coli* SurA, FkpA and Skp/OmpH. *Mol Microbiol* 21, 871-884.

Miticka, H., Rowley, G., Rezuchova, B., Homerova, D., Humphreys, S., Farn, J., Roberts, M. and Kormanec, J. (2003). Transcriptional analysis of the *rpoE* gene encoding extracytoplasmic stress response sigma factor sigmaE in *Salmonella enterica* serovar *Typhimurium*. *FEMS Microbiol Lett* 226, 307-314.

Morita, M., Kanemori, M., Yanagi, H. and Yura, T. (1999). Heat-induced synthesis of sigma32 in *Escherichia coli*: structural and functional dissection of rpoH mRNA secondary structure. *J Bacteriol* 181, 401-410.

Murata, N. and Suzuki, I. (2006). Exploitation of genomic sequences in a systematic analysis to access how cyanobacteria sense environmental stress. *J Exp Bot* 57, 235-247.

Nagai, H., Yano, R., Erickson, J. W. and Yura, T. (1990). Transcriptional regulation of the heat shock regulatory gene *rpoH* in *Escherichia coli*: involvement of a novel catabolite-sensitive promoter. *J Bacteriol* 172, 2710-2715.

Nakayama, S. and Watanabe, H. (1998). Identification of *cpxR* as a positive regulator essential for expression of the *Shigella sonnei virF* gene. *J Bacteriol* 180, 3522-3528.

Nakayama, S., Kushiro, A., Asahara, T., Tanaka, R., Hu, L., Kopecko, D. J. and Watanabe, H. (2003). Activation of *hilA* expression at low pH requires the signal sensor CpxA, but not the cognate response regulator CpxR, in *Salmonella enterica* serovar Typhimurium. *Microbiology* 149, 2809-2817.

Nishino, K., Honda, T. and Yamaguchi, A. (2005). Genome-wide analyses of *Escherichia coli* gene expression responsive to the BaeSR two-component regulatory system. *J Bacteriol* 187, 1763-1772.

Otto, K. and Silhavy, T. J. (2002). Surface sensing and adhesion of *Escherichia coli* controlled by the Cpx-signaling pathway. *Proc Natl Acad Sci U S A* 99, 2287-2292.

Paithoonrangsarid, K., Shoumskaya, M. A., Kanesaki, Y., Satoh, S., Tabata, S., Los, D. A., Zinchenko, V. V., Hayashi, H., Tanticharoen, M., Suzuki, I. and Murata, N. (2004). Five histidine kinases perceive osmotic stress and regulate distinct sets of genes in Synechocystis. *J Biol Chem* 279, 53078-53086.

Parkhill, J., Wren, B. W., Mungall, K., Ketley, J. M., Churcher, C., Basham, D., Chillingworth, T., Davies, R. M., Feltwell, T., Holroyd, S., Jagels, K., Karlyshev, A. V., Moule, S., Pallen, M. J., Penn, C. W., Quail, M. A., Rajandream, M. A., Rutherford, K. M., van Vliet, A. H., Whitehead, S. and Barrell, B. G. (2000). The genome sequence of the food-borne pathogen *Campylobacter jejuni* reveals hypervariable sequences. *Nature* 403, 665-668.

Piper, P. W. (1993). Molecular events associated with acquisition of heat tolerance by the yeast *Saccharomyces cerevisiae. FEMS Microbiol Rev* 11, 339-355.

Piper, P. W., Ortiz-Calderon, C., Holyoak, C., Coote, P. and Cole, M. (1997). Hsp30, the integral plasma membrane heat shock protein of *Saccharomyces cerevisiae*, is a stress-inducible regulator of plasma membrane H(+)-ATPase. *Cell Stress Chaperones* 2, 12-24.

Raffa, R. G. and Raivio, T. L. (2002). A third envelope stress signal transduction pathway in *Escherichia coli. Mol Microbiol* 45, 1599-1611.

Raivio, T. L. (2005). Envelope stress responses and Gram-negative bacterial pathogenesis. *Mol Microbiol* 56, 1119-1128.

Raivio, T. L. and Silhavy, T. J. (2001). Periplasmic stress and ECF sigma factors. *Annu Rev Microbiol* 55, 591-624.

Raivio, T. L., Popkin, D. L. and Silhavy, T. J. (1999). The Cpx envelope stress response is controlled by amplification and feedback inhibition. *J Bacteriol* 181, 5263-5272.

Rhodius, V. A., Suh, W. C., Nonaka, G., West, J. and Gross, C. A. (2006). Conserved and variable functions of the sigmaE stress response in related genomes. *PLoS Biol* 4, 43-59.

Rilfors, L., Lindblom, G., Wieslander, A. and Christiansson, A. (1984). *Lipid bilayer stability in biological membranes*, Kates, M. Manson, L.A. edn. New York: Plenum Press

Rose, J. K. and Rankin, C. H. (2001). Analyses of habituation in *Caenorhabditis elegans. Learn Mem* 8, 63-69.

Rowley, G., Spector, M., Kormanec, J. and Roberts, M. (2006). Pushing the envelope: extracytoplasmic stress responses in bacterial pathogens. *Nat Rev Microbiol* 4, 383-394.

Roy, C. R. (2002). The Dot/lcm transporter of *Legionella pneumophila*: a bacterial conductor of vesicle trafficking that orchestrates the establishment of a replicative organelle in eukaryotic hosts. *Int J Med Microbiol* 291, 463-467.

Russel, N. J. (1989). *Functions of lipids: structural role in membrane function*, Ratledge C. Wilkinson, S. G edn. Toronto: Academic Press

Shoumskaya, M. A., Paithoonrangsarid, K., Kanesaki, Y., Los, D. A., Zinchenko, V. V., Tanticharoen, M., Suzuki, I. and Murata, N. (2005). Identical Hik-Rre systems are involved in perception and transduction of salt signals and hyperosmotic signals but regulate the expression of individual genes to different extents in *synechocystis. J Biol Chem* 280, 21531-21538.

Snyder, W. B., Davis, L. J., Danese, P. N., Cosma, C. L. and Silhavy, T. J. (1995). Overproduction of NlpE, a new outer membrane lipoprotein, suppresses the toxicity of periplasmic LacZ by activation of the Cpx signal transduction pathway. *J Bacteriol* 177, 4216-4223.

Souzu, H. (1986). Fluorescence polarization studies on *Escherichia coli* membrane stability and its relation to the resistance of the cell to freeze-thawing. I. Membrane stability in cells of differing growth phase. *Biochim Biophys Acta* 861, 353-360.

Straus, D. B., Walter, W. A. and Gross, C. A. (1987). The heat shock response of *E. coli* is regulated by changes in the concentration of sigma 32. *Nature* 329, 348-351.

Suzuki, I., Kanesaki, Y., Mikami, K., Kanehisa, M. and Murata, N. (2001). Cold-regulated genes under control of the cold sensor Hik33 in *Synechocystis*. *Mol Microbiol* 40, 235-244.

Tam, C. and Missiakas, D. (2005). Changes in lipopolysaccharide structure induce the sigma(E)-dependent response of *Escherichia coli*. *Mol Microbiol* 55, 1403-1412.

Testerman, T. L., Vazquez-Torres, A., Xu, Y., Jones-Carson, J., Libby, S. J. and Fang, F. C. (2002). The alternative sigma factor sigmaE controls antioxidant defences required for *Salmonella* virulence and stationary-phase survival. *Mol Microbiol* 43, 771-782.

Thanassi, D. G., Saulino, E. T. and Hultgren, S. J. (1998). The chaperone/usher pathway: a major terminal branch of the general secretory pathway. *Curr Opin Microbiol* 1, 223-231.

Tobe, T., Yoshikawa, M., Mizuno, T. and Sasakawa, C. (1993). Transcriptional control of the invasion regulatory gene *virB* of *Shigella flexneri*: activation by *virF* and repression by H-NS. *J Bacteriol* 175, 6142-6149.

Tomoyasu, T., Ogura, T., Tatsuta, T. and Bukau, B. (1998). Levels of DnaK and DnaJ provide tight control of heat shock gene expression and protein repair in *Escherichia coli*. *Mol Microbiol* 30, 567-581.

Tomoyasu, T., Gamer, J., Bukau, B., Kanemori, M., Mori, H., Rutman, A. J., Oppenheim, A. B., Yura, T., Yamanaka, K., Niki, H. and et al. (1995). *Escherichia coli* FtsH is a membrane-bound, ATP-dependent protease which degrades the heat-shock transcription factor sigma 32. *Embo J* 14, 2551-2560.

Urban, S. and Freeman, M. (2002). Intramembrane proteolysis controls diverse signalling pathways throughout evolution. *Curr Opin Genet Dev* 12, 512-518.

Walsh, N. P., Alba, B. M., Bose, B., Gross, C. A. and Sauer, R. T. (2003). OMP peptide signals initiate the envelope-stress response by activating DegS protease via relief of inhibition mediated by its PDZ domain. *Cell* 113, 61-71.

Weber, H., Polen, T., Heuveling, J., Wendisch, V. F. and Hengge, R. (2005). Genome-wide analysis of the general stress response network in *Escherichia coli*: sigmaS-dependent genes, promoters, and sigma factor selectivity. *J Bacteriol* 187, 1591-1603.

Weitzel, G., Pilatus, U. and Rensing, L. (1987). The cytoplasmic pH, ATP content and total protein synthesis rate during heat-shock protein inducing treatments in yeast. *Exp Cell Res* 170, 64-79.

Yu, H., Mudd, M., Boucher, J. C., Schurr, M. J. and Deretic, V. (1997). Identification of the *algZ* gene upstream of the response regulator *algR* and its participation in control of alginate production in *Pseudomonas aeruginosa*. *J Bacteriol* 179, 187-193.

Zheng, M. and Storz, G. (2000). Redox sensing by prokaryotic transcription factors. *Biochem Pharmacol* 59, 1-6.

In: Focus on Signal Transduction Research
Editor: Gavin McAlpine, pp. 145-158
ISBN 1-60021-376-6

Chapter 6

CELL-TO-CELL COMMUNICATION IN PATHOGENIC BACTERIA

Mario Juhas*

Clinical Microbiology and Infectious Diseases, Nuffield Department of Clinical Laboratory Sciences, University of Oxford, UK

ABSTRACT

Many bacteria use small secreted signaling molecules called autoinducers to communicate and to coordinate expression of certain sets of genes in a process known as quorum sensing. Quorum sensing allows bacteria to launch a unified, coordinated response on a population-wide scale to accomplish tasks which would be impossible to achieve for a single bacterial cell. Work over the past years has provided evidence that bacterial quorum sensing signaling is exploited not only for intraspecies, but also for interspecies cell-to-cell communication with other bacteria occupying the same ecological niche and even for cross-talk with their eukaryotic hosts. Quorum sensing is a generic regulatory mechanism used by a broad spectrum of bacteria including human pathogens where it highly influences the ability to establish an infection. In the opportunistic human pathogen *Pseudomonas aeruginosa*, quorum sensing-dependent regulation of gene expression is essential for the synchronization of the production of a battery of virulence factors as well as for biofilm development. Given the essential role of this phenomenon for the pathogenic potential of bacteria, cell-to-cell communication represents an attractive target for the design of novel drugs.

---

* E-mail: juhas_mario@pobox.sk; mario.juhas@ndcls.ox.ac.uk

INTRODUCTION

For many years, bacteria were thought to be primitive single-celled organisms incapable of social interactions typical for eukaryotes.

Work over the last three decades has challenged this view. Considerable evidence has accumulated from the research performed in many laboratories around the world that bacteria can use chemical signals to communicate with each other in order to respond collectively to different environmental stimuli.

First recognized types of bacterial cell-to-cell communication were the regulation of bioluminescence in the marine bacterium *Vibrio fischeri* (Nealson *et al.,* 1970) and the regulation of the genetic competence in *Streptococcus pneumoniae* (Tomasz, 1965). Both, *Vibrio fischeri* and *Streptococcus pneumoniae* were shown to regulate expression of certain sets of genes via the production of small signaling molecules, called autoinducers, in a population density-dependent manner.

In the simplest form, cell-to-cell communication networks of Gram-negative bacteria consist of two major components: of the transcriptional regulator and of the autoinducer synthase. During bacterial growth, autoinducer signaling molecules are synthesized by the autoinducer synthase and are secreted into the surrounding environment. At low cell densities, the concentration of signaling molecules is too low to be detected (Figure 1A); however, when the bacterial population and the resulting autoinducer concentration reaches a critical threshold level, autoinducer molecules bind to and activate the transcriptional regulator, which in turn induces or represses expression of target genes (Figure 1B).

Term "quorum sensing" is usually used for this type of cell-to-cell communication because a certain minimal number of bacteria, or "quorum", of bacteria is required in order to initiate the cooperative gene expression (for recent reviews see Lazdunski *et al.*, 2004; Waters and Bassler, 2005). Quorum sensing allows bacteria to launch a unified, coordinated response to environmental stimuli on a population-wide scale to accomplish tasks which would be impossible to achieve for a single bacterial cell.

Several groups of bacterial signaling molecules have been identified. Gram-negative bacteria use preferentially acylated homoserine lactones (AHL), while Gram-positive bacteria exploit peptide-based signaling molecules (Waters and Bassler, 2005). Autoinducer-2 (AI-2) is the universal signaling molecule detectable in both in Gram-negative as well as in Gram-positive bacteria (Vendeville *et al.,* 2005).

Bacterial cell-to-cell communication was shown to be involved in the regulation of a wide variety of important processes, including antibiotic biosynthesis, biofilm development, bioluminescence, genetic competence, plasmid conjugal transfer, production of virulence factors, swarming, swimming, and twitching motility (Figure 1B).

Work over the past years has provided evidence that bacterial quorum sensing signaling is exploited not only for intraspecies, but also for interspecies cell-to-cell communication with other bacteria occupying the same ecological niche and even for cross-talk with their eukaryotic hosts.

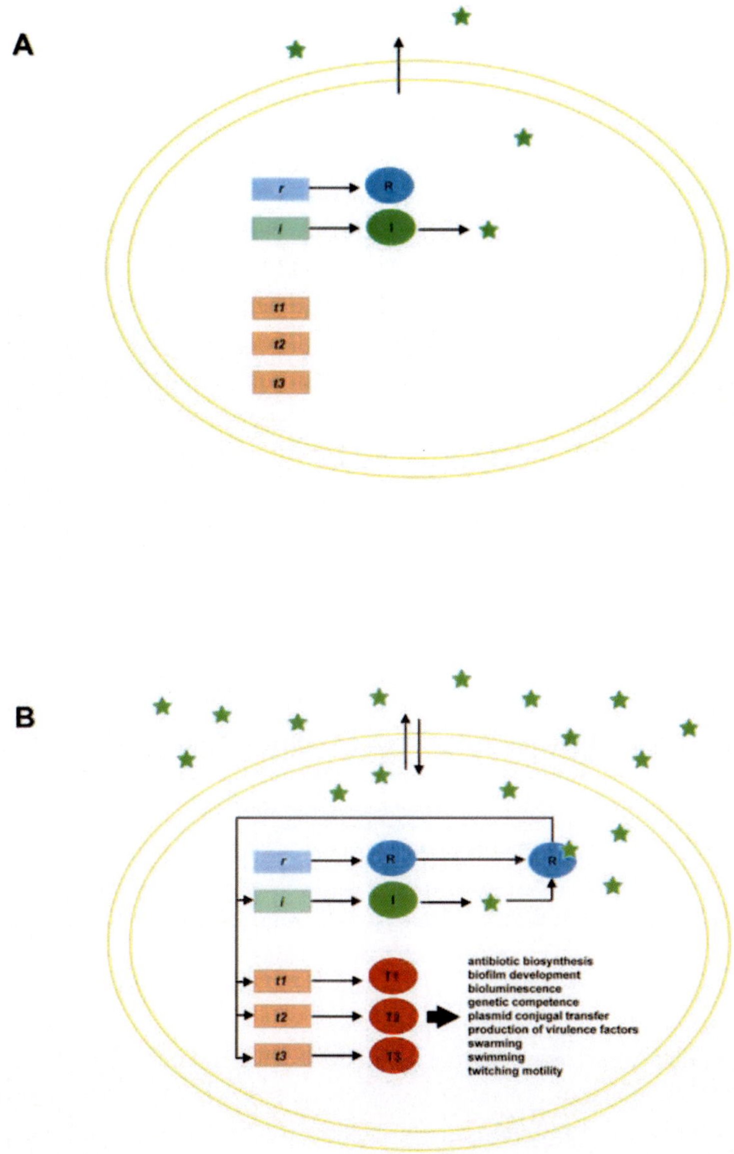

Figure 1. Generic scheme of quorum sensing. In the simplest form, quorum sensing of Gram-negative bacteria consist of two major components: of the transcriptional regulator (R) and of the autoinducer synthase (I). During bacterial growth, autoinducer signaling molecules (represented by green star) are synthesized by the autoinducer synthase (I) and are secreted into the surrounding environment.

(A) At low cell densities, the concentration of signaling molecules is too low to induce expression of target genes (*t1-t3*). (B) When the bacterial population and the resulting autoinducer concentration reach a critical threshold level, autoinducer molecules bind to and activate the transcriptional regulator (R), which in turn induces or represses expression of target genes (*t1-t3*). Products of these genes (T1-T3) are involved in a variety of processes including antibiotic biosynthesis, biofilm development, bioluminescence, genetic competence, plasmid conjugal transfer, production of virulence factors, swarming, swimming, and twitching motility.

The ability to sense and respond to local environmental conditions in a population-wide scale is particularly essential for pathogenic bacteria as it helps them to establish an infection in a wide variety of habitats in host. Moreover, quorum sensing-dependent regulation is hypothesized to provide pathogenic bacteria with the opportunity to appropriately time expression of immune response-activating virulence products. Thus, bacteria can grow undetected without the threat of being eradicated by the host immune system before a sufficient number of invading bacteria has amassed to overwhelm the host defenses with a concerted expression of virulence traits (de Kievit and Iglewski, 2000). A broad spectrum of human pathogens, including *Pseudomonas aeruginosa*, *Burkholderia cepacia*, *Staphylococcus aureus*, *Streptococcus gordonii*, *Porphyromonas gingivalis*, *Veillonella atypica*, *Salmonella enterica*, *Streptococcus pneumoniae*, *Yersinia pseudotuberculosis*, *Escherichia coli*, *Neisseria meningitides*, *Helicobacter pylori*, *Bacillus subtilis*, *Proteus mirabilis*, *Chromobacterium violaceum*, and *Serratia marcescens* were shown to produce quorum sensing signaling molecules (Camara *et al.,* 2002).

The most extensively studied quorum sensing system is the one of the opportunistic human pathogen *P. aeruginosa*.

QUORUM SENSING IN *P. AERUGINOSA*

P. aeruginosa, well known for its resistance against a variety of antibiotics, is the leading source of Gram-negative nosocomial infections. *P. aeruginosa* causes serious and often fatal infections in patients suffering from cystic fibrosis as well as in immunocompromised hosts, such as individuals with cancer, AIDS or severe burn wounds.

Two interlinked quorum sensing systems, *las* system and *rhl* system, are operating in *P. aeruginosa*. The *las* system consists of the transcriptional regulator LasR and the cognate AHL synthase LasI, which produces the cell-to-cell signaling molecule *N*-3-oxododecanoyl-homoserine lactone (3-oxo-C_{12}-HSL), while the *rhl* system consists of the transcriptional regulator RhLR and the AHL synthase RhlI, which directs synthesis of another signaling molecule, *N*-butanoyl-homoserine lactone (C_4-HSL) (Latifi *et al.*, 1995, Pearson *et al.,* 1994; Pearson *et al.,* 1995). Once a threshold concentration is reached, AHL signaling molecules bind to their cognate transcriptional regulators and form a complex which in turn activates or represses expression of target genes. LasR-3-oxo-C_{12}-HSL complex was shown to positively regulate transcription of RhlR and RhlI (Pesci *et al.,* 1997), thus providing evidence that *las* and *rhl* systems of *P. aeruginosa* do not operate independently but rather they are hierarchically arranged with the *las* system being on top of the signaling cascade.

The *P. aeruginosa* quinolone signal (PQS) together with a number of regulators, including AlgR2, DksA, GacA, MvaT, Ppk, RpoN, RpoS, RsaL, and RsmA add additional levels of complexity to quorum sensing circuitry of *P. aeruginosa* (for recent reviews see Juhas *et al.,* 2005; Venturi, 2006, Schuster and Greenberg, 2006). Moreover, besides two AHL signal receptors LasR and RhlR, the genome of *P. aeruginosa* encodes two additional LuxR family members, QscR and VqsR. Inactivation of QscR led to overproduction of virulence factors whose expression is blocked under normal conditions by formation of heterodimers between QscR and LasR and RhlR (Ledgham *et al.,* 2003). In contrary, the inactivation of VqsR abrogated production of both signaling molecules, decreased production of virulence factors and reduced pathogenicity of *P. aeruginosa* (Juhas *et al.*, 2004). These data provided evidence that QscR is the negative regulator of quorum sensing, while VqsR is involved in the positive regulation of the *P. aeruginosa* quorum sensing network.

Production of the battery of virulence factors as well as biofilm development was shown to be regulated by quorum sensing in *P. aeruginosa*. Expression of many virulence factors, including exotoxin A, neuraminidase, catalase, and superoxid dismutase is predominantly regulated by the *las* system, expression of others, including rhamnolipids, pyocyanin, chitinase, lectins A and B, and exoenzyme S is predominantly regulated by the *rhl* system, while expression of elastase, lipase, hydrogen cyanide, as well as swarming and twitching motility are under the control of both quorum sensing systems.

The formation of organized communities enclosed in a slime-like polymeric substance matrix, known as biofilms, has great impact on the ability of *P. aeruginosa* to resist host defenses as well as a broad spectrum of antibiotics. Interestingly, *lasI* mutant failed to form the tower- and mushroom-shaped structures that are typical for *P. aeruginosa* biofilms and was more sensitive to treatment with biocides (Davies *et al.*, 1998; Greenber, 2003). It has been shown that, depending on experimental settings and phase of the biofilm development, both quorum sensing systems of *P. aeruginosa* are involved in the formation of biofilms (Hentzer *et al.*, 2004; Parsek and Greenberg, 2005).

INTERSPECIES CELL-TO-CELL COMMUNICATION IN PATHOGENIC BACTERIA

Work over the past years has provided evidence that quorum sensing signaling is exploited not only for intraspecies, but also for interspecies cell-to-cell communication with other bacteria occupying the same ecological niche.

One of the best established paradigms of the interspecies cell-to-cell communication is that between *P. aeruginosa* and *B. cepacia*. *B. cepacia* is a Gram-negative bacterium often forming mixed biofilms with *P. aeruginosa* in the lungs of cystic fibrosis patients. Similar to *P. aeruginosa*, quorum sensing regulated expression of virulence traits in *B. cepacia* has been shown to rely on AHL signaling molecules, namely *N*-octanoylhomoserine lactone (C_8-HSL) and *N*-hexanoylhomoserine lactone (C_6-HSL) (Lewenza *et al.*, 1999) leading to hypothesis that *P. aeruginosa* and *B.cepacia* can co-operate during infection. Indeed, an exogenous addition of *P. aeruginosa* spent culture supernatant containing AHL signaling molecules to *B. cepacia* increased production of quorum sensing related virulence factors in *B. cepacia* (McKenney *et al.*, 1995). In addition, examination of *P. aeruginosa* and *B. cepacia* mixed biofilms using AHL biosensors revealed that *B. cepacia* is able to perceive AHL signaling molecules produced by *P. aeruginosa* but not vice versa, thus providing evidence of unidirectional signaling between these two pathogens (Riedel *et al.*, 2001; Eberl, 2006). It is speculated that by exploiting *P. aeruginosa* quorum sensing signals, *B. cepacia* is able to enhance its own pathogenicity and increase chances of successful infection.

As mentioned at the beginning, Gram-negative bacteria preferentially use AHLs for cell-to-cell communication, while Gram-positive bacteria exploit peptide-based signaling molecules, thus any cross-talk between Gram-negative and Gram-positive bacteria was considered to be improbable.

However, a very recent study of quorum sensing in *S. aureus* has challenged this view. *S aureus* is a Gram-positive bacterium, often colonizing the lungs of cystic fibrosis patients, which is later in the course of infection often out-competed and replaced by *P. aeruginosa*. In spite of the fact that *S. aureus* does not produce AHLs, it was shown to be forced to repress transcription of its own virulence genes in response to 3-oxo-C_{12}-HSL produced by *P. aeruginosa* (Qazi *et al.*, 2006). This observation suggests that besides the induction of co-operative behavior, the AHL-based cell-to-cell communication signals, as 3-oxo-C_{12}-HSL may also provide an important competitive advantage by inhibition of gene expression in other bacteria occupying the same ecological niche (Keller and Surette, 2006).

AI-2 is the universal signaling molecule detectable in Gram-negative as well as in Gram-positive bacteria. Moreover, *luxS* gene responsible for AI-2 production is present in a broad spectrum of bacteria, thus AI-2 seems to be an ideal candidate for interspecies communication.

Oral pathogens responsible for dental plaque biofilm formation are represented by more than 500 Gram-negative and Gram-positive bacteria, a number of which harbor *luxS* gene and produce AI-2 (Frias *et al.*, 2001). The examination of two of these microorganisms, *S. gordonii* and *P. gingivalis* revealed that mixed *S. gordonii- P. gingivalis* biofilms do not develop in the absence of AI-2 mediated signaling. Moreover, this study showed that the production of AI-2 by either *S. gordonii* or *P. gingivalis* alone is sufficient for the formation of mixed biofilms (McNab *et al.*, 2003).

Diffusible signaling molecules were also demonstrated to be responsible for the formation of *S. gordonii- V. atypica* mixed biofilms (England *et al.*, 2004). *S. gordonii* produces AI-2 molecules; however, no cell-to-cell signaling molecules of any kind were shown to be produced by *V. atypica*. This suggests that some kind of novel, yet uncharacterized signaling molecule is used by *V. atypica* to communicate with *S. gordonii* (England *et al.*, 2004).

Furthermore, *P. aeruginosa*, which does not harbor *luxS* gene was shown to initiate expression of quorum sensing- regulated virulence traits, such as rhamnolipid, phenazine, and elastase in response to AI-2 produced by an avirulent oropharyngeal flora from the lungs of cystic fibrosis patients (Duan *et al.*, 2003). This demonstrates that even non- pathogenic host microflora can play an important role in the pathogenesis of cystic fibrosis by influencing gene expression of *P. aeruginosa* via interspecies cell-to-cell communication. It is hypothesized that AI-2 molecules secreted by non- pathogenic oropharyngeal flora may alert *P. aeruginosa* that it is in the lungs of the host and that a regulated expression of virulence genes should be initiated (Waters and Bassler, 2005).

Another human pathogen, *S. enterica* has an incomplete set of key quorum sensing genes. It harbors the LuxR- type AHL signal receptor, SdiA, but lacks gene encoding protein responsible for the AHL signal synthesis. However, in spite of this apparent disadvantage, *S. enterica* can alter expression of sets of genes encoding proteins involved in the protection against host defenses in response to AHL signaling molecules produced by other bacterial species (Ahmer 2004; Henke and Bassler, 2004).

Taken together, many examples of bacterial interspecies cell-to-cell communication demonstrate that quorum sensing signaling can be exploited for either co-operative or competitive behavior of bacteria colonizing the same ecological niche.

INTERKINGDOM COMMUNICATION: HOST-PATHOGEN INTERACTIONS

By regulation of pathogenic traits expression, quorum sensing highly influences the ability of bacteria to establish an infection.

The substantial amount of work investigating the importance of quorum sensing for pathogenesis has been done in *P. aeruginosa*, where quorum sensing-dependent regulation of gene expression is essential for synchronization of the production of a battery of virulence factors. *P. aeruginosa* strains harboring mutations in the key quorum sensing regulatory genes, *lasR, lasI, rhlR* and *rhlI*, were significantly less pathogenic than the parent strain in a number of different infection models, including the mouse agar bead model, the neonatal mouse model of pneumonia, the burned mouse model, *Arabidopsis thaliana*, *Dictyostelium discoideum*, and *Caenorhabditis elegans* (Rahme *et al.*, 1995; Tang *et al.*, 1996; Rumbaugh *et al.*, 1999; Tan *et al.*, 1999; Cosson *et al.*, 2002). In mammalian experimental infection models, quorum sensing mutants showed reduced ability to spread in host tissues, to cause tissue destruction and to kill the host when compared to the parent strain.

Direct involvement of cell-to-cell communication in the infection process was confirmed by the presence of quorum sensing signaling molecules in the lung tissue of mouse infected with *P. aeruginosa* (Wu *et al.*, 2000) and in indwelling urethral catheters colonized by bacterial biofilms (Stickler *et al.*, 1998). In addition, quorum sensing signaling molecules were also detected in sputa and tissue biopsies from cystic fibrosis patients infected with *P. aeruginosa* and *B. cepacia* (Singh *et al.*, 2000; Erickson *et al.*, 2002; Middleton *et al.*, 2002).

Given the long time of coexistence of bacteria with eukaryotes and a rather intimate relationship between pathogens and their hosts, the discovery of mechanisms of interkingdom communication should not be extremely surprising. The structure of AHL signaling molecules is similar to that of eukaryotic hormones used for communication between eukaryotic cells, thus similar effects were also predicted for bacterial cell-to-cell signals.

Indeed, experiments employing chimeric LuxR-type transcriptional regulators demonstrated that AHLs can enter and retain functionality in eukaryotic cells (Williams *et al.*, 2004). Once inside the host cell, AHLs can induce a wide variety of effects, including apoptosis and immune response.

The long chain AHL, 3-oxo-C_{12}-HSL, produced by *P. aeruginosa* stimulates the expression of the neutrophil chemokine interleukin-8 (IL-8) in respiratory epithelial cells (DiMango *et al.*, 1995). IL-8 subsequently activates the multifunctional transcription factor NFκB involved in regulation of inflammation and apoptosis (Smith *et al.*, 2001). 3-oxo-C_{12}-HSL also influences the expression of two additional macrophage chemokines, macrophage inhibitory protein 2 (MIP-2) and monocyte chemoattractant protein 1 (MCP-1), as well as the expression of the prostaglandin producing enzyme cyclooxygenase-2 (Cox-2) and prostaglandin E2 (PGE2) (Smith *et al.*, 2002; Tateda *et al.*, 2003). The effect of 3-oxo-C_{12}-HSL on the induction of apoptosis has also been demonstrated in a number of experiments (Tateda *et al.*, 2003; Li *et al.*, 2004).

Altogether this can cause host defense cells migration to the site of infection and subsequent induction of inflammation leading to an extensive host tissue damage and more efficient bacterial dissemination. The influence of 3-oxo-C_{12}-HSL on blood vessels smooth muscle contraction might also contribute to rapid spreading of bacteria from the site of infection (Lawrence *et al.*, 1999).

Direct immunomodulatory effect of bacterial cell-to-cell communication was confirmed by influence of 3-oxo-C_{12}-HSL on T and B cells. 3-oxo-C_{12}-HSL significantly inhibited the production of interleukin 12 (IL-12) and tumor necrosis factor α (TNFα) in lipopolysaccharide-activated peritoneal macrophages, thus mediating the Th2 response of T cells instead of the antibacterial Th1 response (Telford *et al.*, 1998). 3-oxo-C_{12}-HSL also stimulated the production of interferon-γ (IFN-γ) in T cells (Smith *et al.*, 2002; Smith and Iglewski, 2003). Besides the antibacterial Th1 response, IFN-γ is also involved in macrophages and B cells activation and therefore plays an important role in the host's defenses against invading bacteria. On the other hand, IFN-γ binds to *P. aeruginosa* outer membrane protein, OprF, thus causing upregulation of bacterial virulence proteins expression (Wu *et al.*, 2005). The ability of *P. aeruginosa* to actively sense IFN-γ suggests the existence of a cyclic network between quorum sensing and T cells used by pathogens to acquire information about the host's immune system activation (Wu *et al.*, 2005; Wagner *et al.*, 2006).

Human airway epithelium was shown to inactivate rapidly *P. aeruginosa* signaling molecule 3-oxo-C_{12}-HSL, but not C_4-HSL (Chun *et al.*, 2004; Hastings, 2004). The ability to inactivate quorum sensing signaling molecules varied substantially between different cell types. Cells from tissues frequently exposed to pathogens, as human lungs or colon were most efficient in 3-oxo-C_{12}-HSL inactivation, whereas cells from other tissues showed reduced ability to inactivate 3-oxo-C_{12}-HSL. This suggests that quorum sensing signaling molecules are most efficiently inactivated by host cells that most likely come in contact with bacteria (Chun *et al.*, 2004).

Sufficient amount of evidence has accumulated that communication between pathogen and host is not strictly unidirectional, but rather that host cells can also interfere with quorum sensing. This interference is mediated by a wide range of chemical compounds, including enzymes, hormones, or molecules structurally related to quorum sensing signaling molecules (QS mimics) (for a more comprehensive overview see Shiner *et al.*, 2005).

Quorum Sensing as a Target of Novel Antimicrobial Drugs

Given the essential role of cell-to-cell communication for the pathogenic potential of bacteria, it represents an attractive target for the design of novel antimicrobial drugs. Potential novel drug capable of interference with cell-to-cell communication could target either the process of the signal generation or the signaling molecule or the signal receptor (for more comprehensive overview see Dong and Zhang, 2005; Rasmussen and Givskov, 2006).

The first strategy to interfere with bacterial cell-to-cell communication is the repression of the signal synthesis. As acyl-acyl carrier protein (acyl-ACP) and *S*-adenosylmethionine (SAM) are AHL precursors, their analogues, namely L/D-*S*-adenosylhomocysteine, butyryl-*S*-adenosylmethionine (butyryl-SAM) and holo-ACP were tested for the ability to inhibit signal generation. According to expectations, these experiments confirmed the inhibitory effect of AHL precursor analogues on the quorum sensing-dependent regulation of virulence traits in *P. aeruginosa* in vitro (Parsek *et al.*, 1999).

The second strategy to interfere with bacterial quorum sensing is the inactivation of signaling molecules. Degradation of AHL signaling molecules can be achieved by lactonolysis: cleavage of the lactone rings of AHLs. Lactonolysis of AHLs was shown to be facilitated in nature by an enzyme, AiiA, which is secreted by several *Bacillus* species (Dong *et al.*, 2000; Dong *et al.*, 2001). As *Bacillus* exploits peptide-based signaling molecules, it is hypothesized that AiiA is used to disrupt AHLs of competing bacteria occupying the same ecological niche. Genes encoding similar quorum sensing signal degrading enzymes were also found to be present in a number of other bacterial species, including *Varvivorax paradoxus*, *Agrobacterium tumefaciens*, *Klebsiella pneumoniae* and *Arthrobacter* sp. (Carlier *et al.*, 2003; Leadbetter and Greenberg, 2000; Huang *et al.*, 2003; Park *et al.*, 2003; Uroz *et al.*, 2003).

The third and most promising strategy to interfere with quorum sensing is the blockage of the signal receptor protein. The halogenated furanones produced by marine alga *Delisea pulchra* were shown to bind to and degrade signal receptor proteins (Givskov *et al.*, 1996; Manefield *et l.*, 2002). *In vivo* studies in the mouse infection model confirmed the potential antimicrobial effect of furanones as the mortality of the *P. aeruginosa* infected mice treated with furanones was significantly lower. Experiments with *P. aeruginosa* genome employing DNA microarrays revealed that the expression of a number of quorum sensing genes, including key virulence factors, has been downregulated by furanone compound 30 (Hentzer *et al.*, 2003).

Moreover, several compounds with the competitive agonistic or inhibitory effect on bacterial cell-to-cell communication have been synthesized in the laboratory (Schaefer *et al.*, 1996; Smith *et al.*, 2003; Rasmussen and Givskov, 2006).

CONCLUSION

The quorum sensing system of the opportunistic human pathogen *P. aeruginosa*, involving hundreds of target genes, represents one of the best studied cell-to-cell communication systems in bacteria. However, work over the past few years has provided evidence that not only *P. aeruginosa* but many other human pathogens use cell-to-cell signaling in order to coordinate expression of their virulence traits. As described in this review, bacterial quorum sensing signaling is exploited not only for intraspecies, but also for interspecies cell-to-cell communication and for interaction with eukaryotic hosts. Given the essential role of cell-to-cell communication for the pathogenic potential of bacteria, novel antimicrobial drugs interfering with bacterial signaling could have an enormous practical impact on the ability of bacteria to establish an infection.

REFERENCES

Ahmer, B.M. (2004). Cell-to-cell signaling in *Escherichia coli* and *Salmonella enterica*. *Mol Microbiology 52:* 933-945.

Camara, M., Williams, P., and Hardman, A. (2002). Controlling infection by tuning in and turning down the volume of bacterial small-talk. *Lancet Infect Dis 2:* 667-676.

Carlier, A., Uroz, S.; Smadja, B., Fray, R., Latour, X., Dessaux, Y., and Faure, D. (2003). The Ti plasmid of *Agrobacterium tumefaciens* harbors an *attM*-paralogous gene, *aiiB*, also encoding *N*-acyl homoserine lactonase activity. *Appl Environ Microbiol 69:* 4989-4993.

Chun, C.K., Ozer, E.A., Welsh, M.J., Zabner, J., and Greenberg, E.P. (2004). Inactivation of a *Pseudomonas aeruginosa* quorum-sensing signal by human airway epithelia. *Proc Natl Acad Sci USA 101:* 3587-3590.

Cosson, P., Zulianello, L., Join-Lambert, O., Faurisson, F., Gebbie, L., Benghezal, M., Van Delden, C., Curty, L.K., and Kohler, T. (2002). *Pseudomonas aeruginosa* virulence analysed in a *Dictyostelium discoideum* host system. *J Bacteriol 184:* 3027-3033.

Davies, D.G., Parsek, M.R., Pearson, J.P., Iglewski, B.H., Costerton, J.W., and Greenberg, E.P. (1998). The involvement of cell-to-cell signals in the development of bacterial biofilm. *Science 280:* 295-298.

DiMango, E., Zar, H.J., Bryan, R., and Prince, A. (1995). Diverse *Pseudomonas aeruginosa* gene products stimulate respiratory epithelial cells to produce interleukin-8. *J Clin Invest 96:* 2204-2210.

Dong, Y.H., Xu, J.L., Li, X.Z., and Zhang, L.H. (2000). AiiA, an enzyme that inactivates the acylhomoserine lactone quorum-sensing signal and attenuates the virulence of *Erwinia carotovora*. *Proc Natl Acad Sci USA 97:* 3526-3531.

Dong, Y.H., Wang, L.H., Xu, J.L., Zhang, H.B., Zhang, X.F., and Zhang, L.H. (2001). Quenching quorum-sensing-dependent bacterial infection by an *N*-acyl homoserine lactonase. *Nature 411:* 813-817.

Dong, Y., and Zhang, L.H. (2005). Quorum sensing and quorum quenching enzymes. *J Microbiol 43:* 101-109.

Duan, K., Dammel, C., Stein, J., Rabin, H., and Surette, M.G. (2003). Modulation of *Pseudomonas aeruginosa* gene expression by host microflora through interspecies communication. *Mol Microbiol 50:* 1477-1491.

Eberl, L. (2006). Quorum sensing in the genus *Burkholderia*. *Int J Med Microbiol 296:* 103-110.

England, P.G., Palmer, R.J.Jr., and Kolenbrander, P.E. (2004). Interspecies communication in *Streptococcus gordonii- Veillonella atypical* biofilms: Signaling in flow conditions requires juxtaposition. *Proc Natl Acad Sci USA 101:* 16917-16922.

Erickson, D.L., Endersby, R., Kirkham, A., Stuber, K., Vollman, D.D., Rabin, H.R., Mitchell, I., and Storey, D.G. (2002). *Pseudomonas aeruginosa* quorum- sensing systems may control virulence factor expression in the lungs of patients with cystic fibrosis. *Infect Immun 70:* 1783-1790.

Frias, J., Olle, E., and Alsina, M. (2001). Periodontal pathogens produce quorum sensing signal molecules. *Infect Immun 69:* 3431-3434.

Givskov, M., de Nys, R., Manefield, M., Gram, L., Maximilien, R., Eberl, L., Molin, S., Steinberg, P.D., and Kjelleberg, S. (1996). Eukaryotic interference with homoserine lactone-mediated prokaryotic signaling. *J Bacteriol 178:* 6618-6622.

Greenberg, E.P. (2003). Tiny teamwork. *Nature 424:* 134.

Hastings, J.W. (2004). Bacterial quorum-sensing signals are inactivated by mammalian cells. *Proc Natl Acad Sci USA 101:* 3993-3994.

Henke, J.M., and Bassler, B.L (2004). Bacterial social engangements. *Trends Cell Biol 14:* 648-656.

Hentzer, M., Wu, H., Andersen, J.B., Riedel, K., Rasmussen, T.B., Bagge, N., Kumar, N., Schembri, M.A., Song, Z., Kristoffersen, P., Manefield, M., Costerton, J.W., Molin, S., Hoiby, N., and Givskov, M. (2003). Attenuation of *Pseudomonas aeruginosa* virulence by quorum sensing inhibitors. *EMBO J 22:* 3803-3815.

Hentzer, M., Givskov, M., and Eberl, L. (2004). Quorum sensing in biofilms: Gossip in slime city. In *Microbial Biofilms*. O'Toole, G., and Ghannoum, M.A. (eds.) Washington, DC, USA: American Society for Microbiology Press, pp. 118-140.

Huang, J.J., Han, J.I., Zhang, L.H., and Leadbetter, J.R. (2003). Utilization of acyl-homoserine lactone quorum signals for growth by a soil pseudomonad and *Pseudomonas aeruginosa* PAO1. *Appl Environ Microbiol 69:* 5941-5949.

Juhas, M., Wiehlmann, L., Huber, B., Jordan, D., Lauber, J., Salunkhe, P., Limpert, A.S., von Gotz, F., Steinmetz, I., Eberl, L., and Tummler, B. (2004). Global regulation of quorum sensing and virulence by VqsR in *Pseudomonas aeruginosa*. *Microbiology 150:* 831-841.

Juhas, M., Eberl, L., and Tummler, B. (2005). Quorum sensing: the power of cooperation in the world of *Pseudomonas*. *Environ Microbiol 7:* 459-471.

Keller, L., and Surette, M.G. (2006). Communication in bacteria: am ecological and evolutionary perspective. *Nature Rev Microbiol 4:* 249-258.

de Kievit, T.R., and Iglewski, B. (2000). Bacterial quorum sensing in pathogenic relationships. *Infect Immun 68:* 4839-4849.

Latifi, A., Winson, M.K., Foglino, M., Bycroft, B.W., Stewart, G.S.A.B., Lazdunski, L., and Williams, P. (1995). Multiple homologues of LuxR and LuxI control expression of virulence and secondary metabolites through quorum sensing in *Pseudomonas aeruginosa* PAO1. *Mol Microbiol 17:* 333-343.

Lawrence, R.N., Dunn, W.R., Bycroft, B., Camara, M., Chhabra, S.R., Williams, P., and Wilson, V.G. (1999). The *Pseudomonas aeruginosa* quorum- sensing signal molecule *N*-(3-oxododecanoyl)-L-homoserine lactone, inhibits porcine arterial smooth muscle contraction. *Br J Pharmacol 128:* 845-848.

Lazdunski, A.M., Ventre, I., and Sturgis, J.N. (2004). Regulatory circuits and communication in gram-negative bacteria. *Nature Rev Microbiol 2:* 581-592.

Leadbetter, J.R., and Greenberg, E.P. (2000). Metabolism of acyl-homoserine lactone quorum-sensing signals by *Varvivorax paradoxus*. *J Bacteriol 182:* 6921-6926.

Ledgham, F., Ventre, I., Soscia, C., Foglino, M., Sturgis, J.N., and Lazdunski, A. (2003). Interactions of the quorum sensing regulator QscR: interaction with itself and the other regulators of *Pseudomonas aeruginosa* LasR and RhlR. *Mol Microbiol 48:* 199-210.

Lewenza, S., Conway, B., Greenberg, E.P., and Sokol, P.A. (1999). Quorum sensing in *Burkholderia cepacia*: identification of the LuxRI homologs CepRI. *J Bacteriol 181:* 748-756.

Li, L., Hooi, D., Chhabra, S.R., Pritchard, D., and Shaw, P.E. (2004). Bacterial *N*-acylhomoserine lactone- induced apoptosis in breast carcinoma cells correlated with down- modulation of STAT32. *Oncogene 23:* 4894-4902.

Manefield, M., Rasmussen, T.B., Hentzer, M., Andersen, J.B., Steinberg, P., Kjelleberg, S., and Givskov, M. (2002). Halogenated furanones inhibit quorum sensing through accelerated LuxR turnover. *Microbiology 148:* 1119-1127.

McKenney, D., Brown, K.E., and Allison, D.G. (1995). Influence of *Pseudomonas aeruginosa* exoproducts on virulence factor production in *Burkholderia cepacia*: evidence of interspecies communication. *J Bacteriol 177:* 6989-6992.

McNab, R., Ford, S.K. El-Sabaeny, A., Barbieri, B., Cook, G.S., and Lamont, R.J. (2003). LuxS-based signaling in *Streptococcus gordonii*: autoinducer 2 controls carbohydrate metabolism and biofilm formation with *Porphyromonas gingivalis*. *J Bacteriol 185:* 274-284.

Middleton, B., Rodgers, H.C., Camara, M., Knox, A.J., Williams, P., and Hardman, A. (2002). Direct detection of *N*- acylhomoserine lactones in cystic fibrosis sputum. *FEMS Microbiol Lett 207:* 1-7.

Nealson, K.H., Platt, T., and Hastings, J.W. (1970). Cellular control of the synthesis and activity of the bacterial luminescent system. *J Bacteriol 104:* 313-322.

Park, S.Y., Lee, S.J., Oh, T.K., Oh, J.W., Koo, B.T., Yum, D.Y., and Lee, J.K. (2003). AhlD, an *N*-acylhomoserine lactonase in *Arthrobacter* sp., and predicted homologues in other bacteria. *Microbiology 149:* 1541-1550.

Parsek, M.R., Val, D.L., Hanzelka, B.L., Cronan, J.E.Jr., and Greenberg, E.P. (1999). Acyl homoserine-lactone quorum-sensing signal generation. *Proc Natl Acad Sci USA 96:* 4360-4365.

Parsek, M.R., and Greenberg, E.P. (2005). Sociomicrobiology: the connections between quorum sensing and biofilms. *Trends Microbiol 13:* 27-33.

Pearson, J.P., Gray, K.M., Passador, L., Tucker, K.D., Eberhard, A., Iglewski, B.H., and Greenberg, E.P. (1994). Structure of the autoinducer required for expression of *Pseudomonas aeruginosa* virulence genes. *Proc Natl Acad Sci USA 91:* 197-201.

Pearson, J.P., Passador, L., Iglewski, B.H., and Greenberg, E.P. (1995). A second *N*-acyl homoserine lactone signal produced by *Pseudomonas aeruginosa. Proc Natl Acad Sci USA 92:* 1490-1494.

Pesci, E.C., Pearson, J.P., Seed, P.C., and Iglewski, B.H. (1997). Regulation of *las* and *rhl* quorum sensing in *Pseudomonas aeruginosa. J Bacteriol 179:* 3127-3232.

Qazi, S., Middleton, B., Muharram, S.H., Cockayne, A., Hill, P., O'Shea, P., Chhabra, S.R., Camara, M., and Williams, P. (2006). *N*-Acylhomoserine lactones antagonize virulence gene expression and quorum sensing in *Staphylococcus aureus. Infect Immun 74:* 910-919.

Rahme, L.G., Stevens, E.J., Wolfort, S.F., Shao, J., Tompkins, R.G., and Ausubel, F.M. (1995). Common virulence factors for bacterial pathogenicity in plants and animals. *Science 268:* 1899-1902.

Rasmussen, T.B., and Givskov, M. (2006). Quorum-sensing inhibitors as anti-pathogenic drugs. *Int J Med Microbiol 296:* 149-161.

Riedel, K., Hentzer, M., Geisenberger, O., Huber, B., Steidle, A., Wu, H., Hoiby, N., Givskov, M., Molin, S., and Eberl, L. (2001). *N*-Acylhomoserine-lactone-mediated communication between *Pseudomonas aeruginosa* and *Burkholderia cepacia* in mixed biofilms. *Microbiology 147:* 3249-3262.

Rumbaugh, K.P., Griswold, J.A., and Hamood, A.N. (1999). Contribution of the regulatory gene lasR to pathogenesis of *Pseudomonas aeruginosa* infection of burned mouse. *J Burn Care Rehabil 20:* 42-49.

Schaefer, A.L., Hanzelka, B., Eberhard, A., and Greenberg, E.P. (1996). Quorum sensing in *Vibrio ficheri*: probing autoinducer-LuxR interactions with autoinducer analogs. *J Bacteriol 178:* 2897-2901.

Schuster, M., and Greenberg, E.P. (2006). A network of networks: quorum sensing gene regulation in *Pseudomonas aeruginosa. Int J Med Microbiol 296:* 73-81.

Shiner, E.K., Rumbaugh, K.P., and Williams, S.C. (2005). Interkingdom signaling: Deciphering the language of acyl homoserine lactones. *FEMS Microbiol Rev 29:* 935-947.

Singh, P.K., Schaefer, A.L., Parsek, M.R., Moninger, T.O., Welsh, M.J., and Greenberg, E.P. (2000). Quorum- sensing signals indicate that cystic fibrosis lungs are infected with bacterial biofilms. *Nature 407:* 762-764.

Smith, R.S., Fedyk, E.R., Springer, T.A., Mukaida, N., Iglewski, B.H., and Phipps, R.P. (2001). IL-8 production in human lung fibroblasts and epithelial cells activated by the *Pseudomonas* autoinducer *N*- 3- oxododecanoyl homoserine lactone is transcriptionally regulated by NF- kappa B and activator protein- 2. *J Immunol 167:* 366-374.

Smith, R.S., Kelly, R., Iglewski, B.H., and Phipps, R.P. (2002). The *Pseudomonas* autoinducer *N*-(3-oxododecanoyl) homoserine lactone induces cyclooxygenase-2 and prostaglandin E2 production in human lung fibroblasts: implications for inflammation. *J Immunol 169:* 2636-2642.

Smith, R.S., and Iglewski, B.H. (2003). *Pseudomonas aeruginosa* quorum-sensing systems and virulence. *Curr Opin Microbiol 6:* 56-60.

Smith, K.M., Bu, Y., and Suga, H. (2003). Induction and inhibition of *Pseudomonas aeruginosa* quorum sensing by synthetic autoinducer analogs. *Chem Biol 10:* 81-89.

Stickler, D.J., Morris, N.S., McLean, N.J.C., and Fuqua, C. (1998). Biofilms on indwelling urethral catheters produce quorum- sensing signal molecules in situ and in vitro. *Appl Environ Microbiol 64:* 3486-3490.

Tan, M.W., Rahme, L.G., Sternberg, J.A., Tompkins, R.G., and Ausubel, F.M. (1999). *Pseudomonas aeruginosa* killing of *Caenorhabditis elegans* used to identify *P. aeruginosa* virulence factors. *Proc Natl Acad Sci USA 96:* 2408-2413.

Tang, H.B., DiMango, E., Bryan, R., Gambello, M., Iglewski, B.H., Goldberg, J.B., and Prince, A. (1996). Contribution of specific *Pseudomonas aeruginosa* virulence factors to pathogenesis of pneumonia in a neonatal mouse model of infection. *Infect Immun 64:* 37-43.

Tateda, K., Ishii, Y., Horikawa, M., Matsumoto, T., Miyairi, S., Pechere, J.C., Standiford, T.J., Ishiguro, M., and Yamaguchi, K. (2003). The *Pseudomonas aeruginosa* autoinducer *N*-3-oxododecanoyl homoserine lactone accelerates apoptosis in macrophages and neutrophils. *Infect Immun 71:* 5785-5793.

Telford, G., Wheeler, D., Williams, P., Tompkins, P.T., Appleby, P., Sewell, H., Stewart, G.S., Bycroft, B.W., and Pritchard, D.I. (1998). The *Pseudomonas aeruginosa* quorum-sensing signal molecule *N*-(3-oxododecanoyl)-L-homoserine lactone has immunomodulatory activity. *Infect Immun 66:* 36-42.

Tomasz, A. (1965). A control of the competent state in *Pneumococcus* by a hormone-like cell product: an example of a new type of regulatory mechanism in bacteria. *Nature 208:* 155-159.

Uroz, S., D'Angelo-Piçard, C., Carlier, A., Elasri, M., Sicot, C., Petit, A., Oger, P., Faure, D., and Dessaux, Y. (2003). Novel bacteria degrading *N*-acylhomoserine lactones and their use as quenchers of quorum-sensing-regulated functions of plant-pathogenic bacteria. *Microbiology 149:* 1981-1989.

Vendeville, A., Winzer, K., Heurlier, K., Tang, C.M., and Hardie, K.R. (2005). Making „sense" of metabolism: autoinducer-2, LuxS and pathogenic bacteria. *Nature Rev Microbiol 3:* 383-396.

Venturi, V. (2006). Regulation of quorum sensing in *Pseudomonas*. *FEMS Microbiol Rev 30:* 274-291.

Wagner, V.E., Frelinger, J.G., Barth, R.K., and Iglewski, B.H. (2006). Quorum sensing: dynamic response of *Pseudomonas aeruginosa* to external signals. *Trends Microbiol 14:* 55-58.

Waters, C.M., and Bassler, B.L. (2005). Quorum sensing: cell-to-cell communication in bacteria. *Annu Rev Cell Dev Biol 21:* 319-346.

Williams, S.C., Patterson, E.K., Carty, N.L. Griswold, J.A., Hamood, A.N., and Rumbaugh, K.P. (2004). *Pseudomonas aeruginosa* autoinducer enters and functions in mammalian cells. *J Bacteriol 186:* 2281-2287.

Wu, H., Song, Z., Hentzer, M., Andersen, J.B., Heydorn A., Mathee, K., Moser, C., Eberl, L., Molin, S., Hoiby, N., and Givskov, M. (2000). Detection of *N*- acylhomoserine lactones in lung tissues of mice infected with *Pseudomonas aeruginosa*. *Microbiology 146:* 2481-2493.

Wu, L., Estrada, O., Zaborina, O., Bains, M., Shen, L., Kohler, J.E., Patel, N., Musch, M.W., Chang, E.B., Fu, Y.X., Jacobs, M.A., Nishimura, M.I., Hancock, R.E., Turner, J.R., and Alverdy, J.C. (2005). Recognition of host immune activation by *Pseudomonas aeruginosa*. *Science 309:* 774-777.

In: Focus on Signal Transduction Research
Editor: Gavin McAlpine, pp. 159-171
ISBN 1-60021-376-6

Chapter 7

ROLES OF P38 MAPK SIGNALING PATHWAY IN VIRAL INFECTION

Tetsuya Mizutani*
Virology 1, National Institute of Infectious Diseases, Musashimurayama, Tokyo, Japan

ABSTRACT

Mitogen-activated protein kinases (MAPKs) function as serine/threonine kinases that regulate intracellular signaling cascades. The MAPK, p38, is stimulated by environmental stress, proinflammatory cytokines, oxidative stimuli, and UV irradiation. p38 MAPK is known to phosphorylate its downstream targets, mitogen and stress-activated protein kinase 1 (MSK1), MAP kinase-interacting kinase 1 (MNK1), and MAPK-activated protein kinase 2, 3 (MAPKAPK 2, 3). MAPKAPK 2 activates heat shock protein 27 (HSP27), cAMP response element-binding protein (CREB), and transcription factor-1 (ATF-1). Infection of cells by viruses has been shown to result in phosphorylation of p38 MAPK. In virus-infected cells, viral proteins have been shown to be key mediators of up- and down-regulation of p38 MAPK by experiments using specific inhibitors and over-expression of kinase-defective mutants. p38 MAPK is known to promote both cell death and survival, depending on the cell types and stimuli. The alterations in p38 MAPK activities by viral proteins influence apoptotic events and cytokine secretion in vivo and *in vitro*. Therefore, p38 MAPK may be a therapeutic target for suppressing viral replication and reducing host responses in virus-infected cells. This review will highlight recent finding on the role of p38 MAPK signaling pathway including its up- and down-stream components by viral infection.

---

* Phone +81-425-61-0771; Fax +81-425-65-3315; E-mail tmizutan@nih.go.jp

Introduction

Mitogen-activated protein kinases (MAPKs), which are a superfamily of protein kinases, have a number of important roles in living cells. These molecules are highly conserved in a wide range of species from yeast to mammals. Generally, MAPK is activated by MAPK kinase (MAPKK), and MAPKK is activated by MAPKK kinase (MAPKKK). Three major MAPKs, NH_2-terminal kinase (JNK), extracellular signal-regulated kinase (ERK), and p38 MAPK, have been characterized in mammals.

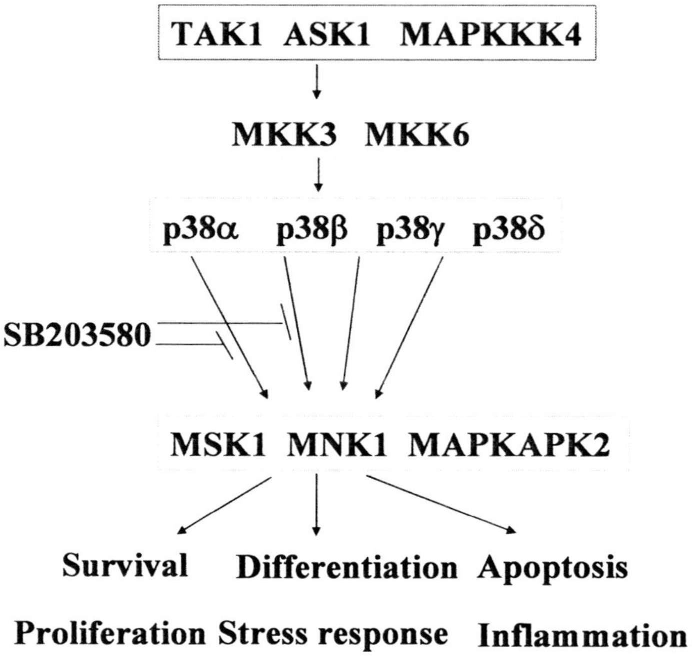

Figure 1. Upstream and downstream components of p38 MAPK pathway in mammalian cells.

The MAPK, p38, is stimulated by environmental stress, proinflammatory cytokines, oxidative stimuli, and UV irradiation. At least four isoforms of p38 have been identified to date: p38α, p38β, p38γ, and p38δ [3, 24, 43] (Figure 1). Of these, p38α and p38β have more than 70% similarity at the amino acid level, while p38γ and p38δ have 60% similarity to p38α. Both p38α and p38β are widely expressed in tissues, whereas p38γ and p38δ show tissue-specific expression. Interestingly, although activities of p38α and p38β/β2 are inhibited by the pyridinylimidazole compound, SB203580 [4-(4-fluorophenyl)-2-(4-methylsulfinylphenyl)-5-(4-pyridyl) imidazole], p38γ and p38δ are not [24, 43]. MKK3 and 6 are MAPKKs for p38, and TGF-β-activating kinase (TAK1), apoptosis signal-regulatory kinase (ASK1), and MAPKKK4 are MAPKKKs of MKK3 and 6. These MAPKKKs are also capable of activating the JNK signaling pathway. On the other hand, mitogen and stress-activated protein kinase 1 (MSK1), MAP kinase-interacting kinase 1 (MNK1), and MAPK-activated protein kinase 2, 3 (MAPKAPK 2, 3) are kinase substrates of p38 MAPK. MAPKAPK2 activates heat shock protein 27 (HSP27), cAMP response element-binding protein (CREB), and transcription factor-1 (ATF-1). MNK1 activates eukaryotic initiation factor of translation 4E (eIF4E). As p38 MAPK is activated by pro-inflammatory cytokines, as described above, chemical inhibitors of p38 MAPK may be of therapeutic value. For example, AMG-548 (Amgen) is a potent selective p38a inhibitor, which has potent inhibitory effect on LPS-stimulated tumor necrosis factor alpha (TNF-α) and interleukin-1 beta (IL-1β), TNF-α-induced IL-8, and IL-1β-induced IL-6 in whole blood [20, 27].

Several viruses have been shown to induce activation and phosphorylation of the p38 MAPK signaling pathway in infected cells. However, the precise role of p38 MAPK in viral replication remains unclear. This review highlights progress regarding p38 MAPK signal transduction in virus-infected cells.

Regulation of p38 MAPK Activation by Viral Infection

As p38 MAPK is expressed in response to stressors, viral infection generally induces activation of p38 MAPK (Figure 2). However, viral proteins sometimes negatively regulate p38 MAPK. Varicella-Zoster virus (VZV), which is a neurotropic alphaherpesvirus, transcribes immediate early (IE) genes following entry into the host cell. Among these IE genes, open reading frame (ORF) 61 encodes a phosphoprotein of 62–66 kDa [22, 54, 41], and this protein enhances the expression of other VZV immediate-early and putative early promoters. Furthermore, this protein increases infectivity of VZV DNA [39]. When ORF61 was deleted in the genome of VZV, viral replication was impaired [7]. Levels of transcription of AP1 components and ATF-2, which are downstream of JNK and p38 MAPK, are increased in VZV-infected cells [46]. On the other hand, ORF61 is known to repress phosphorylation of p38 MAPK [45]. Following infection and expression of other viral proteins, the repression of phosphorylation of p38 MAPK by ORF61 is removed and the level of phosphorylated p38 MAPK is increased, followed by induction of transcription factors that respond to proinflammatory mediators. Thus, ORF61 may be a negative regulator of cellular proinflammatory responses *via* repression of p38 MAPK activation.

Recent studies demonstrated the relationship between p38 MAPK and the interferon (IFN) system [20, 43]. p38 MAPK is activated by treatment with IFN-α, but shows no response to IFN-γ. Thus, type I IFN-mediated signaling pathways include activation of p38 MAPK. Furthermore, a recent study indicated that PKR interacts with and activates p38 MAPK [50]. Interestingly, MKK6, which is a MAPKK of p38 MAPK, has affinity to PKR in the presence of double-stranded RNA (dsRNA), poly(rI;rC), but not MKK3, which is also MAPKK of p38 MAPK. PKR is capable of activating MKK6. Hepatitis C virus (HCV) core protein expression in cells increases activity of cyclin B1-CDK1 complex through JNK and p38 MAPK signaling pathways [52]. Furthermore, the core protein promotes nuclear import of cyclin B1. PKR interacts strongly with p38 MAPK in HCV core-expressing cells. These findings suggest that core protein induces deregulation of the mitotic checkpoint at the boundary of the G2 phase of the cell cycle and mitosis by interaction between p38 MAPK and PKR.

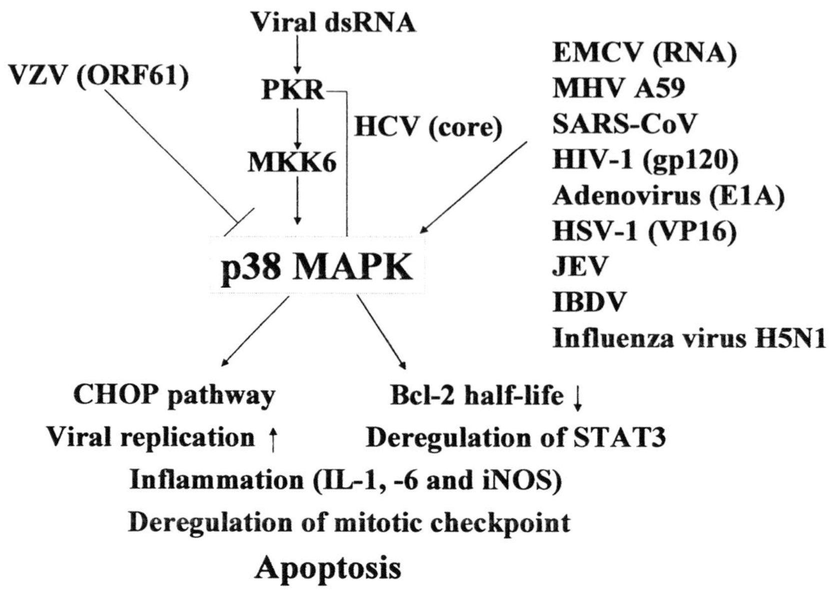

Figure 2. Response of p38 MAPK following viral infection.

Viral RNA also regulates activation of p38 MAPK. Encephalomyocarditis virus (EMCV), which is a member of the *Picornaviridae* and a positive-stranded RNA virus, causes type 1 diabetes, encephalitis, and myocarditis in mice [8]. p38 MAPK and JNK were shown to be activated in EMCV-infected cells [17]. p38 MAPK inhibitors suppress viral replication and translation of viral RNA transcripts. p38 inhibitors can inhibit viral transcription, but this may be a secondary effect of inhibition of viral protein synthesis. p38 MAPK plays a critical role in the replication of EMCV, probably in the translation of viral RNA. Iordanov *et al.* demonstrated that activation of p38 MAPK in mouse fibroblasts occurred in response to infection of EMCV, which replicates through a dsRNA intermediate [18]. IL-6 production from virus-infected fibroblasts is inhibited by the p38-selective inhibitor, SB203580. They suggested that activation of p38 MAPK by viral dsRNA is independent of 2-5A [44, 51] and the PKR [53] system, which activate JNK.

As describe above, activation of p38 MAPK sometimes influences viral replication. Mouse hepatitis virus (MHV)-3 is known to induce fulminant hepatitis, which depends on macrophage activation and expression of the specific prothrombinase fgl-2 [10], in mice. The prothrombinase is central to the pathogenesis of hepatitis caused by MHV-3. MHV-3 induces activation of ERK and p38 MAPK in macrophages, and fgl-2 mRNA level is increased by activation of p38 MAPK [31]. p38 inhibitors may be useful for modulation of fulminant viral hepatitis. Further analyses were reported using MHV A59 strain and the macrophage-derived cell line J774.1 [1]. p38 MAPK activation was shown to increase production of IL-6 in an experiment using SB203580. Interestingly, p38 MAPK inhibitors also suppress viral mRNA and protein production. Phosphorylation of translation initiation factor 4E (eIF4E), which enhances translation rates of cap-containing mRNAs [14], is increased in virus-infected cells. The phosphorylation of eIF4E is p38 MAPK-dependent in the cells. Therefore, MHV may utilize eIF4E phosphorylation by activation of p38 MAPK to promote virus-specific protein synthesis. On the other hand, human immunodeficiency virus (HIV)-1 gp120 induces neuronal dysfunction and death by activation of p38 MAPK [42]. The p38 MAPK inhibitor, RWJ67657 (4-[4-(4-fluorophenyl)-1-(3-phenylpropyl)-5-(4-pyridyl)-1H-imidazol-2-yl]-3-butyn-1-ol), inhibits HIV-1 replication in both T-cell and monocyte cell lines, suppresses both reverse transcriptase and protease-resistant escape mutant viruses, and inhibits HIV-induced T-cell apoptosis [40]. RWJ67657 is a p38/JNK2 inhibitor, which potently inhibits lipopolysaccharide (LPS)-induced tumor necrosis factor (TNF)-α production from human PBMC. Therefore, p38 inhibitors may be good candidates as anti-HIV drugs that target host cellular proteins.

PI3K/Akt is well-characterized anti-apoptotic signaling pathway. Sensitization to apoptosis of adenoviral early region 1A (E1A) protein is mediated by inactivation of Akt and activation of p38 MAPK [28]. Inactivation of Akt induces activation of p38 MAPK by activation of ASK1 and MEKK3, which act upstream of p38. A correlation was found between enhanced Akt phosphorylation and reduced p38 phosphorylation in human tumor tissues *in vivo* and in human cancer cell lines, suggesting that p38 activity is repressed in human cancer, which is associated with enhanced Akt activation. The conserved domain (CR2) of E1A is known to be required for down-regulation of Akt phosphorylation. Recently, we also found that cell death in severe acute respiratory syndrome coronavirus (SARS-CoV)-infected cells is induced by inactivation of Akt and activation of p38 MAPK, as discussed below. Furthermore, the anti-apoptotic protein, Bcl-2, is regulated by p38 MAPK. In the case of herpes simplex virus type 1 (HSV-1), the virion transactivator protein VP16 stimulates p38/JNK, but no immediate-early, early, or late viral genes are involved [61]. In addition, using mutant viruses *ts*k (lacking ICP4 function) and 27lacZ (lacking ICP27 function)[62], ICP4 and 27 were found to have anti-apoptotic effects. In these virus-infected cells, activation of p38 MAPK decreases the half-life of Bcl-2 and inhibition with p38 MAPK increases the levels of Bcl-2 protein. However, p38 MAPK is activated independently of caspases during apoptosis. Thus, caspases and p38 MAPK act independently, and both act upstream of Bcl-2 protein.

p38 MAPK is known to enhance transcription of CHOP, which is a protein homologous to C/EBP as growth arrest- and DNA damage-inducible gene 153 (GADD153) [59]. CHOP induces apoptosis and cell cycle arrest in response to endoplasmic reticulum (ER) stress [59]. Japanese encephalitis virus (JEV) induces apoptosis under conditions of ER stress, and treatment with a p38-specific inhibitor, SB203580, partially blocks this apoptosis [55]. Thus, activated p38 MAPK induces apoptosis *via* activation of the CHOP pathway.

The activation of p38 MAPK by viral infection enhances inflammation. For example, the proinflammatory cytokine interleukin (IL)-1 promotes HIV-1 replication *in vitro*. p38 MAPK inhibitors suppress IL-1β-induced virus production and -induced IL-8 production, and also decrease constitutive HIV production in freshly infected peripheral blood mononuclear cells [49]. Infectious bursal disease virus (IBDV), which belongs to the family *Birnaviridae*, causes acute, highly contagious, and immunosuppressive disease in chickens. The p38 MAPK inhibitor, SB203580, inhibits expression of iNOS, COX-2, and IL-8 in virus-infected cells [21]. Therefore, p38 MAPK is thought to be a good candidate for IBDV-induced macrophage activation.

Activation of p38 MAPK may be related to the pathogenesis of influenza virus. Apoptosis signal-regulating kinase1 (ASK1) is known to be a MAPKKK of JNK and p38 MAPK. Influenza virus strain A/Udon/307/72 (H3N2) activates ASK1, JNK, and p38 MAPK in human bronchial epithelial cells [30]. Virus-induced apoptosis is depressed in embryonic fibroblasts derived from ASK1 knockout mice. Human influenza subtype H5N1, isolated in 1997 and 2003, induces high levels of tumor necrosis factor alpha (TNF-α) as compared with other subtypes [5, 16]. On the other hand, in influenza virus-infected bronchial epithelial cells, both JNK and p38 MAPK regulate RANTES expression [23]. Viral proteins, such as hemagglutinin and nucleocapsid, activate inhibitor of NF-κB (IκB) kinase (IKK), resulting in activation of the NF-κB signaling pathway [11]. Davy *et al.* compared the activation status of NF-κB and p38 MAPK in H5N1 and H1N1 subtypes of influenza virus *in vitro* [26]. Although NF-κB signaling pathways are activated in both viruses, only H5N1 virus induces high levels of p38 MAPK. TNF-α expression is inhibited by treatment with SB203580. Infectivity by H5N1 virus is not affected by p38 MAPK inhibitor. Thus, p38 MAPK induced by H5N1 viral infection plays important regulatory roles in deregulation of cytokines. This phenomenon may be reflected in the pathogenesis of disease caused by H5N1.

SARS-CoV and p38 MAPK

SARS is a newly found infectious disease caused by a novel coronavirus, SARS-CoV [29, 48]. SARS spread to more than 30 countries and caused severe atypical pneumonia in 2002/2003. As the virulence and mortality rate of SARS patients are very high, it is important to understand the pathological mechanisms of SARS to prevent further outbreaks. Recently, the mechanisms of SARS pathogenesis have been clarified based on molecular pathological data, including signaling pathways in SARS-CoV-infected cells. As described above, recent studies have clarified that activation of MAPKs in virus-infected cells plays important roles in up-regulation of cytokine expression and apoptosis. The monkey kidney cell line, Vero E6, is widely used in SARS-CoV research because these cells are highly sensitive to infection with the virus. SARS-CoV infection inhibits cell proliferation [32]. SARS-CoV-infected Vero E6 cells undergo apoptosis, detected by DNA fragmentation and caspase activation [35, 60]. Phosphorylation of p38 MAPK is significantly up-regulated at 18 h post-infection (h.p.i.) in SARS-CoV-infected Vero E6 cells [35]. Other MAPKs, ERK1/2 and JNK, are also up-regulated in SARS-CoV-infected Vero E6 cells [34]. The p38 MAPK-specific inhibitor, SB203580, can partially prevent the cytopathic effects of virus infection in Vero E6 cells [35]. Thus, the p38 MAPK signaling pathway may be involved in cell death in SARS-CoV-infected Vero E6 cells. On the other hand, activation of anti-apoptotic Akt is too weak to prevent apoptotic cell death [36]. Downstream targets of p38 MAPK are phosphorylated in SARS-CoV-infected cells by SB203580 [35]. MAPK-activated protein kinase (MAPKAPK)-2 is phosphorylated in virus-infected cells. Hsp-27, cAMP response element-binding protein (CREB), and activation of transcription factor (ATF)-1 are also phosphorylated in virus-infected cells. Nucleocapsid (N) protein can induce phosphorylation of Hsp-27 and CREB in transfected cells, as described below [56]. eIF4E is known to enhance translation rates of cap-containing mRNAs [14]. As described above, phosphorylation of eIF4E promotes virus-specific protein synthesis in the case of MHV [1]. Although the level of eIF4E phosphorylation is increased by SARS-CoV infection, similar kinetics of viral protein accumulation are observed in infected Vero E6 cells in the presence and absence of SB

203580 [35]. In addition to such *in vitro* studies, p38 MAPK activation in CD14 monocytes was observed in SARS patients [25].

The p90 ribosomal S6 kinase (RSK) family, which are serine/threonine kinases, consist of four isoforms (RSK1, 2, 3, and 4) and two structurally related RSK-like protein kinases (RLPK/MSK1) and RSK-B (MSK2) in humans [12]. As the C-terminal of p90RSK has an ERK docking site [13], p90RSK1 is phosphorylated at Thr573 in the activation loop of the C-terminal kinase domain by ERK. Autophosphorylation at Ser380 in the linker region is induced by activation of the C-terminal kinase domain [57], and then PDK1 phosphorylates at Ser221 in the activation loop of the N-terminal kinase domain [19]. Thus, p90RSK is believed to be a substrate of ERK and PDK-1. As p90RSK phosphorylates Bad [58] and C/EBPβ [4] to protect cells against apoptosis, p90RSK plays an important role in apoptosis. Vero E6 cells mainly express p90RSK1 and show weak expression of p90RSK2 [38]. Ser221 in the N-terminal kinase domain is phosphorylated constitutively in mock-infected confluent and subconfluent cells. Ser220 is phosphorylated constitutively despite infection by SARS-CoV. Thr573 and Ser380 are not phosphorylated in confluent cells. Ser380 is phosphorylated in confluent SARS-CoV-infected cells. Interestingly, this Ser380 phosphorylation in SARS-CoV-infected cells is inhibited by the p38MAPK inhibitor, SB203580. Thus, phosphorylation of p90RSK is regulated by p38 MAPK in SARS-CoV-infected Vero E6 cells. Furthermore, phosphorylation of Thr573 is not up-regulated in SARS-CoV-infected cells, indicating that phosphorylation of Thr573 is not necessary to induce phosphorylation of Ser380. Several reports have suggested that p90RSK activation results in phosphorylation of CREB [2, 12]. Phosphorylation of ERKs is partially down-regulated by treatment with SB203580 in SARS-CoV-infected cells. There may be cross-talk between signaling pathways of p38 MAPK and ERKs in Vero E6 cells. The signaling cascade of p38 MAPK in SARS-CoV-infected Vero E6 cells is p38 MAPK > (ERK >) p90RSK > CREB, which suggests an anti-apoptotic role of the p38 MAPK signaling pathway.

As described above, phosphorylation of Hsp-27, p90RSK, CREB, and ATF-1 may induce an anti-apoptotic environment in SARS-CoV-infected cells, suggesting that activated p38 MAPK in SARS-CoV-infected cells acts as an anti-cell death signaling pathway. However, SB203580 partially prevents the cytopathic effects of virus infection in Vero E6 cells [35]. Therefore, we searched for other downstream targets of p38 MAPK capable of inducing apoptosis. Signal transducer and activator of transcription (STAT) 3 is constitutively phosphorylated at Tyr705 and is slightly phosphorylated at Ser727 in Vero E6 cells [34]. STAT3 is a major transcription factor activated in response to cytokines, such as interleukin-6 (IL-6) and IL-10, and inhibition of the STAT3 signaling pathway by dominant negative and antisense oligonucleotides against STAT3 decreases cell viability and induces apoptosis [47, 15]. Thus, STAT3 has anti-apoptotic effects. SARS-CoV replication induces dephosphorylation of STAT3 at Tyr705. Although inhibitors of MEK and JNK had no effect on the phosphorylation status of STAT3 in virus-infected cells, two inhibitors of p38 MAPK (SB203580 and SB202190) partially inhibited dephosphorylation of STAT3 at Tyr705. Thus, the p38 MAPK signaling pathway acts upstream of Tyr705 dephosphorylation of STAT3 in Vero E6 cells. In contrast, the level of STAT3 phosphorylation at Ser727 is increased in virus-infected cells. Although the effect of Ser727 phosphorylation on the function of STAT3 remains unclear, it was reported that phosphorylation of Ser727 of STAT3 negatively modulates its tyrosine phosphorylation [6]. Tyr705 dephosphorylation and Ser727 phosphorylation show almost the same timing in SARS-CoV-infected cells. STAT3 phosphorylated at Tyr705 is localized mainly in the nucleus in mock-infected cells, whereas STAT3 disappears from the nucleus after viral infection. As STAT3 acts as an activator of transcription in the nucleus, STAT3 does not act as a transcription factor in SARS-CoV-infected Vero E6 cells. Therefore, STAT3 decreases anti-apoptotic activity in the cells as a downstream target of p38 MAPK.

Although the majority of SARS-CoV-infected cells die by apoptosis within 2 days post-infection, we found that a small population of virus-infected cells survived and these cells grew with production of virus [33, 37]. JNK and PI3K/Akt signaling pathways are important for establishment of persistent infection. p38 MAPK, JNK, and Akt are strongly phosphorylated in surviving cells at 50 h.p.i. We also found that the protein levels of Bcl-2 and Bcl-xL are increased in these cells. Dent et al. reported that p38 MAPK can promote both cell death and survival [9]. Generally, both pro-apoptotic and pro-survival signaling pathways are activated during viral replication. Many pro- and anti-apoptotic proteins are involved in these pathways in SARS-CoV-infected cells. The balance between activation of pro-apoptotic and anti-apoptotic signaling pathways may determine cell death or survival. In acute SARS-CoV-infected Vero E6 cells, pro-apoptotic (cell death) signaling pathways, such as p38 MAPK, may be stronger than anti-apoptotic (cell death) signaling pathways. However, anti-apoptotic proteins, such as Bcl-2 and Bcl-xL, may be stronger than p38 MAPK in persistently SARS-CoV-infected cells. Interestingly, we found that all persistently SARS-CoV-infected cells die by superinfection with *Mycoplasma fermentans* (paper submitted). We showed that *M. fermentans* induces inhibition of proliferation in Vero E6 cells, but the cells remained alive. Only the hypophosphorylated form of retinoblastoma (Rb), which is largely found in early G1 phase, is detected in *M. fermentans*-infected cells at 2 d.p.i. Apoptosis is induced easily in *M. fermentans*-infected cells, indicating that they are primed for apoptosis. Thus, *M. fermentans* enhances apoptosis in persistently SARS-CoV-infected cells, possibly due to p38 MAPK activation by SARS-CoV.

Further Studies of the Role of p38 MAPK in Viral Infection

p38 MAPK is important for regulating apoptosis and cell death in virus-infected cells. Many parts of the signaling pathways between viral proteins and p38 MAPK activation are still unclear. p38 MAPK inhibitors will be valuable tools against the pathogenesis of viral infection. Treatment with p38 MAPK inhibitors and siRNA *in vivo* may result in cell survival and may inhibit viral replication. Although they are good candidates for anti-viral drugs, activation of p38 MAPK has different effects on different cell types. Therefore, it is necessary to carefully check the side effects of inhibitors *in vivo*. Further studies are required to generate specific inhibitors of p38 MAPK.

ACKNOWLEDGMENTS

I am grateful to Drs. S. Fukushi, M. Saijo, M. Ogata, K. Sakai, I. Kurane, and S. Morikawa for comments on this manuscript.

REFERENCES

[1] Banerjee, S., Narayanan, K., Mizutani, T., S. Makino, S. (2002) Murine coronavirus replication-induced p38 mitogen-activated protein kinase activation promotes interleukin-6 production and virus replication in cultured cells. *J. Virol.* 76. 5937-5948.

[2] Böhm, M., Moellmann, G., Cheng, E., Alvarez-Franco, MSW.,. Sassone-Corsi, PRH. (1995) Identification of p90RSK as the probable CREB-Ser133 kinase in human melanocytes, *Cell Growth Differ.* 6. 291–302.

[3] Bonni, A., Brunet, A., West, AE., Datta, SR., Takasu, MA., Greenberg, ME. (1999) Cell survival promoted by the Ras-MAPK signaling pathway by transcription-dependent and -independent mechanisms, *Science* 286. 1358–1362.

[4] Buck, M., Poli, V., Hunter, T., Chojkier, M. (2001) C/EBPβ phosphorylation by RSK creates a functional XEXD caspase inhibitory box critical for cell survival, *Mol. Cell* 8. 807–816.

[5] Cheung, CY., Poon, LL., Lau, AS., Luk, W., Lau, YL., Shortridge, KF., Gordon, S., Guan, Y., Peiris, JS. (2002) Induction of proinflammatory cytokines in human macrophages by influenza A (H5N1) viruses: a mechanism for the unusual severity of human disease? *Lancet* 360. 1831-1837.

[6] Chung, J., E. Uchida, T. C. Grammer and J. Blenis. (1997) STAT3 serine phosphorylation by ERK-dependent and -independent pathways negatively modulates its tyrosine phosphorylation. *Mol. Cell. Biol.* 17. 6508-6516.

[7] Cohen, JI., Nguyen, H. (1998) Varicella–zoster virus ORF61 deletion mutants replicate in cell culture, but a mutant with stop codons in ORF61 reverts to wild-type virus. *Virology* 246. 306–316.

[8] Craighead, JE., McLane, MF. (1968) Diabetes mellitus: induction in mice by encephalomyocarditis virus. *Science* 162. 913-914.

[9] Dent, P., Yacoub, A., Fisher, PB., Hagan, PMP, Grant, S. (2002) MAPK pathways in radiation responses. *Oncogene* 22. 5885-5896.

[10] Fingerote, RJ., Abecassis, M., Phillips, MJ., Rao, YS., Cole, EH., Leibowitz, J., Levy, GA. (1996) Loss of resistance to murine hepatitis virus strain 3 infection after treatment with corticosteroids is associated with induction of macrophage procoagulant activity. *J. Virol.* 70, 275-4282.

[11] Flory, E., Kunz, M., Scheller, C., Jassoy, C., Stauber, R., Rapp, UR., Ludwig, S. (2000) Influenza virus-induced NF-κB-dependent gene expression is mediated by overexpression of viral proteins and involves oxidative radicals and activation of IκB kinase. *J. Biol. Chem.* 275. 8307-8314

[12] Frodin, M. Gammeltoft, S. (1999) Role and regulation of 90 kDa ribosomal S6 kinase (RSK) in signal transduction. Mol. Cell. Endocrinol. 151. 65–77.

[13] Gavin, AC., Nebreda, AR. (1999) A MAP kinase docking site is required for phosphorylation and activation of p90rsk/MAPKAP kinase-1. *Curr. Biol.* 9. 281–284.

[14] Gingras, AC., Raught, B., Sonenberg, N. (1999) eIF4 initiation factors: effectors of mRNA recruitment to ribosomes and regulators of translation. *Annu. Rev. Biochem.* 68. 913-963.

[15] Grandis, JR., Drenning, SD., Zeng, Q., Watkins, SC., Melhem, MF., Endo, S., Johnson, DE., Huang, L., He, Y., Kim, JD. (2000) Constitutive activation of Stat3 signaling abrogates apoptosis in squamous cell carcinogenesis in vivo. *Proc. Natl. Acad. Sci.* USA 97. 4227-4232.

[16] Guan, Y., Poon, LLM., Cheung, CY., Ellis, TM., Lim, W., Lipatov, AS., Chan, KH., Strum-Ramirez, KM., Cheung, CL., Leung, YHC., Yuen, KY., Webster, RG., Peiris, JS. 2004. H5N1 influenza: a protean pandemic threat. *Proc. Natl. Acad. Sci.* USA 101. 8156-8161

[17] Hirasawa, K., Kim, A., Han, HS., Han, J., Jun, HS., Yoon, JW. (2003) Effect of p38 Mitogen-Activated Protein Kinase on the Replication of Encephalomyocarditis Virus. *J. Virol.* 77. 5649-5656.

[18] Iordanov, MS., Paranjape, JM., Zhou, A.,Wong, J.,Williams, BR., Meurs, EF., Silverman, RH., Magun, BE. (2000) Activation of p38 mitogenactivated protein kinase and c-Jun NH(2)-terminal kinase by double-stranded RNA and encephalomyocarditis virus: Involvement of RNase L, protein kinase R, and alternative pathways. *Mol. Cell Biol* 20. 617–627.

[19] Jensen, CJ., Buch, MB, Krag, TO., Hemmings, BA., Gammeltoft, S., Frodin, M. (1999) 90-kDa ribosomal S6 kinase is phosphorylated and activated by 3-phosphoinositide-dependent protein kinase-1. *J. Biol. Chem.* 274. 27168–27176.

[20] Katsoulidis, E., Li, Y., Mears, H., Platanias, LC. (2005) The p38 mitogen-activated protein kinase pathway in interferon signal transduction. *J. Interferon and Cytokine Res.* 25. 749-756.

[21] Khatri, M., Sharma, JM.. (2006) Infectious bursal disease virus infection induces macrophage activation via p38 MAPK and NF-kappaB pathways. *Virus Res.* 118. 70-77.

[22] Kinchington, PR., Vergnes, JP., Turse, SE. (1995) Transcriptional mapping of varicella–zoster virus regulatory proteins. *Neurology* 45. S33–S35.

[23] Kujime, K., Hashimoto, S., Gon, Y., Shimizu, K., Horie, T. (2000) p38 mitogen-activated protein kinase and c-jun-NH2-terminal kinase regulate RANTES production by influenza virus-infected human bronchial epithelial cells. *J. Immunol.* 164. 3222-3228.

[24] Kumar, S., Boehm, J., Lee, JC. (2003) p38 MAP kinases: key signaling molecules as therapeutic targets for inflammatory diseases. *Nat. Rev. Drug Discov.* 2. 717-726.

[25] Lee CH., Chen, RF., Liu, JW., Yeh, WT., Chang, JC., Liu, PM., Eng, HL., Lin, MC., Yang, KD. (2004) Altered p38 mitogen-activated protein kinase expression in different leukocytes with increment of immunosuppressive mediators in patients with severe acute respiratory syndrome. *J Immunol.* 172. 7841-7847.

[26] Lee, DC., Cheung, CY., Law, AH., Mok, CK., Peiris, M., Lau, AS.(2005) p38 Mitogen-Activated Protein Kinase-Dependent Hyperinduction of Tumor Necrosis Factor Alpha Expression in Response to Avian Influenza Virus H5N1. *J. Virol.* 79. 10147-10154.

[27] Lee, JC., Kumar, S., Griswold, DE., Underwood, DC., Votta, BJ., Adams, JL. (2000) Inhibition of p38 MAP kinase as a therapeutic strategy. *Immunopharmacology* 47. 185-201.

[28] Liao, Y., Hung, MC. (2003) Regulation of the Activity of p38 Mitogen-Activated Protein Kinase by Akt in Cancer and Adenoviral Protein E1A-Mediated Sensitization to Apoptosis. *Mol. Cell. Biol.* 23. 6836-6848.

[29] Marra MA, Jones SJM, Astell CR, Holt RA, Brooks-Wilson A, Butterfield YSN, Khattra J, Asano JK, Barber SA, Chan SY et al. (2003) The genome sequence of the SARS associated coronavirus. *Science* 300: 1399–1404.

[30] Maruoka, S., Hashimoto, S., Gon, Y., Nishitoh, H., Takeshita, I., Asai, Y., Mizumura, K., Shimizu, K., Ichijo, H., Horie, T. (2003) ASK1 regulates influenza virus infection-induced apoptotic cell death. *Biochem. Biophys. Res. Commun.* 307. 870-876.

[31] McGilvray, ID., Lu, Z., Wei, AC., Dackiw, AP., Marshall, JC., Kapus, A., Levy, G., Rotstein, OD. (1998) Murine Hepatitis Virus Strain 3 Induces the Macrophage Prothrombinase fgl-2 through p38 Mitogen-activated Protein Kinase Activation. *J Biol Chem.* 273. 32222-32229,

[32] Mizutani T, Fukushi S, Iizuka D, Inanami O, Kuwabara M, Takashima H, Yanagawa H, Saijo M, Kurane I, Morikawa S. (2006) Inhibition of cell proliferation by SARS-CoV infection in Vero E6 cells. *FEMS Immunol. Microbiol.* 46: 236-243.

[33] Mizutani T, Fukushi S, Ishii K, Sasaki Y, Kenri T, Saijo M, Kanaji Y, Shirota K, Kurane I, Morikawa S. Mechanisms of establishment of persistent SARS-CoV-infected cells. *Biochem Biophys Res Commun.* (in press)

[34] Mizutani T, Fukushi S, Murakami M, Hirano T, Saijo M, Kurane I, Morikawa S. (2004) Tyrosine dephosphorylation of STAT3 in SARS coronavirus-infected Vero E6 cells. *FEBS Lett.* 577: 187-192.

[35] Mizutani T, Fukushi S, Saijo M, Kurane I, Morikawa S. (2004) Phosphorylation of p38 MAPK and its downstream targets in SARS coronavirus-infected cells. *Biochem Biophys Res Commun.* 319: 1228-1234.

[36] Mizutani T, Fukushi S, Saijo M, Kurane I, Morikawa S. (2004) Importance of Akt signaling pathway for apoptosis in SARS-CoV-infected Vero E6 cells. Virology 327: 169-174.

[37] Mizutani T, Fukushi S, Saijo M, Kurane I, Morikawa S. (2005) JNK and PI3k/Akt signaling pathways are required for establishing persistent SARS-CoV infection in Vero E6 cells. Biochem. Biophys. *Acta.* 1741: 4-10.

[38] Mizutani T, Fukushi S, Saijo M, Kurane I, Morikawa S. (2006) Regulation of p90RSK phosphorylation by SARS-CoV infection in Vero E6 cells. *FEBS Lett.* 580: 1417-1424.

[39] Moriuchi, H., Moriuchi, M., Straus, SE., Cohen, JI. (1993) Varicella–zoster virus (VZV) open reading frame 61 protein transactivates VZV gene promoters and enhances the infectivity of VZV DNA. *J Virol* 67. 4290–4295.

[40] Muthumani, K., Wadsworth, SA., Dayes, NS., Hwang, DS., Choo, AY., Abeysinghe, HR., Siekierka, JJ., Weiner, DB. (2004) Suppression of HIV-1 viral replication and cellular pathogenesis by a novel p38/JNK kinase inhibitor. *AIDS.* 18. 739-748.

[41] Ng, TI., Keenan, L., Kinchington, PR., Grose, C. (1994) Phosphorylation of varicella–zoster virus open reading frame (ORF) 62 regulatory product by viral ORF47-associated protein kinase. *J. Virol* 68. 1350–1359.

[42] Platanias, LC. (2003) p38 mitogen-activated protein kinase pathway and its role in interferon signaling. *Pharmacol. Ther*. 98. 129-142.

[43] Player, MR., Torrence, PF. (1998) The 2-5A system: modulation of viral and cellular processes through acceleration of RNA degradation. *Pharmacol. Ther.* 78. 55-113

[44] Rahaus, M., Desloges, N., Wolff, MH. (2005) ORF61 Protein of Varicella–Zoster Virus Influences JNK/SAPK and p38/MAPK Phosphorylation. *J. Med. Virol.* 76. 424-433.

[45] Rahaus, M., Wolff, MH. (2003) Reciprocal effects of varicella-zoster virus (VZV) and AP1: Activation of jun, fos and ATF-2 after VZV infection and their importance for the regulation of viral genes. *Virus Res.* 92. 9–21.

[46] Rajan, P., McKay, RD. (1998) Multiple routes to astrocytic differentiation in the CNS. *J. Neurosci.* 18. 3620-3629.

[47] Rota PA, Oberste MS, Monroe SS, Nix WA, Campagnoli R, Icenogle JP, Penaranda S, Bankamp B, Maher K, Chen MH et al. (2003) Characterization of a novel coronavirus associated with severe acute respiratory syndrome. *Science* 300: 1394–1399

[48] Shapiro, L., Heidenreich, KA., Meintzer, MK., Dinarello, CA. (1998) Role of p38 mitogen-activated protein kinase in HIV type 1 production in vitro. *Proc. Natl. Acad. Sci.* USA 95. 7422-7426

[49] Silva, AM., Whitmore, M., Xu, Z., Jiang, Z., Li, X., Williams, BR. (2004) Protein kinase R (PKR) interacts with and activates mitogen-activated protein kinase kinase 6 (MKK6) in response to double-stranded RNA stimulation. *J Biol Chem.* 279. 37670-37676.

[50] Silverman, RH. (1997) 2-5A dependent RNase L: a regulated endoribonuclease in the interferon system, p515-551. In J. F. Riordan, and G. D'Alessio (ed.), *Ribonucleases: structures and functions.* Academic Press, Inc., San Diego, Calif

[51] Singh, IN., El-Hage, N., Campbell, ME., Lutz, SE., Knapp, PE., Nath, A., Hauser, KF. (2005) Differential involvement of p38 and JNK MAP kinases in HIV-1 Tat and gp120-induced apoptosis and neurite degeneration in striatal neurons. *Neuroscience* 135. 781-790.

[52] Spaziani, A., Alisi, A., Sanna, D., Balsano, C. (2006) Role of p38 MAPK and RNA-dependent Protein Kinase (PKR) in Hepatitis C Virus Core-dependent Nuclear Delocalization of Cyclin B1. *J. Biol. Chem.* 281. 10983-10989.

[53] Stark, GR., Kerr, IM., Williams, BR., Silverman, RH., Schreiber, RD. (1998) How cells respond to interferons. *Annu. Rev. Biochem*. 67. 227-264

[54] Stevenson, D., Colman, KL., Davison, AJ. (1992) Characterization of the varicella–zoster virus gene 61 protein. *J Gen Virol* 73. 521–530.

[55] Su, HL., Liao, CL., Lin, YL. (2002) Japanese Encephalitis Virus Infection Initiates Endoplasmic Reticulum Stress and an Unfolded Protein Response. *J. Virol.* 76. 4162-4171.

[56] Surjit, M., Liu, B., Jameel, S., Chow, VTK., Lal, SK. (2004) The SARS coronavirus nucleocapsid (N) protein induces actin reorganization and apoptosis in COS-1 cells. *Biochem. J.* 383. 13-18.

[57] Vik, TA., Sweet, LJ., Erikson, RL. (1990) Coinfection of insect cells with recombinant baculovirus expressing pp60v-src results in the activation of a serine-specific protein kinase pp90rsk. *Proc. Natl. Acad. Sci.* USA 87. 2685–2689.

[58] Wada, T., Penninger, JM. (2004) Mitogen-activated protein kinases in apoptosis regulation. *Oncogene* 23. 2838-2849.

[59] Wang, XZ., Lawson, B., Brewer, JW., Zinszner, H., Sanjay, A., Mi, LJ., Boorstein, R., Kreibich, G., Hendershot, LM., Ron, D.. (1996) Signals from the stressed endoplasmic reticulum induce C/EBP-homologous protein (CHOP/GADD153). *Mol Cell Biol.* 16. 4273-4280.

[60] Yan, H., Xiao, G., Zhang, J., Hu, Y., Yuan, F., Cole, DK., Zheng, C., Gao, GF. (2004) SARS Coronavirus Induces Apoptosis in Vero E6 Cells. *J. Med. Virol.* 73. 323–331.

[61] Zachos, G., Clements, B., Conner, J. (1999) Herpes simplex virus type 1 infection stimulates p38/c-Jun N-terminal mitogen-activated protein kinase pathways and activates transcription factor AP-1. *J. Biol. Chem.* 274. 5097-5103.

[62] Zachos, G., Koffa, M., Preston, CM., Clements, JB., Conner, J. (2001) Herpes simplex virus type 1 blocks the apoptotic host cell defense mechanisms that target Bcl-2 and manipulates activation of p38 mitogen-activated protein kinase to improve viral replication. *J. Virol.* 75. 2710-2728.

In: Focus on Signal Transduction Research
Editor: Gavin McAlpine, pp. 173-188

ISBN 1-60021-376-6

Chapter 8

INNATE IMMUNE SIGNALLING PATHWAYS AND VIRAL RECOGNITION: NEW PERSPECTIVES

Ismar R. Haga* and Andrew G. Bowie
School of Biochemistry and Immunology, Trinity College Dublin, Dublin, Ireland

ABSTRACT

It has been known for many years that the innate immune response triggered by viral infections ultimately leads to the activation of a number of transcription factors involved in anti-viral mechanisms, such as NFκB and the interferon regulatory factor (IRF) family, via different signalling pathways. Recently, molecular details of some of the receptors responsible for viral recognition have been intensely studied, including the toll-like receptors (TLRs), and cytoplasmic helicase-domain containing proteins. The TLRs participate in the first step of host defence by triggering innate immunity in response to the recognition of pathogen-associated molecular patterns (PAMPs). TLRs signal via homotypic interactions between their intracellular Toll/Interleukin-1/Resistance (TIR) domains and those on TIR adaptor molecules such as MyD88, Mal/TIRAP and TRIF. Helicase-domain containing proteins, such as RIG-I and Mda5, lack TIR domains but signal via another protein-interacting domain, named caspase activation and recruitment domain (CARD). TIR- and CARD-mediated signalling pathways lead to the activation of NFκB and IRFs.

Both viral proteins and nucleic acids can work as PAMPs. For instance, both TLRs and, more recently, RIG-I and Mda5 have been shown to sense dsRNA produced during viral infections. The specific signalling pathways triggered by each of these PAMPs have also been the focus of intense scrutiny.

Here, we review current advances in viral recognition by cellular components such as TLRs and CARD-domain containing proteins and the increasingly complex signalling pathways triggered.

---

* Corresponding author: Ismar R. Haga, School of Biochemistry and Immunology, Trinity College, Dublin 2, Ireland. Telephone: +353 1 896 3047, fax: +353 1 677 2400, email: ismar.haga@tcd.ie

INTRODUCTION

The immune system detects and eliminates invading pathogenic microorganisms by discriminating between self and non-self. In mammals the immune system can be divided into two branches: innate and adaptive immunity. Adaptive immunity detects non-self through recognition of peptide antigens using antigen receptors expressed on the surface of B and T cells. In order to respond to a wide range of potential antigens, B and T cells rearrange their immunoglobulin and T cell receptor (TCR) genes to generate over 10^{11} different species of antigen receptors. Engagement of antigen receptors by cognate antigen triggers clonal expansion of the lymphocyte and further production of antigen-specific Abs. This highly sophisticated system is observed only in vertebrates and is a potent defense against microbial infection (Takeda *et al.*, 2003).

In contrast, the innate immune system is phylogenetically conserved and is present in almost all multicellular organisms (Hoffmann *et al.*, 1999). Whereas the system of adaptive immunity has been the subject of considerable study in the past century, innate immunity has been less well appreciated. Therefore, the mechanism by which the innate immune system recognises non-self remained unknown until fairly recently. The innate immune response is not completely non-specific, but rather able to discriminate between self and non-self. Recognition of non-self by the innate immune system is due to a limited (when compared to that of the adaptive immune response) number of germline-encoded pattern recognition receptors (PRRs) (Akira *et al.*, 2006).

PRRs recognise microbial components (known as pathogen-associated molecular patterns, PAMPs) that are essential for the survival of the microorganism, are expressed constitutively in the host and are expressed on all cells of a given type, independently of immunological memory. Different PRRs interact with specific PAMPs, show distinct expression patterns, activate specific signalling pathways and lead to distinct anti-pathogen responses. Examples of PRR/PAMP interactions include TLR3/dsRNA and TLR4/LPS (Akira *et al.*, 2006).

Here we review recent advances on the recognition of pathogens by the innate immune system and the respective signalling pathways triggered, focusing on viral recognition by TLRs and novel cytoplasmic receptors.

VIRAL RECOGNITION BY TLRS

The signalling pathway of *Drosophila* Toll (the founding member of the TLR family) shows remarkable similarity to the mammalian IL-1 pathway, which leads to the activation of NFκB, a transcription factor responsible for many aspects of inflammatory and immune responses. Indeed, the cytoplasmic domains of *Drosophila* Toll and the mammalian IL-1 receptor are highly conserved (Gay and Keith, 1991) and are referred to as the Toll/IL-1 receptor (TIR) domain.

Based on the similarity between the TIR domains of *Drosophila* Toll and the mammalian IL-1R, it was proposed that the Toll-mediated pathway might be involved in regulating immune responses. This was subsequently demonstrated in a study in which mutant flies lacking individual components of this pathway, i.e., Toll, Spaetzle, Tube or Pelle, had an impaired response to fungal infection (Lemaitre *et al.*, 1996).

A year after the discovery of the role of *Drosophila* Toll in the fly immune response, a mammalian counterpart of *Drosophila* Toll was identified (Medzhitov *et al.*, 1997). Subsequently, a family of proteins structurally related to *Drosophila* Toll was identified, collectively referred to as the Toll-like receptors (TLRs) (Rock *et al.*, 1998). TLR family members are characterised structurally by the presence of a leucine-rich repeat (LRR) in their extracellular domain and a TIR domain in their intracellular domain.

The TLR family signalling pathway is similar to that of the IL-1R family. Both IL-1R and TLRs (with the exception of TLR3) interact with adaptor protein myeloid differentiation factor-88 (MyD88), which has a TIR domain in its C-terminal portion and a death domain in its N-terminal portion. Apart from MyD88, there are at least other four mammalian TIR domain-containing adaptors: MyD88-adaptor like (Mal), TIR domain-containing adaptor inducing IFN (TRIF), TRIF-related adaptor molecule (TRAM) and sterile alpha and HEAT–Armadillo motifs (SARM) (reviewed in (O'Neill *et al.*, 2003)).

Upon stimulation, MyD88 associates with either the TLRs or the IL-1R complex via TIR domain interactions. MyD88 recruits a death domain-containing serine/threonine kinase, the IL-1R-associated kinase (IRAK) (Martin *et al.*, 1994). There are four IRAKs in the human and murine genomes: two with demonstrated kinase activity, IRAK1 and IRAK4, and two inactive kinases, IRAK2 and IRAKM (Janssens and Beyaert, 2003). The exact physiological function of IRAK2 has not been clarified, although it is likely to have a role in NFκB activation (Fitzgerald *et al.*, 2001; Harte *et al.*, 2003). IRAK4 is an upstream kinase for IRAK1. Phosphorylation of IRAK1 by IRAK4 leads to the dissociation of IRAK1 from the IL-1R complex (Janssens and Beyaert, 2003). IRAK1 in turn interacts with the tumour necrosis factor receptor-associated factor (TRAF) family member TRAF6 (initially identified as a signal transducer for IL-1 (Cao *et al.*, 1996)), eventually leading to activation of NFκB.

NFκB is a transcription factor that has crucial roles in inflammation, immunity, cell proliferation and apoptosis. The activity of NFκB is regulated by association with IκB, which sequesters NFκB in the cytoplasm until phosphorylated on serine residues by the IκB kinase (IKK) complex (Viatour *et al.*, 2005). This phosphorylation targets IκB for degradation, allowing nuclear translocation of NFκB. The IKK complex contains two catalytic subunits, IKKα and IKKβ, as well as a scaffolding protein, IKKγ. Different pathways, not only IL-1- and TLR-dependent, converge to phosphorylation and subsequent activation of the IKK complex. One mammalian protein complex that activates IKK was purified and analysed and found to be composed of two subunits: TGFβ-activated kinase 1 (TAK1) and a ubiquitin conjugating enzyme complex (Deng *et al.*, 2000). A complex consisting of TAK1 and the TAK1 binding proteins TAB1, TAB2 and TAB3 is recruited to TRAF6. Once activated by its association with TRAF6, TAK1 mediates phosphorylation of the IKK complex (Wang *et al.*, 2001) and of MAP kinases kinases. Recent data have demonstrated that TAB1 is capable of inducing p38 activity and also has a role in modulating downstream effects (Lu *et al.*, 2006).

NFκB was identified as a nuclear factor necessary for the transcription of immunoglobulin light chain in B cells. Subsequently, it was shown to be expressed in a variety of cells types. The NFκB family of transcription factors is structurally related and evolutionary conserved, with five members in mammals: Rel (c-Rel), RelA (p65), RelB, NFκB1 (p50 and its precursor p105) and NFκB2 (p52 and its precursor p100) (Ghosh *et al.*,

1998). NFκB regulates a wide variety of genes, including those encoding cytokines (e.g., IL-1, IL-2, IL-6, IL-12, TNFα, LTα, LTβ and GM-CSF), chemokines (e.g., IL-8, MIP-1α, MCP1, RANTES and eotaxin), adhesion molecules (e.g., ICAM, VCAM and E-selectin), acute phase proteins (e.g., SAA), and inducible effector enzymes (e.g., iNOS and COX-2) (Ghosh and Karin, 2002). In addition, genes encoding evolutionary conserved antimicrobial peptides, such as β-defensins, are also regulated by NFκB (Zhang and Ghosh, 2001). NFκB also plays a role in expression of genes encoding molecules important for the adaptive immune response, such as MHC proteins, costimulatory molecules such as CD80, and cytokines such as IL-2, IL-12 and IFNβ (Kopp and Medzhitov, 1999). Additionally, some of the cytokines and chemokines produced in response to NFκB activation can stimulate the migration and maturation of lymphocytes. NFκB also activates genes coding for regulators of apoptosis and cell proliferation (Karin and Lin, 2002).

NFκB is not the only transcription factor activated by TLRs. Members of the interferon regulatory factor (IRF) family of transcription factors (composed of nine members that are critical regulators of innate immune responses) (Taniguchi *et al.*, 2001), namely IRF3, IRF5 and IRF7, are also activated by TLRs. Specifically, TLR3 and TLR4, via TRIF, activate IRF3, while TLR7 and TLR8, via MyD88, activate IRF5 and IRF7 (Wietek *et al.*, 2003; Takaoka *et al.*, 2005; Uematsu *et al.*, 2005). IRFs are transcriptional activators for IFNα, IFNβ and IFN-stimulated genes and can act as direct transducers of virus-mediated signaling pathways activating IFNα and IFNβ in infected cells (Taniguchi *et al.*, 2001). Genes targeted by IRF3 include not only the classical IRF-responsive genes IFN-α1 and β but also IL-15 and RANTES (Servant *et al.*, 2002). Two IKK-related kinases are involved in the TRIF-mediated activation of IRF3 by TLR3 and TLR4: IKKε (also called IKKi) and TRAF family member-associated NFκB activator (TANK)-binding kinase 1 (TBK1), forming a TRIF-IKKε/TBK1-IRF3 axis (Fitzgerald *et al.*, 2003), probably with some degree of redundancy between the IKKε and TBK1 (for a detailed review, please see (Moynagh, 2005)).

Both viral proteins and nucleic acids are recognised as PAMPs by PRRs, including TLRs. TLR2, TLR3, TLR4, TLR7, TLR8 and TLR9 have been involved in the recognition of viral PAMPs. Here we present a brief overview of viral recognition of TLRs (Bowie and Haga, 2005; Akira *et al.*, 2006; Kawai and Akira, 2006).

Cell-surface TLR2 and TLR4 recognise some viral glycoproteins present on virions. The first described viral PAMP recognised by TLRs was the F protein of respiratory syncytial virus (RSV), which induced IL-6 production in a TLR4-dependent manner (Kurt-Jones *et al.*, 2000). Despite similarities in structure between the F proteins of RSV and Sendai virus, TLR4 was not involved in host defence against Sendai virus (van der Sluijs *et al.*, 2003). The functional significance of TLR4 for RSV remains, however, unspecified, with different studies reporting somewhat conflicting results (Haeberle *et al.*, 2002; Monick *et al.*, 2003; Ehl *et al.*, 2004). TLR4 was also activated by the envelope (env) protein of mouse mammary tumour virus (MMTV), leading to maturation of bone marrow derived dendritic cells (DCs). Also, TLR4 has a role in MMTV-induced activation of B cells (which abrogated in TLR4 -/- cells) (Burzyn *et al.*, 2004). Measles virus and human cytomegalovirus (HCMV) can mediate cellular effects, such as the induction of IL-6, via TLR2 (Bieback *et al.*, 2002; Compton *et al.*, 2003).

Double-stranded RNA (dsRNA) is a PAMP that can be generated during the life cycle of a large number of viruses. Alexopoulou *et al.* showed that TLR3 mediates responses to poly(I:C), a synthetic analog to dsRNA, including activation of NFκB and the production of type I IFNs. Further studies, however, have shown that TLR3 is not required for the initial, cell-autonomous recognition of viral infection (Lopez *et al.*, 2004) and that TLR3 -/- mice fail to show increased susceptibility to a number of viral infections, including murine cytomegalovirus virus (MCMV), vesicular stomatitis virus (VSV), lymphocytic choriomeningitis virus (LCMV) and reovirus (Edelmann *et al.*, 2004). Interestingly, West Nile virus benefits from its interaction with TLR3, which ultimately leads to a disruption in the blood-brain barrier, allowing the virus access to the brain. TLR3 -/- mice are, expectedly, more resistant to lethal West Nile virus infection (Wang *et al.*, 2004). Overall, the role of TLR3 in antiviral responses remains unclear, although it is possible that TLR3 has a more prominent role in recognising viruses that possess successful IFN-evading strategies (Schroder and Bowie, 2005).

TLR9 was originally identified as the PRR for bacterial DNA sequences containing unmethylated CpG dinucleotides (Hemmi *et al.*, 2000). Such motifs can also be found in some viral genomes, including those of herpes simplex virus 1 (HSV-1), HSV-2 and MCMV. Both HSV-1 and HSV-2 have been shown to activate plasmacytoid dendritic cells (pDCs) to produce type I IFNs through TLR9 (Lund *et al.*, 2003; Krug *et al.*, 2004).

TLR7 and TLR8 were the most recent TLRs to be implicated in recognition of viral PAMPs. Mouse TLR7 and human TLR8 recognise synthetic antiviral imidazoquinoline components (imiquimod and resiquimod) and guanine nucleotide analogues (such as loxoribine). Some of these analogues were shown to activate TLR7 in a process requiring endosomal maturation (Lee *et al.*, 2003). Further studies revealed that human TLR8 and mouse TLR7 recognise G- and U- rich ssRNA oligonucleotides derived from HIV-1 (Heil *et al.*, 2004) and that, in addition, mouse TLR7 is also involved in recognising viral (influenza) and non-viral (polyU) ssRNAs (Diebold *et al.*, 2004), as well as ssRNA from VSV (Lund *et al.*, 2004).

In the past months, novel TLR-independent viral recognition mechanisms have been described, after a series of studies found that a number of immune responses were maintained or only slightly affected in the absence of TLRs or TLR adaptors (Kawai and Akira, 2006). As discussed in the next section, these findings might eventually help elucidating how cells distinguish between self and non-self nucleic acids.

Viral Recognition by Intracellular CARD Domain Containing Proteins

Helicases are enzymes that catalyse the unwinding of double-stranded nucleic acids in an energy-dependent manner. They utilise the free energy change of binding and hydrolysing a nucleotide triphosphate to dissociate duplexes or disband bound proteins. Some helicases, when purified, processively translocate along nucleic acids *in vitro* and unwind double-stranded regions. However, most helicases show little or no processivity when assayed (Cordin *et al.*, 2006). Further classification depends on whether the helicase can bind single-stranded nucleic acid, unwind DNA or RNA or both and the direction of unwinding (Kwong *et al.*, 2005). Helicases have been divided into three super-families (SF1, SF2 and SF3) and two smaller families based on sequence comparisons and conserved motifs (Gorbalenya and Koonin, 1993).

Helicases containing a DEAD (Asp-Glu-Ala-Asp) box (Linder *et al.*, 1989) and the related DEAH, DExH and DExD boxes, which are commonly referred as the DExD/H helicase family, are the most numerous members of SF2 (Kwong *et al.*, 2005; Cordin *et al.*, 2006). Members of this family include two proteins recently involved in intracellular viral recognition, retinoic acid inducible gene I (RIG-I) and melanoma differentiation-associated gene 5 (Mda5). Both proteins contain at least one copy of a homotypic interaction motif called the caspase recruitment domain (CARD) (Andrejeva *et al.*, 2004; Yoneyama *et al.*, 2004).

The CARD module was first identified in a subset of caspases and their adaptor molecules, being implicated in proteolytic processing of cytokines (Bouchier-Hayes and Martin, 2002). However, it has become apparent that some of the CARD domain-containing proteins do not recruit caspases or promote caspase activation. Instead, many (such as RIG-I and Mda5) participate in signalling pathways leading to NFκB activation (Bouchier-Hayes and Martin, 2002) or are involved in pathways regulating apoptosis (Dufner and Mak, 2006). As an example, caspase-1 was recently proposed to contribute to inflammation by activating a distinct pathway in B cells. This pathway involves an interaction of the CARD domain of procaspase-1 with the CARD-containing kinase receptor-induced protein RIP2, which, in turn, recruits the IKK complex to activate NFκB and p38 (Lamkanfi *et al.*, 2004).

RIG-I has two copies of the CARD domain at its N terminus in addition to its C-terminal helicase domain (Yoneyama *et al.*, 2004). Mda5 has a single CARD domain in its N terminus (Andrejeva *et al.*, 2004).

The initial report on RIG-I-mediated viral recognition identified the protein as an essential regulator for dsRNA-induced signalling. The overexpression of RIG-I blocks the replication of VSV and encephalomyocarditis virus and the knockdown of RIG-I by small interfering RNA (siRNA) reduces type I IFN production after infection with Newcastle disease virus (NDV), Sendai virus and encephalomyocarditis virus (Yoneyama *et al.*, 2004). Additionally, conventional dendritic cells (cDCs) and fibroblasts derived from RIG-I -/- mice were almost completely unable to produce cytokines and type I IFN following infection with NDV, Sendai virus or VSV, indicating that RIG-I is essential for the induction of type I IFNs in those cells. However, pDCs, which produce large amounts of IFNα, used the TLR system, rather than RIG-I, for IFN induction in response to viruses (Kato *et al.*, 2005). Specifically, TLR7 and, to a lesser extent, TLR9 played a role in NDV recognition. Considering the expression patterns and endosomal localisation of these receptors, the authors suggested that pDCs might recognise the viral ssRNA upon endosomal uptake of virions, unlike cDCs and fibroblasts, indicating that the involvement of RIG-I in viral recognition might be cell type-specific (Kato *et al.*, 2005). In a different study, nonimmune cells primarily responded to intracellular accumulation of dsRNA via RIG-I while myeloid cells sensed Sendai virus via TLR7- and TLR8-dependent but RIG-I and PKR-independent mechanisms, also suggesting the existence of cell-specific mechanisms of viral recognition (Melchjorsen *et al.*, 2005).

Mda5 is an IFN-inducible protein and its exogenous expression activates the IFNβ promoter and enhances its activation by dsRNA. It also possesses growth-suppressive properties, as proven by suppression of colony formation in HO-1 human melanoma cells following its ectopic expression (Kang *et al.*, 2002). The finding that Mda5 binds to a highly conserved cysteine-rich domain of V proteins of a wide variety of paramyxoviruses, including Sendai virus, mumps and simian virus 5 (SV5) (Andrejeva *et al.*, 2004) was the first report of the involvement of this protein in intracellular viral recognition.

The recently reported generation of Mda5 knockout mice has pointed towards differential roles for RIG-I and Mda5 in recognition of viruses. Mda5 is essential for pIC-induced IFNα production and TLR3 signalling is critical for IL-12 production, whereas both Mda5 and TLR3 regulate IL-6 production. Response to pIC in bone marrow-derived dendritic cells (BMDCs) derived from Mda5 -/- mice, but not RIG-I -/- or TRIF -/- mice, was severely impaired, indicating that pIC is primarily recognised by Mda5 in these cells. IFN responses to a number of negative-sense ssRNA viruses (including NDV, Sendai virus and influenza virus) were severely impaired in mouse embryonic fibroblasts (MEFs) derived from RIG-I -/- mice, but not in cells derived from Mda5 -/- mice (Kato *et al.*, 2006). Interestingly, IFNβ responses were abrogated in Mda5 -/- MEFs following infection with encephalomyocarditis virus, a positive sense ssRNA virus belonging to the picornavirus family, showing that RIG-I and Mda5 have roles in the recognition of different groups of viruses. The authors suggested that RIG-I probably recognises dsRNA generated over the course of RNA virus replication. However, how RIG-I and Mda5 differentially recognise RNAs remains undetermined and other not yet identified cellular proteins might also function as PRRs for viral RNAs (Kato *et al.*, 2006). It should also be noted that at least one cellular protein (Lgp2, which contains a DExD/H box highly homologous to that of RIG-I but lacks a CARD domain), acts as a repressor of RIG-I signalling (Rothenfusser *et al.*, 2005).

Upon dsRNA recognition and biding by its RNA helicase activity, RIG-I dimerises and undergoes conformational alterations that enable the N-terminal CARD domain to interact with another downstream adaptor protein (s). RIG-I signalling ultimately engages the IKK complex and stress-activated kinases, as well as the IKK-related kinases TBK1 and IKKε (see previous section), leading to the phosphorylation and activation of NFκB, ATF-2/c-jun and IRF3 transcription factors. Coordinated activation of these factors results in the formation of a transcriptionally competent enhanceosome that triggers IFNβ production (Lin *et al.*, 2006).

Four different groups have reported the adaptor molecule that provides the link between RIG-I and the downstream activation events (Kawai *et al.*, 2005; Meylan *et al.*, 2005; Seth *et al.*, 2005; Xu *et al.*, 2005). The name IFNβ promoter stimulator 1 (IPS-1) was given to a protein that activates IFNα, IFNβ and NFκB promoters in a TBK-dependent manner. Similar to RIG-I, IPS-1 has an N-terminal CARD domain and a C-terminal effector domain that recruits the Fas-associated death domain protein FADD and the kinase receptor interacting protein 1 (RIP1) (Kawai *et al.*, 2005). The same protein, named mitochondrial antiviral signalling (MAVS), was shown to have a C-terminal transmembrane domain and to localise in the mitochondria (this location being essential for signalling) (Seth *et al.*, 2005). In a third report, the protein was named virus-induced signalling adaptor (VISA) and shown to interact with TRIF, TRAF2 and TRAF6 through a proline-rich domain, suggesting a role for the protein in both TLR3 and RIG-I virus-triggered pathways (Xu *et al.*, 2005). The final report named the protein CARD adaptor inducing IFNβ (Cardif) and demonstrated that it interacts with RIG-I and recruits IKKα, IKKβ and IKKε through its C-terminal region (Meylan *et al.*, 2005). This study also showed that Cardif is cleaved by the NS3/4A protease encoded by hepatitis C virus (HCV). The NS3/4A cleavage of IPS-1/MAVS/VISA/Cardif results in its dissociation from the mitochondrial membrane and in the disruption of the downstream signalling. Virus-induced IKKε, but not TBK1, colocalised with IPS-1/MAVS/VISA/Cardif at the mitochondrial membrane, and the localisation of both molecules was disrupted by the expression of NS3/4A (Lin *et al.*, 2006). Furthermore, early activation of IRF3 and induction of antiviral genes are reversed by the cleavage of IPS-1/MAVS/VISA/Cardif by NS3/4A (Loo *et al.*, 2006).

Recently, the generation of IPS-1/MAVS/VISA/Cardif knockout mice has showed that the loss of the protein abolished viral induction of interferons and prevented the activation of IRF3 and NFκB in different cell types, except pDCs. The protein, however, was not required for IFN induction by cytosolic DNA or *Listeria monocytogenes*, suggesting that the protein is specifically required for innate immune responses against viral infection (Sun *et al.*, 2006). A different experimental approach also proved that *Listeria monocytogenes* stimulates IFN synthesis in an IPS-1/MAVS/VISA/Cardif–independent manner (Soulat *et al.*, 2006). The protein has also been directly implicated in anti-viral responses triggered by RIG-I and Mda5 (Kumar *et al.*, 2006). IPS-1/MAVS/VISA/Cardif is required for both RIG-I- and Mda5-mediated responses triggered by a number of ssRNA viruses (NDV, Sendai virus and VSV for RIG-I-dependent responses, EMCV for Mda5-dependent ones). Moreover, IPS-1/MAVS/VISA/Cardif -/- mice are more susceptible to VSV and EMCV infection than wild type mice, indicating an important *in vivo* role for the protein. The same study showed that IPS-1/MAVS/VISA/Cardif is not essential for DNA virus-mediated responses (Kumar *et al.*, 2006). Undoubtedly, IPS-1/MAVS/VISA/Cardif will be the focus of a large number of studies in the near future, as several points involving its regulation and function remain inconclusive (Hiscott *et al.*, 2006).

Perspectives on Viral DNA Recognition

As mentioned in the last section, there has been a lot of progress in the understanding of how viral RNA is sensed intracellularly. Most of the studies of intracellular recognition of viral DNA have been focused on TLRs (as mentioned in Viral recognition by TLRs) but recent work has pointed to the existence of a TLR-independent recognition mechanism.

Production of type I IFNs in response to mammalian DNA formulated in cationic lipids has been shown to be partially TLR-independent (Yasuda *et al.*, 2005). Also, TLR9 -/- and MyD88 -/- cells are still able to produce IFNs in response to HSV-1 infection (Hochrein *et al.*, 2004), suggesting this response to be at least partially TLR-independent. *Listeria monocytogenes* DNA also activated a potent type I IFN response in a TLR-independent manner (Stetson and Medzhitov, 2006). Another indication of the existence of a TLR-independent sensing mechanism was given in a study of deoxyribonuclease (DNAse) II-deficient mice. In macrophages, DNAse II cleaves the DNA of engulfed apoptotic cells and of nuclei expelled from erythroid precursor cells. DNAse II -/- mouse embryos accumulate undigested DNA in macrophages and die due to the activation of the IFNβ gene (Kawane *et al.*, 2003). Macrophages from the DNAse II -/- foetal livers produce IFNβ, TNFα and CXCL10, but the activation of these genes was not due to TLR3, TLR9 or MyD88, as seen by backcrossing DNAse II deficient mice onto mouse strains lacking TLR3, TRIF, TLR9 and MyD88 (Okabe *et al.*, 2005), indicating the existence of a TLR-independent sensing mechanism for the endogenous DNA that escapes lysosomal degradation (Okabe *et al.*, 2005).

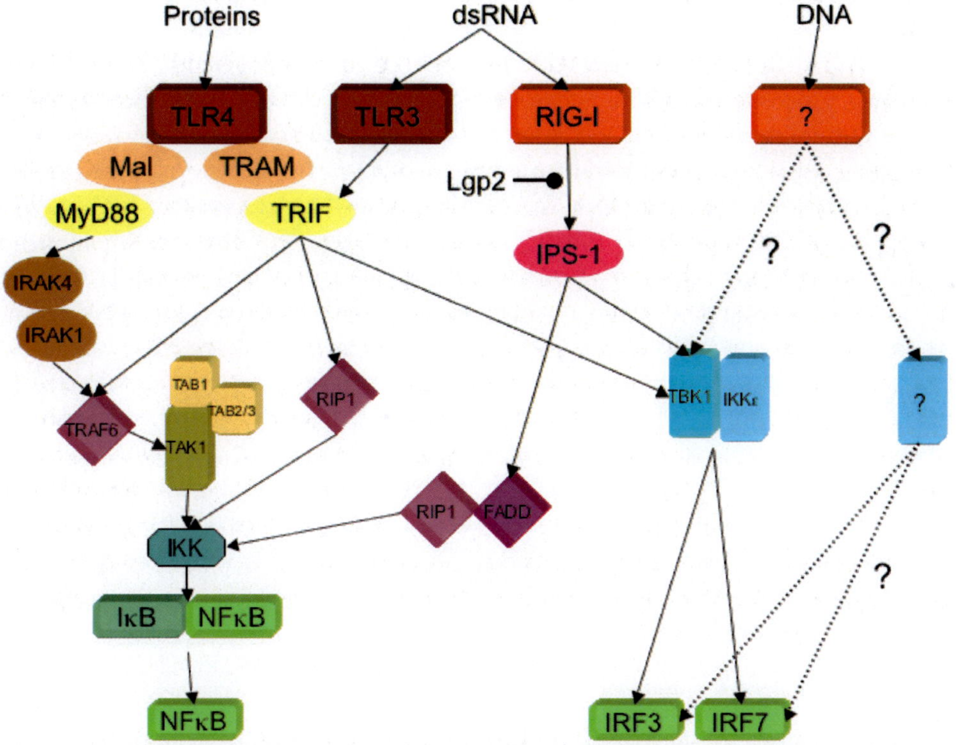

Figure 1. Schematic representation of some of the signalling pathways triggered by viral components. Stimulation of TLR4 by viral proteins triggers MyD88- dependent and independent pathways resulting in the phosphorylation of IκB and activation of NFκB. MyD88 is recruited via the adaptor Mal, with the subsequent recruitment of IRAK4, IRAK1 and TRAF6. TRAF6 activates TAK1 in a ubiquitination–dependent mechanism. TAK1, together with TAB1, TAB2 and TAB3, activates NFκB via IKK-mediated IκB degradation. Freed NFκB translocates into the nucleus and initiates the expression of proinflammatory cytokine genes. The MyD88-independent pathway uses TRIF, through the adaptor TRAM, to activate NFκB in either a TRAF6- dependent or independent manner (possibly involving RIP1). TRIF also associates with TBK1 and IKKε, which in turn phosphorylate IRF3 and IRF7, leading to their nuclear translocation and induction of type I IFN genes and co-stimulatory molecules. Upon dsRNA stimulation, TLR3 interacts directly with TRIF (which interacts with TBK1, RIP1 and TRAF6) to trigger these MyD88-independent pathways. RIG-I and Mda5 recognize dsRNA to initiate antiviral signalling. IPS-1 interacts with RIG-I and Mda5 via the CARD-like domain, followed by the activation of IRF3 and IRF7 via TBK1- and IKKε-dependent phosphorylation. IPS-1 also activates NFκB via FADD and RIP1. Synthetic dsDNA was also reported to activate an IPS-1-dependent pathway, although a DNA receptor has not been identified.

The intracellular administration of B-form dsDNA (B-DNA, the conformation most commonly found in physiological conditions) but not Z-form DNA (Z-DNA) stimulated mouse and human stromal and DCs, resulting in the production of type I IFNs and chemokines. B-DNA activated IRF3 and the IFNβ promoter via both TBK1 and IKKε (a noncanonical IKK family member), whereas it activated NFκB independently of both TBK1 and IKKε. Interestingly, both these pathways require IPS-1/MAVS/VISA/Cardif, but not TLRs or RIG-I, suggesting that the signalling pathways induced by dsRNA- and B-DNA are closely related but probably possess different receptor molecules (Ishii *et al.*, 2006).

These data point towards the existence of a cytosolic DNA receptor. Apart from sensing foreign DNA, a mechanism also capable of sensing endogenous DNA would be advantageous, for instance, when large quantities of self-DNA are released during tissue apoptosis to promote cytokine-driven tissue repair (Wagner and Bauer, 2006). So far, TLR9 has been involved in the cytoplasmic recognition of viral DNA (see Viral recognition by TLRs) but the still unidentified DNA receptor might also be responsible for sensing viral DNA. The replication of dsDNA viruses occurs mostly in the nucleus, with the notable exception of poxviruses, which replicate entirely in the cytoplasm. Poxvirus DNA would therefore be a major target for the putative intracellular DNA receptor. This is a rather interesting scenario, considering the plethora of immunemodulators encoded by various poxviruses (Haga and Bowie, 2005). Which immune evasion mechanism (s) would poxviruses use to escape recognition of their DNA? A virus-encoded protein could conceivably interact with the receptor or with a component of its signalling pathway, therefore suppressing the detection of viral DNA or blocking the signalling cascade triggered by the receptor. Here we might have another elegant example of viral subversion of the host immune response.

Conclusion

The identification of the cellular DNA receptor will help not only elucidating the mechanism of viral DNA detection but also clarifying the intricacies of the triggered signalling pathways. Together with the RNA detection, it will provide a much better understanding of how cells sense invading viruses, how viruses counteract the innate immune response and ultimately lead to new therapeutic strategies against viral infections.

Acknowledgements

This work was supported by Science Foundation Ireland (SFI), Investigator Programme Grant number 02/IN.1/B192.

References

Akira, S., Uematsu, S. and Takeuchi, O. (2006). Pathogen recognition and innate immunity. *Cell* 124, 783-801.

Andrejeva, J., Childs, K.S., Young, D.F., Carlos, T.S., Stock, N., Goodbourn, S. and Randall, R.E. (2004). The V proteins of paramyxoviruses bind the IFN-inducible RNA helicase, mda-5, and inhibit its activation of the IFN-beta promoter. *Proc Natl Acad Sci U S A* 101, 17264-17269.

Bieback, K., Lien, E., Klagge, I.M., Avota, E., Schneider-Schaulies, J., Duprex, W.P., Wagner, H., Kirschning, C.J., Ter Meulen, V. and Schneider-Schaulies, S. (2002). Hemagglutinin protein of wild-type measles virus activates toll-like receptor 2 signaling. *J Virol* 76, 8729-8736.

Bouchier-Hayes, L. and Martin, S.J. (2002). CARD games in apoptosis and immunity. *EMBO Rep* 3, 616-621.

Bowie, A.G. and Haga, I.R. (2005). The role of Toll-like receptors in the host response to viruses. *Mol Immunol* 42, 859-867.

Burzyn, D., Rassa, J.C., Kim, D., Nepomnaschy, I., Ross, S.R. and Piazzon, I. (2004). Toll-like receptor 4-dependent activation of dendritic cells by a retrovirus. *J Virol* 78, 576-584.

Cao, Z., Xiong, J., Takeuchi, M., Kurama, T. and Goeddel, D.V. (1996). TRAF6 is a signal transducer for interleukin-1. *Nature* 383, 443-446.

Compton, T., Kurt-Jones, E.A., Boehme, K.W., Belko, J., Latz, E., Golenbock, D.T. and Finberg, R.W. (2003). Human cytomegalovirus activates inflammatory cytokine responses via CD14 and Toll-like receptor 2. *J Virol* 77, 4588-4596.

Cordin, O., Banroques, J., Tanner, N.K. and Linder, P. (2006). The DEAD-box protein family of RNA helicases. *Gene* 367, 17-37.

Deng, L., Wang, C., Spencer, E., Yang, L., Braun, A., You, J., Slaughter, C., Pickart, C. and Chen, Z.J. (2000). Activation of the IkappaB kinase complex by TRAF6 requires a dimeric ubiquitin-conjugating enzyme complex and a unique polyubiquitin chain. *Cell* 103, 351-361.

Diebold, S.S., Kaisho, T., Hemmi, H., Akira, S. and Reis e Sousa, C. (2004). Innate antiviral responses by means of TLR7-mediated recognition of single-stranded RNA. *Science* 303, 1529-1531.

Dufner, A. and Mak, T.W. (2006). CARD Tricks: Controlling the Interactions of CARD6 with RICK and Microtubules. *Cell Cycle* 5, 797-800.

Edelmann, K.H., Richardson-Burns, S., Alexopoulou, L., Tyler, K.L., Flavell, R.A. and Oldstone, M.B. (2004). Does Toll-like receptor 3 play a biological role in virus infections? *Virology* 322, 231-238.

Ehl, S., Bischoff, R., Ostler, T., Vallbracht, S., Schulte-Monting, J., Poltorak, A. and Freudenberg, M. (2004). The role of Toll-like receptor 4 versus interleukin-12 in immunity to respiratory syncytial virus. *Eur J Immunol* 34, 1146-1153.

Fitzgerald, K.A., Palsson-McDermott, E.M., Bowie, A.G., Jefferies, C.A., Mansell, A.S., Brady, G., Brint, E., Dunne, A., Gray, P., Harte, M.T., McMurray, D., Smith, D.E., Sims, J.E., Bird, T.A. and O'Neill, L.A. (2001). Mal (MyD88-adapter-like) is required for Toll-like receptor-4 signal transduction. *Nature* 413, 78-83.

Fitzgerald, K.A., McWhirter, S.M., Faia, K.L., Rowe, D.C., Latz, E., Golenbock, D.T., Coyle, A.J., Liao, S.M. and Maniatis, T. (2003). IKKepsilon and TBK1 are essential components of the IRF3 signaling pathway. *Nat Immunol* 4, 491-496.

Gay, N.J. and Keith, F.J. (1991). Drosophila Toll and IL-1 receptor. *Nature* 351, 355-356.

Ghosh, S., May, M.J. and Kopp, E.B. (1998). NF-kappa B and Rel proteins: evolutionarily conserved mediators of immune responses. *Annu Rev Immunol* 16, 225-260.

Ghosh, S. and Karin, M. (2002). Missing pieces in the NF-kappaB puzzle. *Cell* 109 Suppl, S81-96.

Gorbalenya, A.E. and Koonin, E.V. (1993). Helicases: amino acid sequence comparisons and structure-function relationships. *Curr Opin Struc Biol* 3, 419-429.

Haeberle, H.A., Takizawa, R., Casola, A., Brasier, A.R., Dieterich, H.J., Van Rooijen, N., Gatalica, Z. and Garofalo, R.P. (2002). Respiratory syncytial virus-induced activation of nuclear factor-kappaB in the lung involves alveolar macrophages and toll-like receptor 4-dependent pathways. *J Infect Dis* 186, 1199-1206.

Haga, I.R. and Bowie, A.G. (2005). Evasion of innate immunity by vaccinia virus. *Parasitology* 130 Suppl, S11-25.

Harte, M.T., Haga, I.R., Maloney, G., Gray, P., Reading, P.C., Bartlett, N.W., Smith, G.L., Bowie, A. and O'Neill, L.A. (2003). The poxvirus protein A52R targets Toll-like receptor signaling complexes to suppress host defense. *J Exp Med* 197, 343-351.

Heil, F., Hemmi, H., Hochrein, H., Ampenberger, F., Kirschning, C., Akira, S., Lipford, G., Wagner, H. and Bauer, S. (2004). Species-specific recognition of single-stranded RNA via toll-like receptor 7 and 8. *Science* 303, 1526-1529.

Hemmi, H., Takeuchi, O., Kawai, T., Kaisho, T., Sato, S., Sanjo, H., Matsumoto, M., Hoshino, K., Wagner, H., Takeda, K. and Akira, S. (2000). A Toll-like receptor recognizes bacterial DNA. *Nature* 408, 740-745.

Hiscott, J., Lin, R., Nakhaei, P. and Paz, S. (2006). MasterCARD: a priceless link to innate immunity. *Trends Mol Med* 12, 53-56.

Hochrein, H., Schlatter, B., O'Keeffe, M., Wagner, C., Schmitz, F., Schiemann, M., Bauer, S., Suter, M. and Wagner, H. (2004). Herpes simplex virus type-1 induces IFN-alpha production via Toll-like receptor 9-dependent and -independent pathways. *Proc Natl Acad Sci U S A* 101, 11416-11421.

Hoffmann, J.A., Kafatos, F.C., Janeway, C.A. and Ezekowitz, R.A. (1999). Phylogenetic perspectives in innate immunity. *Science* 284, 1313-1318.

Ishii, K.J., Coban, C., Kato, H., Takahashi, K., Torii, Y., Takeshita, F., Ludwig, H., Sutter, G., Suzuki, K., Hemmi, H., Sato, S., Yamamoto, M., Uematsu, S., Kawai, T., Takeuchi, O. and Akira, S. (2006). A Toll-like receptor-independent antiviral response induced by double-stranded B-form DNA. *Nat Immunol* 7, 40-48.

Janssens, S. and Beyaert, R. (2003). Functional diversity and regulation of different interleukin-1 receptor-associated kinase (IRAK) family members. *Mol Cell* 11, 293-302.

Kang, D.C., Gopalkrishnan, R.V., Wu, Q., Jankowsky, E., Pyle, A.M. and Fisher, P.B. (2002). mda-5: An interferon-inducible putative RNA helicase with double-stranded RNA-dependent ATPase activity and melanoma growth-suppressive properties. *Proc Natl Acad Sci U S A* 99, 637-642.

Karin, M. and Lin, A. (2002). NF-kappaB at the crossroads of life and death. *Nat Immunol* 3, 221-227.

Kato, H., Sato, S., Yoneyama, M., Yamamoto, M., Uematsu, S., Matsui, K., Tsujimura, T., Takeda, K., Fujita, T., Takeuchi, O. and Akira, S. (2005). Cell type-specific involvement of RIG-I in antiviral response. *Immunity* 23, 19-28.

Kato, H., Takeuchi, O., Sato, S., Yoneyama, M., Yamamoto, M., Matsui, K., Uematsu, S., Jung, A., Kawai, T., Ishii, K.J., Yamaguchi, O., Otsu, K., Tsujimura, T., Koh, C.S., Reis e Sousa, C., Matsuura, Y., Fujita, T. and Akira, S. (2006). Differential roles of MDA5 and RIG-I helicases in the recognition of RNA viruses. *Nature* 441, 101-105.

Kawai, T., Takahashi, K., Sato, S., Coban, C., Kumar, H., Kato, H., Ishii, K.J., Takeuchi, O. and Akira, S. (2005). IPS-1, an adaptor triggering RIG-I- and Mda5-mediated type I interferon induction. *Nat Immunol* 6, 981-988.

Kawai, T. and Akira, S. (2006). Innate immune recognition of viral infection. *Nat Immunol* 7, 131-137.

Kawane, K., Fukuyama, H., Yoshida, H., Nagase, H., Ohsawa, Y., Uchiyama, Y., Okada, K., Iida, T. and Nagata, S. (2003). Impaired thymic development in mouse embryos deficient in apoptotic DNA degradation. *Nat Immunol* 4, 138-144.

Kopp, E.B. and Medzhitov, R. (1999). The Toll-receptor family and control of innate immunity. *Curr Opin Immunol* 11, 13-18.

Krug, A., Luker, G.D., Barchet, W., Leib, D.A., Akira, S. and Colonna, M. (2004). Herpes simplex virus type 1 activates murine natural interferon-producing cells through toll-like receptor 9. *Blood* 103, 1433-1437.

Kumar, H., Kawai, T., Kato, H., Sato, S., Takahashi, K., Coban, C., Yamamoto, M., Uematsu, S., Ishii, K.J., Takeuchi, O. and Akira, S. (2006). Essential role of IPS-1 in innate immune responses against RNA viruses. *J Exp Med* 10.1084/jem.20060792.

Kurt-Jones, E.A., Popova, L., Kwinn, L., Haynes, L.M., Jones, L.P., Tripp, R.A., Walsh, E.E., Freeman, M.W., Golenbock, D.T., Anderson, L.J. and Finberg, R.W. (2000). Pattern recognition receptors TLR4 and CD14 mediate response to respiratory syncytial virus. *Nat Immunol* 1, 398-401.

Kwong, A.D., Rao, B.G. and Jeang, K.T. (2005). Viral and cellular RNA helicases as antiviral targets. *Nat Rev Drug Discov* 4, 845-853.

Lamkanfi, M., Kalai, M., Saelens, X., Declercq, W. and Vandenabeele, P. (2004). Caspase-1 activates nuclear factor of the kappa-enhancer in B cells independently of its enzymatic activity. *J Biol Chem* 279, 24785-24793.

Lee, J., Chuang, T.H., Redecke, V., She, L., Pitha, P.M., Carson, D.A., Raz, E. and Cottam, H.B. (2003). Molecular basis for the immunostimulatory activity of guanine nucleoside analogs: activation of Toll-like receptor 7. *Proc Natl Acad Sci U S A* 100, 6646-6651.

Lemaitre, B., Nicolas, E., Michaut, L., Reichhart, J.M. and Hoffmann, J.A. (1996). The dorsoventral regulatory gene cassette spatzle/Toll/cactus controls the potent antifungal response in Drosophila adults. *Cell* 86, 973-983.

Lin, R., Lacoste, J., Nakhaei, P., Sun, Q., Yang, L., Paz, S., Wilkinson, P., Julkunen, I., Vitour, D., Meurs, E. and Hiscott, J. (2006). Dissociation of a MAVS/IPS-1/VISA/Cardif-IKKepsilon molecular complex from the mitochondrial outer membrane by hepatitis C virus NS3-4A proteolytic cleavage. *J Virol* 80, 6072-6083.

Linder, P., Lasko, P.F., Ashburner, M., Leroy, P., Nielsen, P.J., Nishi, K., Schnier, J. and Slonimski, P.P. (1989). Birth of the D-E-A-D box. *Nature* 337, 121-122.

Loo, Y.M., Owen, D.M., Li, K., Erickson, A.K., Johnson, C.L., Fish, P.M., Carney, D.S., Wang, T., Ishida, H., Yoneyama, M., Fujita, T., Saito, T., Lee, W.M., Hagedorn, C.H., Lau, D.T., Weinman, S.A., Lemon, S.M. and Gale, M., Jr. (2006). Viral and therapeutic control of IFN-beta promoter stimulator 1 during hepatitis C virus infection. *Proc Natl Acad Sci U S A* 103, 6001-6006.

Lopez, C.B., Moltedo, B., Alexopoulou, L., Bonifaz, L., Flavell, R.A. and Moran, T.M. (2004). TLR-independent induction of dendritic cell maturation and adaptive immunity by negative-strand RNA viruses. *J Immunol* 173, 6882-6889.

Lu, G., Kang, Y.J., Han, J., Herschman, H.R., Stefani, E. and Wang, Y. (2006). TAB-1 modulates intracellular localization of p38 MAP kinase and downstream signaling. *J Biol Chem* 281, 6087-6095.

Lund, J., Sato, A., Akira, S., Medzhitov, R. and Iwasaki, A. (2003). Toll-like receptor 9-mediated recognition of Herpes simplex virus-2 by plasmacytoid dendritic cells. *J Exp Med* 198, 513-520.

Lund, J.M., Alexopoulou, L., Sato, A., Karow, M., Adams, N.C., Gale, N.W., Iwasaki, A. and Flavell, R.A. (2004). Recognition of single-stranded RNA viruses by Toll-like receptor 7. *Proc Natl Acad Sci U S A* 101, 5598-5603.

Martin, M., Bol, G.F., Eriksson, A., Resch, K. and Brigelius-Flohe, R. (1994). Interleukin-1-induced activation of a protein kinase co-precipitating with the type I interleukin-1 receptor in T cells. *Eur J Immunol* 24, 1566-1571.

Medzhitov, R., Preston-Hurlburt, P. and Janeway, C.A., Jr. (1997). A human homologue of the Drosophila Toll protein signals activation of adaptive immunity. *Nature* 388, 394-397.

Melchjorsen, J., Jensen, S.B., Malmgaard, L., Rasmussen, S.B., Weber, F., Bowie, A.G., Matikainen, S. and Paludan, S.R. (2005). Activation of innate defense against a paramyxovirus is mediated by RIG-I and TLR7 and TLR8 in a cell-type-specific manner. *J Virol* 79, 12944-12951.

Meylan, E., Curran, J., Hofmann, K., Moradpour, D., Binder, M., Bartenschlager, R. and Tschopp, J. (2005). Cardif is an adaptor protein in the RIG-I antiviral pathway and is targeted by hepatitis C virus. *Nature* 437, 1167-1172.

Monick, M.M., Yarovinsky, T.O., Powers, L.S., Butler, N.S., Carter, A.B., Gudmundsson, G. and Hunninghake, G.W. (2003). Respiratory syncytial virus up-regulates TLR4 and sensitizes airway epithelial cells to endotoxin. *J Biol Chem* 278, 53035-53044.

Moynagh, P.N. (2005). TLR signalling and activation of IRFs: revisiting old friends from the NF-kappaB pathway. *Trends Immunol* 26, 469-476.

O'Neill, L.A., Fitzgerald, K.A. and Bowie, A.G. (2003). The Toll-IL-1 receptor adaptor family grows to five members. *Trends Immunol* 24, 286-290.

Okabe, Y., Kawane, K., Akira, S., Taniguchi, T. and Nagata, S. (2005). Toll-like receptor-independent gene induction program activated by mammalian DNA escaped from apoptotic DNA degradation. *J Exp Med* 202, 1333-1339.

Rock, F.L., Hardiman, G., Timans, J.C., Kastelein, R.A. and Bazan, J.F. (1998). A family of human receptors structurally related to Drosophila Toll. *Proc Natl Acad Sci U S A* 95, 588-593.

Rothenfusser, S., Goutagny, N., DiPerna, G., Gong, M., Monks, B.G., Schoenemeyer, A., Yamamoto, M., Akira, S. and Fitzgerald, K.A. (2005). The RNA helicase Lgp2 inhibits TLR-independent sensing of viral replication by retinoic acid-inducible gene-I. *J Immunol* 175, 5260-5268.

Schroder, M. and Bowie, A.G. (2005). TLR3 in antiviral immunity: key player or bystander? *Trends Immunol* 26, 462-468.

Servant, M.J., Grandvaux, N. and Hiscott, J. (2002). Multiple signaling pathways leading to the activation of interferon regulatory factor 3. *Biochem Pharmacol* 64, 985-992.

Seth, R.B., Sun, L., Ea, C.K. and Chen, Z.J. (2005). Identification and characterization of MAVS, a mitochondrial antiviral signaling protein that activates NF-kappaB and IRF 3. *Cell* 122, 669-682.

Soulat, D., Bauch, A., Stockinger, S., Superti-Furga, G. and Decker, T. (2006). Cytoplasmic Listeria monocytogenes stimulates IFN-beta synthesis without requiring the adapter protein MAVS. *FEBS Lett* 580, 2341-2346.

Stetson, D.B. and Medzhitov, R. (2006). Recognition of cytosolic DNA activates an IRF3-dependent innate immune response. *Immunity* 24, 93-103.

Sun, Q., Sun, L., Liu, H.H., Chen, X., Seth, R.B., Forman, J. and Chen, Z.J. (2006). The Specific and Essential Role of MAVS in Antiviral Innate Immune Responses. *Immunity* 24, 633-642.

Takaoka, A., Yanai, H., Kondo, S., Duncan, G., Negishi, H., Mizutani, T., Kano, S., Honda, K., Ohba, Y., Mak, T.W. and Taniguchi, T. (2005). Integral role of IRF-5 in the gene induction programme activated by Toll-like receptors. *Nature* 434, 243-249.

Takeda, K., Kaisho, T. and Akira, S. (2003). Toll-like receptors. *Annu Rev Immunol* 21, 335-376.

Taniguchi, T., Ogasawara, K., Takaoka, A. and Tanaka, N. (2001). IRF family of transcription factors as regulators of host defense. *Annu Rev Immunol* 19, 623-655.

Uematsu, S., Sato, S., Yamamoto, M., Hirotani, T., Kato, H., Takeshita, F., Matsuda, M., Coban, C., Ishii, K.J., Kawai, T., Takeuchi, O. and Akira, S. (2005). Interleukin-1 receptor-associated kinase-1 plays an essential role for Toll-like receptor (TLR)7- and TLR9-mediated interferon-{alpha} induction. *J Exp Med* 201, 915-923.

van der Sluijs, K.F., van Elden, L., Nijhuis, M., Schuurman, R., Florquin, S., Jansen, H.M., Lutter, R. and van der Poll, T. (2003). Toll-like receptor 4 is not involved in host defense against respiratory tract infection with Sendai virus. *Immunol Lett* 89, 201-206.

Viatour, P., Merville, M.P., Bours, V. and Chariot, A. (2005). Phosphorylation of NF-kappaB and IkappaB proteins: implications in cancer and inflammation. *Trends Biochem Sci* 30, 43-52.

Wagner, H. and Bauer, S. (2006). All is not Toll: new pathways in DNA recognition. *J Exp Med* 203, 265-268.

Wang, C., Deng, L., Hong, M., Akkaraju, G.R., Inoue, J. and Chen, Z.J. (2001). TAK1 is a ubiquitin-dependent kinase of MKK and IKK. *Nature* 412, 346-351.

Wang, T., Town, T., Alexopoulou, L., Anderson, J.F., Fikrig, E. and Flavell, R.A. (2004). Toll-like receptor 3 mediates West Nile virus entry into the brain causing lethal encephalitis. *Nat Med* 10, 1366-1373.

Wietek, C., Miggin, S.M., Jefferies, C.A. and O'Neill, L.A. (2003). Interferon regulatory factor-3-mediated activation of the interferon-sensitive response element by Toll-like receptor (TLR) 4 but not TLR3 requires the p65 subunit of NF-kappa. *J Biol Chem* 278, 50923-50931.

Xu, L.G., Wang, Y.Y., Han, K.J., Li, L.Y., Zhai, Z. and Shu, H.B. (2005). VISA is an adapter protein required for virus-triggered IFN-beta signaling. *Mol Cell* 19, 727-740.

Yasuda, K., Yu, P., Kirschning, C.J., Schlatter, B., Schmitz, F., Heit, A., Bauer, S., Hochrein, H. and Wagner, H. (2005). Endosomal translocation of vertebrate DNA activates dendritic cells via TLR9-dependent and -independent pathways. *J Immunol* 174, 6129-6136.

Yoneyama, M., Kikuchi, M., Natsukawa, T., Shinobu, N., Imaizumi, T., Miyagishi, M., Taira, K., Akira, S. and Fujita, T. (2004). The RNA helicase RIG-I has an essential function in double-stranded RNA-induced innate antiviral responses. *Nat Immunol* 5, 730-737.

Zhang, G. and Ghosh, S. (2001). Toll-like receptor-mediated NF-kappaB activation: a phylogenetically conserved paradigm in innate immunity. *J Clin Invest* 107, 13-19.

In: Focus on Signal Transduction Research
Editor: Gavin McAlpine, pp. 189-224
ISBN 1-60021-376-6

Chapter 9

EXTRINSIC REGULATION OF METABOLIC AND ENERGETIC MITOCHONDRIAL FUNCTIONS

V. I. Kulinsky,* and L. S. Kolesnichenko
Irkutsk State Medical University, Irkutsk, Russia

ABSTRACT

The discovery of the complex extrinsic regulation of mitochondrial functions by hormones and signal transduction systems is a new and important achievement of a mitochondriology. A number of hormones of all the chemical classes and with different mechanisms of activity stimulate a great plenty of processes; namely, Krebs cycle, respiratory chain, oxidative phosphorylation, energy dependent syntheses, and other mitochondrial enzymes. The realization of these effects are reproduced by signal transduction systems, including receptors, the second messengers (especially by cAMP and Ca^{2+}), protein and tyrosine kinases, protein and tyrosine phosphatases, anchor proteins, and transcription factors. The receptors of lipophylic hormones and some kinase receptors of protein hormones (growth factors, neurotrophic factors, growth hormone) are revealed in mitochondria. All the main cellular kinases are found in mitochondria. Protein kinases phosphorylate the protein 18 kDa from complex I, cytochrom *c*-oxidase, ATP synthase, cAMP/Ca^{2+} response element binding protein, voltage dependent anione channel, steroidogenic acute regulator, pro- and antiapoptotic proteins etc. Tyrosine kinases phosphorylate a number of mitochondrial proteins, including cytochrom *c*-oxidase and ATP synthase. The pleiotropy of calcium regulation of mitochondrial functions is proved. Mitochondria as a whole realize the integrative function in cellular signaling. The main unsolved problems are noted. The present review is the first one, in which all the problem are considered as a whole and in a development.

Keywords: hormones, receptors, signal transduction systems, mitochondria.

* Tel.: (395)224-0826, E-mail : kulinsky@pp.irkutsk.ru

MAIN ABBREVIATION

ACAP – A-kinase anchor protein,
CcO – cytochrome *c*-oxidase,
CREB – cAMP response element binding protein,
GPCR – G-protein coupled protein,
ICDH – isocitrate dehydrogenase,
IP_3 – inositol 1,4,5-trisphosphate,
KGDH – α-ketoglutarate dehydrogenase,
M_i – inner mitochondrial membrane,
M_o – outer mitochondrial membrane,
OxPhos – oxidative phosphorylation,
PDH – pyruvate dehydrogenase,
PK – protein kinase,
PKA – cAMP-dependent PK,
SDH – succinate dehydrogenase,
StAR – steroidogenic acute regulatory protein,
STS – signal transduction system,
TAC – Krebs tricarboxylic acid cycle,
TK – tyrosine kinase.

INTRODUCTION

Classic mitochondriology is considered as a mature field from 90s. Within the last 15 years important observations have been made showing "an extensive regulatory role for mitochondria in both normal and pathological cell function" [1]. To our mind the complex regulation of mitochondria function by hormones and signal transduction systems (STSs) have been discovered and has been working out at present.

At 60s and early 70s many thought that a mitochondrion is self-sufficient in regulatory relation and does not need in an extrinsic control. The start of the regulation of mitochondrial functions by STSs was the revelation in 1969-1974 in R.C. Haynes laboratory of the short-term regulation of pyruvate metabolism in mitochondria by glucagon and cAMP [2] and P.J. Randle and R.M. Denton – pyruvate dehydrogenase (PDH) activation by Ca^{2+} [3]. In 1975 year Haynes et al. revealed the stimulation of citrate and malate production from pyruvate by glucagon, catecholamines and cAMP [4] and mitochondrial respiration by glucagon [5] and our collective determined the activation by catecholamines and cAMP of NAD-dependent isocitrate dehydrogenase (ICDH) [6] and in a year the stimulation of mitochondrial respiration [7, 8]. In 1976 one from us suggested that the any biologically important process, including biooxidation, necessary regulated not only by metabolites, but hormones and second messengers [9]. These laboratories and A.P. Hallestrap and M. Crompton ones investigated actively the short-term mitochondrial regulation up to middle of 90s, their results have been analyzed in a number of reviews [10-16].

The present review is the first one, in which all the problem is considered in a development and as a whole. Its publication does quite not mean that all the problems in this field have been solved. An purpose of this review is a description and generalization of available results for helping next investigation. Therefore we also point out the unsettled questions. The references available up to 16.05.2006 have been analyzed.

1. Hormonal Regulations of Mitochondrial Metabolic and Energetic Functions

1.1. Short-term Regulation

In this section short-term (seconds-minutes) effects of five hormones: catecholamines (epinephrine and norepinephrine), glucagon, vasopressin, and angiotensin II are considered. They are distinguished chemically: catecholamines are monoamines, the rest are protein-peptide hormones, but they are similar by activity mechanisms. These hormones have also long-term effects, for example, catecholamines and angiotensin II induce a remodulation (structural-functional changes) in the heart and vessels. However for mitochondria such data are absent. More early references are cited in review [10].

Catecholamines in a number of tissues and glucagon in the liver activate key enzymes of Krebs tricarboxylic acid cycle (TAC): ICDH [6, 10, 14, 15, 17], α-ketoglutarate dehydrogenase (KGDH) [14, 15], succinate dehydrogenase (SDH) [18] and citrate synthase [19] *(Figure 1)*. Vasopressin activates ICDH in the liver [14]. PDH is stimulated by glucagon, vasopressin and angiotensin II in hepatocytes and by catecholamines in the heart and white fat [14, 20]. The opposed inhibitory effect of fat acids, released by catecholamines induced lypolysis, is observed in experiments in vivo in the liver and kidney [21]. Catecholamines activate glyceropol-3-phosphate dehydrogenase [22]. Activities of malate dehydrogenase, NADP-dependent ICDH and glutamate dehydrogenase are not changed by these hormones [10, 21]. Catecholamines and glucagon activate in the inner mitochondrial membrane (M_i) energy-unlinked transhydrogenase in the direction NADPH → NAD^+ (but not reverse one) [17, 23]. ICDH, SDH and transhydrogenase activation by catecholamines is also maintained in mitoplasts and submitochondrial particles [17, 18, 21, 23]. The activation of KGDH and glycerol-3-phosphate dehydrogenase is realized by K_M mechanism, PDH and transhydrogenase by V_{max} mechanism and ICDH by both. These data conform the increase by catecholamines and glucagon of the metabolite flow through TAC in the liver [4, 24, 25] and heart [26]. Glucagon increases acetyl-CoA and malate concentration and decreases isocitrate, α-ketoglutarate and glutamate (cit. by [10]), that is a crossover point is between citrate and isocitrate – ketoglutarate. All four hormones stimulate glutaminase activity [27, 28]. Glucagon raises the oxidative glycine metabolism [14, 29]. Catecholamines increase NADH in mitochondria and in parallel decrease it in cytosol [22] and glucagon increases in both compartments [30].

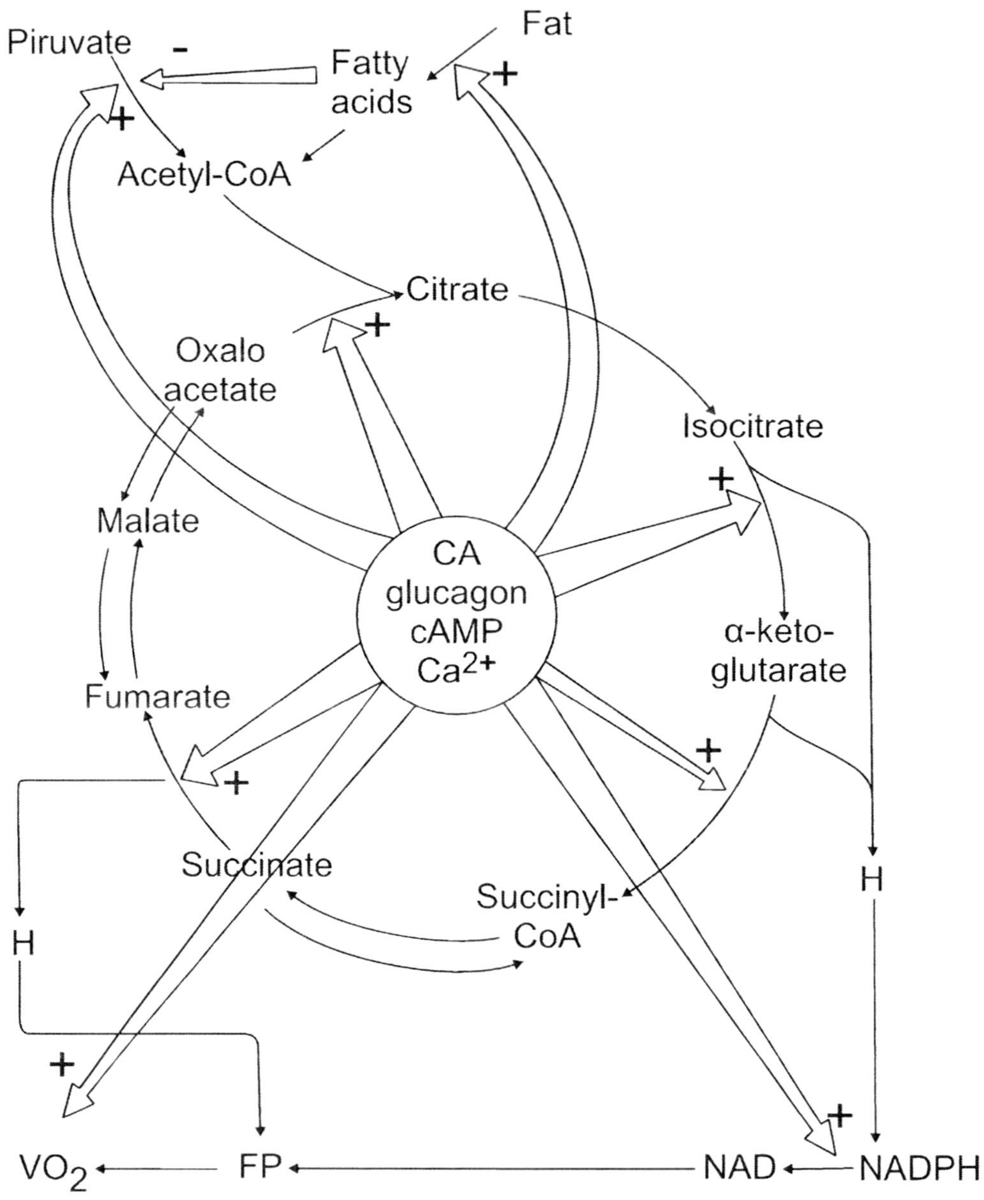

Figure 1. Short-term regulation of Krebs cycle. FP – flavoprotein, VO2 – O2 consumption. + - increase, – - decrease.

Glucagon [5, 14-16, 24], catecholamines [7, 8, 14, 21, 31-33], vasopressin and angiotensin II [15, 32, 33] raise respiration of cells and mitochondria in state 3 and uncoupled state at using a number of NAD-dependent substrates of complex I, substrates of complex II (succinate, acyl carnitine and octanoate) and duroquinol, oxidized in complex III. Effect as a rule is absent with substrates of complex IV. The meaning of respiratory control was more often risen. Glucagon, catecholamine and vasopressin increase steadily the reduction of cytochromes *b*, c_1 and *c*, some flavoproteins, especially an electron-transferring protein, participating in transfer of atom H from acyl-CoA to Q [16]. These and other data show an activation of respiratory chain [16]. It is evident that the plural activation of pathways reducing complexes I-III, but not IV (*Figure 2)*.

Glucagon and catecholamines increase pH gradient on M_i [12, 13, 24], ATP concentration and ratio ATP/ADP in mitochondria [11, 14, 15], the latter is shown for vasopressin [11]. These effects are retained even at the augmentation of citrulline synthesis and pyruvate carboxylation in the liver and at intense work in the heart [15]. Catecholamines stimulate ATP synthesis [34]. Glucagon and catecholamines increase ATPase activity of ATP synthase [11]. Though direct data on ATP synthase are absent there are facts testifing in favor of oxidative phosphorylation (OxPhos) stimulation. Glucagon, catecholamines and vasopressin increase the mitochondrial matrix volume [16], M_i stability [16, 23] and decrease adenine nucleotide leakage [16].

Catecholamines, glucagon, vasopressin and angiotensin II augment the consumption of converted energy in liver mitochondria: K^+, Mg^{2+}, P_i and pyrophosphate accumulation [11, 16], pyruvate carboxylation and gluconeogenesis [11, 15, 31, 32], ketogenesis [32, 35], citrulline synthesis and ureogenesis [11, 27, 32] are risen; glucagons activate mitochondrial β-hydroxy-β-methylglutaryl-CoA synthase [36].

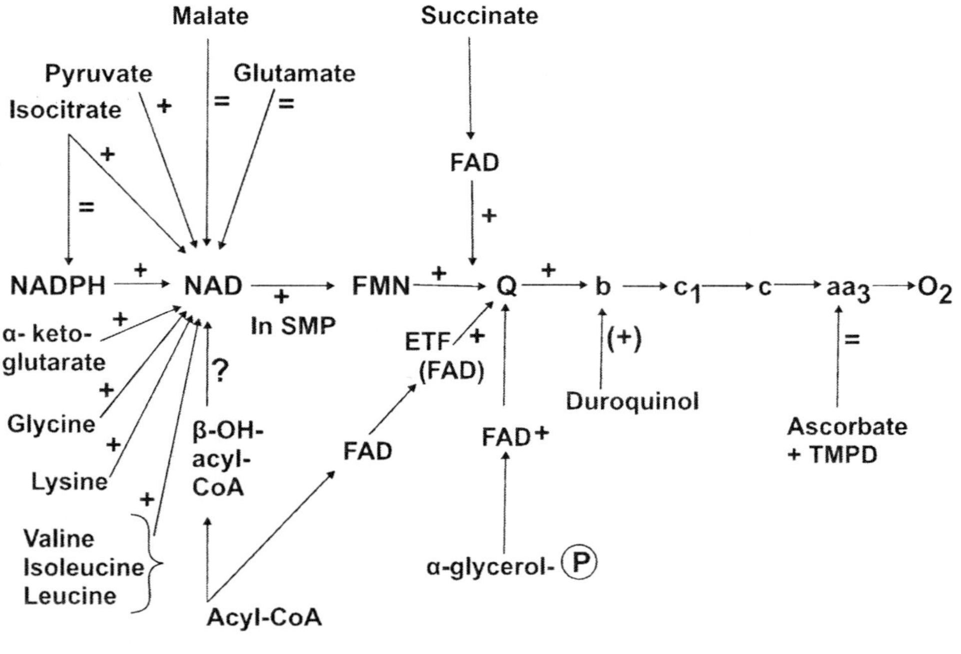

Figure 2. Short-term regulation of respiratory chain. SMP – submitochondrial particles, ETF – electron transferring flavoprotein TMPD – tetramethyl-p-fenylendiamine, + - - increase, = - absence of changes.

The speed of development and the pleiotropy of hormones action are evident: described changes are realized for seconds-minutes and their total quantity reaches sixty [10]. The interconnection of metabolic and energetic hormonal effects and their interpretation are shown on *Figure 3*. Mainly key and regulatory enzymes catalyzing nonreversible reactions are activated: PDH and three dehydrogenases of TAC – ICDH, KGDH, SDH (SDH reaction is reversible in vitro, but practically not reversible in vivo). This is the cause both of the augmentation of general cycle activity with rising metabolite flux and changes of substrate concentration. A crossover point corresponds with the cycle part where hormone activated enzymes (ICDH, KGDH, SDH) are found. The activation of both these enzymes and PDH, β-oxidation (in vivo catecholamine induced lipolysis promotes this change), amino acids catabolism (and in a stress condition a transhydrogenase stimulation) leads to NAD^+ reduction. The stimulation of β-oxidation and PDH induces acetyl-CoA accumulation, that together with the activation of β-hydroxy-β-methylglutaryl-CoA synthase increase ketogenesis. Glutaminase stimulation promotes citrulline synthesis.

The second block of hormone effects is connected with the activation of complexes I-III (but not IV), that augments mitochondrial respiration with substrates of three complexes in state 3 and uncoupled one. It is not connected with catecholamine [8] and glucagon action [12, 16] on other metabolic processes (glycogenolysis, lipolysis). The stimulation of respiratory chain can accelerate substrate oxidation, including substrates, accumulation of which are not increased by hormones. The appearing stimulation of mitochondrial respiration is steady in difference from dehydrogenase activation [15]. The stimulation of respiratory chain augments an output of converted energy forms – $\Delta\mu_H$ and then ATP, "redox flow pushes OxPhos" [14]. $\Delta\mu_H$ augmentation promotes ion accumulation in mitochondria and ATP production favors to mitochondrial syntheses as a result of their key enzymes activation. The mechanism of increasing matrix volume by hormones [16] remains unexplained.

Main effects of hormones on mitochondria are reproduced by mobilization of endogenous catecholamines in stress [37, 38], including cooling [39], and increase of glucagons secretion by high-protein diet [28] and mannoheptulosa introduction [40]. This proves that hormone effects are of physiological, but not pharmacological type.

Thus the stimulation of oxidative processes by hormones of short-term action is not the secondary but is a result of influence on mitochondria. It is proved also by identity of main effects at a simplification of studied system – from a hormone introduction in the body to an addition to slices, cells and homogenates.

Norepinephrine in mitochondria of the rodent brown fat tissue greatly (up to 10 times) activates an expression of unique *unc*oupled *p*rotein UCP-1 (earlier thermogenin) – "inner body heater". Fat acids at an interaction with UCP-1 stimulate H^+ leakage and thermogenesis, the latter decreases 30 times at knock out of gene UCP-1. In newborn children but not in adults UCP-1 is active and necessary [41]. Now proteins $UCP_{2\text{-}5}$ are known; UCP-2 is in many organs, UCP_3 is in the muscles; they do not increase heat production and their physiological function is vague, but they are supposed to induce mild uncoupling [41, 42].

In ischemia-reperfusion of the heart a melatonin has strong protective effect on mitochondrial complexes I and III and cardiolipin [43].

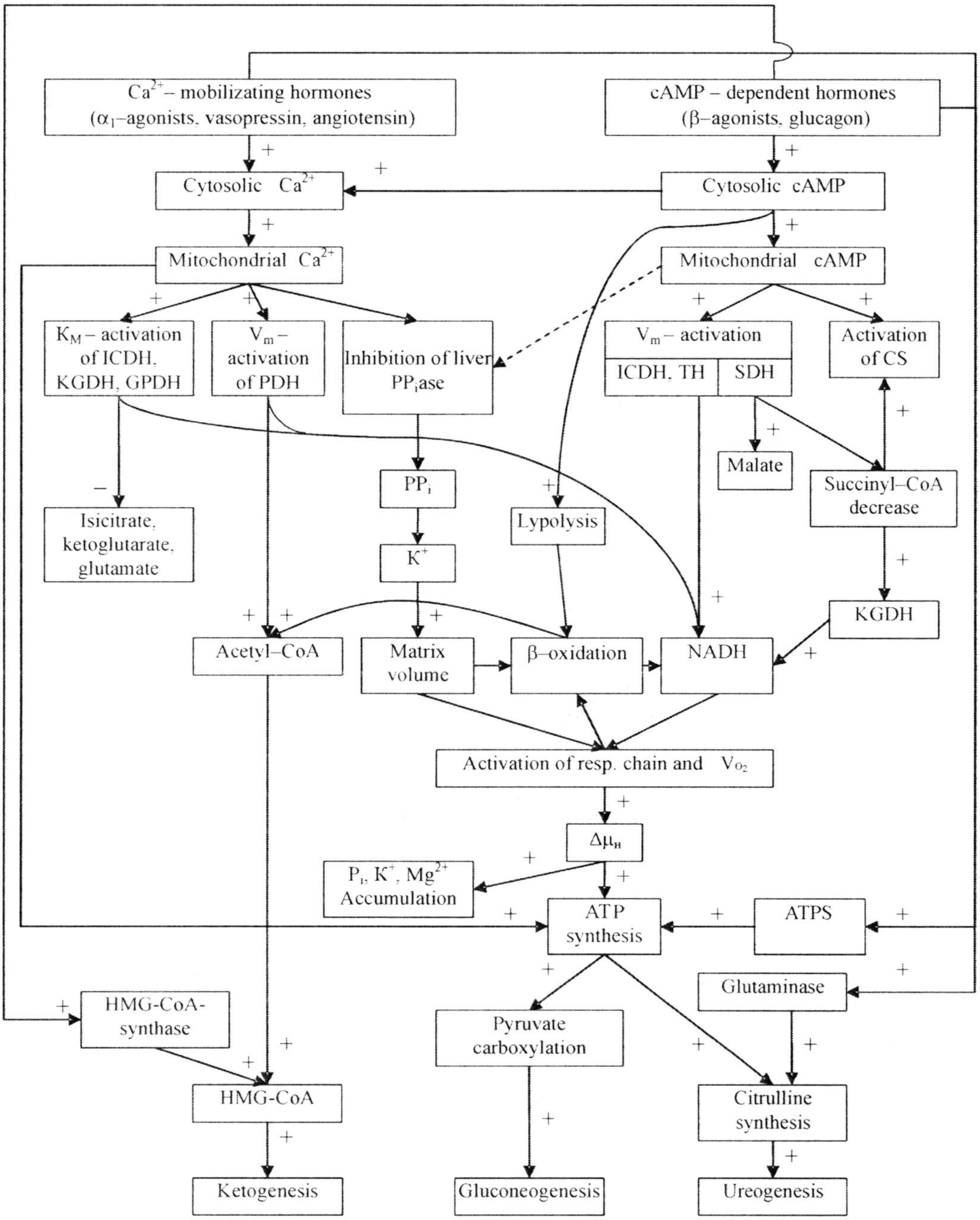

Figure 3. Interconnections of metabolic effects of hormones on mitochondria functions. ICDH – NAD+-isocitrate dehydrogenase , KGDH – α-ketoglutarate dehydrogenase, GPDH – glycerophosphate dehydrogenase, PDH - pyruvate dehydrogenase, PPi-ase – pyrophosphatase, TH – transhydrogenase, SDH – succinate dehydrogenase, CS – citrate syntase, PPi – pyrophosphate, resp. – respiratory, VO2 – O2 consumption, ATPS – ATP-syntase, HMG CoA – β-hydroxy-β-methylglutaryl-CoA, + - increase, – - decrease.

1.2. Mainly Long-term Regulation

In this section we consider hormones possessing both long-term (hours-days) and short-term effects: mainly hydrophobic molecules (iodothyronines and steroids) and proteins (insulin, leptin etc.). They are different chemically but activity mechanisms similar inside both pairs but differ between them.

Iodothyronines belong to the most important regulators of mitochondrial functions. Three stages are distinguished: 1) the revelation of uncoupling respiration and OxPhos in 60s, 2) the discovery of a transcription stimulation by triiodothyronine (60s-80s), 3) a revelation of short-term mitochondrial effects (from middle of 80s) [44]. They are apparent already in 2-30 minutes. Short-term effects are the activation of *c*ytochrome *c*-*o*xidase (C*c*O) and adenine nucleotide translocase, the increase of $\Delta\mu_H$, mitochondrial respiration, ATP synthesis [45-47], and the stimulation of fat acids oxidation [48]. Long-term effects realizing in hours are the stimulation of mitochondrial respiration, oxidative stress with H^+ leakage and energy dissipation, provoking OxPhos uncoupling, the induction of mitochondrial glycerol-3-phosphate dehydrogenase [49-52], cytochromes *c* and aa_3 [53], gluconeogenesis [54] and ketogenesis [55]; the augmentation of UCP-1 expression in brown fat tissue and UCP-3 in skeletal muscles [56-58], thermogenesis [44], biosynthesis and stability of mRNA, synthesis of proteins and phospholipids [45-47], mitochondriogenesis [59]; also normal embryogenesis, growth, development of the brain and intellect [60], increase of cell proliferation and differentiation [45-47, 61, 62]. These effects are realized in the level of transcription. Hypothyroidism decreases in M_i an activity and expression of tricarboxylate carrier that is need for synthesis of fatty acids and cholesterol [63]; contributes in inactivation of mitochondrial complex I by translocation of neuronal nitric oxide synthase [64]. Short- and long-term effects of iodothyronines are astonishingly synergic and coordinated [47] and they function in synergy with catecholamines [51, 65]. Jointly these hormones increase thrice UCP-1 level in brown fat tissue; provoked by a cold noradrenaline release induces (10-50 times) short-living isoenzyme of deiodinase D_2 ($T_{0.5}$ = 1 hour), converting prohormone thyroxine to the more active and true hormone triiodothyronine [66].

Steroid hormones are the only whose synthesis is impossible without mitochondria participation. The letter ensures cholesterol transport to limiting enzyme of synthesis and its fuctioning. In all the steroidogenic cells protein hormones (tropic ones, angiotensin II, insulin-like growth factor I, interleukin-1α) induce an expression of the *st*eroidogenic *a*cute *r*egulatory protein (StAR), that is activated by cAMP and protein kinase A (PKA) and proteolysis from proprotein 37 kDa to active regulator 30 kDa [67-72]. With a help of disposed in an outer mitochondrial membrane (M_o) benzodiazepine receptor of peripheral type (PBR) StAR is importing in mitochondria [73-75] (these stages demand 3-4 hours [75]). In these processes the tyrosine phosphatase and the release of arachidonic acid take an important part [72]. StAR at an active mitochondrial respiration, intact $\Delta\psi$, ΔpH, ATP synthesis, polarization and positive regulation by *e*xtracellular signal *r*egulated *k*inase (ERK) [70, 71] quickly (within minutes) transfers cholesterol from M_o to M_i. where specific limiting enzyme CYP11 (earlier cytochrome P450scc) cleavages side chain of cholesterol and converts it in pregnenolone [76]. This mechanism is also used in neurosteroid synthesis in glia. Further synthesis of specific steroid hormones is continued in reticulum [77]. Totally 30 genes participate in steroid hormones synthesis.

The influence of *glucocorticoids* on mitochondria function is still vague. Cortisol introduction during 6 days decreases mitochondrial respiration and ratio P/O in the liver, uncouples OxPhos and lessens mitochondria number, however in consequence of their size augmentation an mitochondrial volume does not change (cit. by [78]). Dexamethasone administration during 5 days decreases mitochondria respiration in state 3 (but not state 4) and respiratory chain activity, H^+ permeability rises, but authors think that it is compensated by better control of OxPhos [79]. The acute and chronic corticosterone introduction in developing rats also diminishes the cerebral mitochondrial respiration in state 3 with different substrates, ratio ADP/O, OxPhos rate and induces energetic deficiency [80]. Glucocorticoids decreases C*c*O activity in the kidney (cit. by [78]). It is reported that dexamethasone increases the hepatic mitochondria respiration in state 3 and uncoupled state [81]. In all the body, including human, energy expenditure grows [78, 79]. It is difficult to connect these vague data with the fact that glucocorticoids are stress hormones, for whose action the necessity of an energy (in the most cases in ATP, and in a cooling in a heat) increases. At the same time there is a number of publications about an augmentation by glucocorticoids in the nucleus and mitochondria of a transcription of genes which code subunits of complexes I, II, IV and ATP synthase, accessibility and stability of mRNA transcripts and at a high necessity of energy also mitochondriogenesis (see review [78]) . An addition of dexamethasone to human skeletal muscle from 501 investigated genes connected with mitochondria has most activates the gene of monoamine oxidase A [82]. The dissociation of more frequent decrease of mitochondrial respiration and the activation of template synthesis is a challenge to a further investigation. Glucocorticoids introduction even in 9 minutes [11] augments gluconeogenesis by the way of the increase of a flux through pyruvate carboxylase but not through TAC [11, 83].

Estrogenes have big neuroprotective effect [84]. They increase OxPhos and lessen ATPase activity in cerebral neurons, and promote neuroprotection; authors suppose that mitochondria are a therapeutic target of estrogenes [85]. Mitochondria play the central role in protection by estrogenes of neurons [86] and other cells [87], in decrease of membrane peroxidation. Vasoprotective activity of estrogenes is known [88]. It is connected with a modulation of mitochondrial functions in vessels of the brain, an increase in specific proteins level, which coded both by nuclear genome (estrogen receptor α, cytochrome *c,* subunit IV of complex IV, Mn-superoxide dismutase) and mitochondrial one (subunit I of complex IV), an activation of enzymes (citrate synthase, complex IV) limiting energy production and a decrease of reactive oxygen species (ROS) [89]. In hepatocytes these hormones stimulate the formation of DNA transcripts in mitochondria [90]. The cell proliferation induced by estrogenes can be explained by the augmentation of OxPhos and transcription of mitochondrial genes with the following signal transduction in a nucleus with helping ROS [91]. However the inhibition of ATP synthase by physiological estrogen concentrations has also been described [92]. These interesting but diverse data for the present do not permit to concretize clearly a conclusion on control of mitochondrial functions by estrogenes.

Not rarely different steroid hormones induce short-term (minutes) effects, nonsensitive to inhibitors of transcription and translation [78, 93].

The activation by *insulin* of PDH in aerobic carbohydrate metabolism and acetyl-CoA carboxylase in fat acids synthesis was known long ago. Insulin influence over TAC has been studed since 80s. S.P. Bessman et al. on tissue pieces, hepatocytes and homogenates with help of marked ^{14}C substrates was established that insulin stimulate oxidation of C_2 and C_3 atoms of pyruvate and succinate and both C of acetate to CO_2. An augmentation of aliphatic but not carboxyl C atoms shows that: a) only mitochondrial enzymes of TAC are included but not their cytosolic analogs, b) these substrate must cross the whole cycle more than one time. Only these $C_{2,3}$ atoms include in glutamate residues of proteins and in lipids [94]. These elegant works proved that insulin activates TAC indeed. The second big step was made in the laboratory of P. Fernyhough. In sensory neurons the physiological concentration of insulin (1-10 nM) and neurotrophic factors (insulin-like growth factor I, nerve growth factor, and neurotrophin-3, 10-50 ng/ml) do not influence on short-term (20 min), but on long-term action (6-24 hours) in increase the membrane mitochondrial potential 1.5-2 time, ATP concentration (+30-40%), NAD(P) reduction, hexokinase activity and in parallel the neurite growth [95]. Intravenous insulin infusion (8 hours) augments in mitochondria of human skeletal muscle ATP synthesis, NADH dehydrogenase mRNA (subunit IV), mitochondrial protein synthesis, and activities of C*c*O and citrate synthase [96]. The article commentator underlines an insulin role as important regulator of OxPhos in muscle [97]. Insulin stimulates OxPhos enzyme, raises and accelerates expression of subunits I and IV of C*c*O and increases myogenesis [98]. It is revealed the acute insulin activation of $\Delta\mu_H$ in mitochondria of pancreatic β cell [99]. In diabetes mellitus type II ATP synthesis does not increased [96] and what is more a gene set activity involving in OxPhos is decreased [100]. In streptozocin diabetes insulin hyper-stimulates glutamate dehydrogenase and succinate reductase activities [101]. Usually in diabetes mellitus and obesity high-fat diet diminishes mitochondrial mass and functions, inhibits OxPhos genes and mitochondriogenesis [102]. Thus it is getting clear that insulin is not only anabolic hormone, that it stimulates also the main catabolic cycle, OxPhos and as a result provides itself by an necessary energy for synthesis of proteins and lipids. In its turn insulin in β cells generates and integrates metabolic signal, controlling insulin secretion [103]. Mitochondria are implicated in insulin secretion and action [104].

Leptin stimulates directly thermogenesis in muscle [105] and its concentration in blood plasma in food restriction or a glucocorticoids introduction correlates with hepatic mitochondrial respiration [79]. In endothelium cells leptin activates carnitine palmitoyltransferase-1 and fat acids oxidation and inhibits acetyl-CoA carboxylase [106].

However in innate absence of leptin in muscle the activities of enzymes of β-oxidation, citrate synthase, C*c*O, transcription factors of mitochondriogenesis and mRNA UCP-3 do not changed [56, 107]; in the liver H^+ leakage increases but additional leptin introduction normalizes it [108]. At the same genome UCP-1 and 3 decreases in the brown fat tissue and on this background leptin increases these mRNA [56, 109]. At leptin overexpression the white fat tissue approximates to brown one by the augmentation of mitochondria quantity, contents of C*c*O and UCPs [110]. Further investigations are evidently necessary as the picture received is rather complicated and vague.

Growth hormone can stimulate O_2 consumption by cells CHO and mitochondria at transfection of cDNA coding hormone receptors [111].

Tumor necrosis factor-α (TNF-α) in intravenous injection stimulates considerably an expression of UCP-3 and especially UCP-2 in skeletal muscle that conform to a termogenesis increase in the brown fat tissue [112]. TNF-α and interleukin 1β decrease complex I activity and ATP production, induce a mitochondrial depolarization in chondrocyte cells [113]. IL-1β increases $Ca^{2+}{}_i$ in epithelial cells [114].

Thus iodothyronines and insulin concern to the number of main regulators of mitochondrial functions. They possess plural and various effects. An action of other hormones on mitochondria is revealed but demands further intensive investigations. It is principally important that an energy release is typical not only for catabolic hormones but also for insulin. An energy is necessary to all the hormones, but for different purposes: catabolic hormones expend it for a decision of external problems (work, stress), anabolic ones for synthesis and a housekeeping as a whole. Therefore both that and other hormones need to activate energetics.

One from interesting and non-explaining differences between short-term and mainly long-term acting hormones is an absence of regulation complex IV in the first group and a presence such regulation in second group.

2. Realization of Hormone Effects by Signal Transduction Systems

Mitochondria are incapable to synthesize completely any hormon they is always formed (steroid hormones are completed) extramitochondrially. It is possible a rudiment of the remote past of mitochondria as free prokaryotes, not needing in hormones. For regulation of mitochondrial functions a hormonal signal owes to penetrate through proteolipid membranes both into a cell and mitochondrion.

Three variants are possible: 1) receptor is on plasmatic membrane of the cell; 2) receptor localizes in the nucleus or penetrates in it; 3) a resident receptor is present in mitochondrion and waits the hormone penetration. They majority of hormones is hydrophilic or even charged (especially amines of catecholamine type), therefore is needed in special mechanisms. Since proteins are main workers in the cell, they carry out also this task. It is evident a necessity of the special STSs, the first component of that are receptors of plasmatic membrane. In this case "personal presence" of hormone inside a cell is not necessary: not hormone but its signal penetrates. The second variant is typical for lipophylic hormones, but they can use specific protein transporters [115] for a penetration into the cell and nucleus (probably in mitochondria too). The third variant is an important discovery of recent years – mitochondrial receptors for lipophylic and protein hormones (see below).

2.1. Functions and Importance of Receptors

Molecular receptors are specific proteins transferring any extrinsic signals inside the cell. All hormonal receptors are proteins. They distinguished 5 types and 8 subtype of receptors; hormones, circumscribed in section 1, operate through 4 types and 5 subtypes [115]. *R*eceptors *c*oupled with *G*TP-activated *p*roteins (GPCR) realize effects of the first group of hormones (section 1.1). Glucagon has only one type of receptors, vasopressin acts on the liver through V_1-receptor, angiotensin II – through AT_1-receptor. Catecholamines stimulate the cell through $\beta_{1\text{-}3}$ and α_1-receptors [10, 37, 38, 115]. Hormone-receptor complex transfers their signals to G_s (*s*timulated *G*-protein) → adenylyl cyclase (catecholamines through β-receptors; glucagon) or G_q-protein → phospholipase C (catecholamine through α_1-receptor; angiotensin II, vasopressin). It is necessary to take into account that there are 30 G_s-dependent hormones (45 GPCR) and 38 G_q-dependent hormones (59 GPCR), but an influence of majority on mitochondria is not studied. The activation by catecholamine both of mitochondrial enzymes

(ICDH, SDH, transhydrogenase) and substrate oxidation and mitochondria respiration is realized through β_1-adrenoceptors [17, 18, 21, 22]. In the whole body catecholamines possess not only activating, but also more weak inhibiting effect – decrease of O_2 consumption through α_2-adrenoceptors [116], that transfer an inhibiting signal to G_i-protein → adenylyl cyclase.

Other group of hormones acts through *t*yrosine *k*inase (TK): membrane receptor TKs (insulin, growth factors), non-receptor (cytosolic) TK (leptin, growth hormone), and non-receptor PK (tumor necrosis factor-α) [117]. It has been recently revealed the penetration of growth hormone into mitochondria (by the caveola pathway) and the activation of its mitochondrial receptors are realizing the stimulation of mitochondrial respiration [111]. There are neurothrophin receptors in mitochondria (see section 2.2.1). In all these cases hormone or its signal penetrates in the cell. Next stages of signal transduction will be consider further.

Lipophylic hormones act through intracellular receptors that bind hormons in cytosol (steroids) or nucleus (iodothyronines, retinoate, calcitriol). The formed hormone receptor complex itself is a transcription factor that stimulates directly the expression of mitochondrial or nuclear genome. Nuclear receptors of estrogens (ERα and β) and iodothyronines (TRα and β) realize their classic genomic effects. The possible localization of their receptors in other parts of cell have been not known of long duration. However for recent years the presence of fast (during 2-30 min) effects of lipophylic hormones has been determined, this cannot by explained by action to nucleus. The inert genomic mechanism, including an expression of genes with the following import of formed proteins in mitochondria, is not able to act during so short time. In addition inhibitors of transcription and translation do not prevent these effects [45, 117, 118]. In 1986-2001 years C. Wrutniak-Cabelo et al. discovered the truncated isoforms of TR in M_i (p28) and mitochondrial matrix (p43), coded by the same gene *c-erb Aa* and realizing the stimulation of mitochondrial genome by triiodothyronine; the notion about direct pathway of regulation in mitochondria is argued (in nucleus these receptors are absent) and effective coordination transcription of mitochondrial genome (rapid responses) and nuclear genes, coding mitochondrial proteins (mitochondriogenesis), were elucidated [45]. The presence of truncated isoforms of TRα and β is confirmed in mitochondria of Xenopus and rat and showed that they during 10 minutes through inositol 1,4,5-trisphosphate (IP_3) provokes Ca^{2+} wave in the cell and increases O_2 consumption [119]. The rapid not genomic effects are also described for exogenous diiodothyronine [44, 47].

In human cells of hepatoma and osteosarcoma glucocorticoid receptors GRβ and estrogen receptors ERα are revealed in the nucleus, mainly in a nucleolus, and GRα and ERβ – in mitochondria [120, 121]. In cells of prostate cancer the androgen receptor AR is localized in nucleus and mitochondria and ERβ and AR in mitochondria of human sperm; it is supposed a role of these receptors for physiology of sperm cells [122]. High affinity ER is present and functions in mitochondria of breast cancer and endothelium cells [87]. Negative data for presence of ERβ in the liver has also been described [123]. Glucocorticoid receptors (GR) are revealed in mitochodria of the liver and brain, they disappear after adrenalectomy but are induced by dexamethasone [74]. GR translocating in mitochondria occurs in glucocorticoid sensitive but not insensitive T-cells [124]. It is remained vague why glucocorticoids do not induce regular stimulation O_2 consumption (see section 1.2). In plasmatic membrane and mitochondria the truncated variant of ER46 (nuclear receptor has M = 66 kDa) was revealed [125]. P.Thomas et al. discovered GPCRs of progesterone [126] and estrogens (early receptor-orphan GPR30) [127]. Estrogens through GPCR activate adenylyl cyclase and increase a release of heparin-bounded epidermal growth factor (EGF), that explains a presence of EGF-like effects at estrogens [128]. Membrane estrogen GPCRs are also described [84]. A proof of bounding lipophylic hormones not only with nuclear receptors, but with truncated mitochondrial ones and GPCRs is an important break-through. Two latter forms, especially GPCRs, secure a realization of short-term (minutes) effects of these hormones not blocked by inhibitors of transcription and translation. This is important superiority of GPCRs especially for mobile and energetic animals.

2.2. Signal-transductor Systems of Mitochondria

Signal-transductor systems (STSs) are sequences of proteins and sometimes small molecules that perceive, transduct, sharply strengthen and transfer a signal to all the parts of the cell [115, 117]. Basic components of STS are receptors, G-proteins, effector enzymes, second messengers, ion channels, PKs and TKs, anchor proteins. Signal transfer to a template is realized by transcription and translation factors.

2.2.1. cAMP, Protein Kinases, Tyrosine Kinases

cAMP – protein kinase A (PKA) system transfers a hormonal signal by the chain GPCR → G_s or G_i protein → adenylyl cyclase of plasmatic membrane → cAMP → PKA → effector protein. The membrane adenylyl cyclase (120 kDa) is absent in mitochondria, but the truncated (50 kDa) HCO_3-activated "solving" enzyme has been revealed [129]. Its functional importance is vague. An inhibition of system is realized by phosphodiesterases, inhibiting cAMP, and protein phosphatases, hydrolyzing phosphorylated proteins. In mitochondria there is cAMP – 0.30 pmol/mg protein [130] that corresponding to 0.3 μM, it is approximately 3 times lower than in cytosol. We revealed that at incubation with hepatic mitochondria during 10 min at 30°C [8-^{3}H]cAMP or [U-^{14}C-adenine]cAMP in a close correlation one with another penetrate into all the mitochondrial subcompartments [131]. The specific activity of fractions of matrix and M_o with an intermembranous space is 2.7 times higher than in M_i; 2/3 of radioactive mark is in mitoplasts. A considerable part of marked cAMP is intact one [132]. [2.8-^{3}H]cAMP specifically and reversibly is bounded by mitochondria with 50% saturation at 0.3-0.6 μM [132]. Consequently cAMP penetrates into mitochondria and can regulate their functions.

Catecholamines [7, 8, 23] and glucagon [11] do not act to isolated mitochondria, but glucagon and agonists of β_1 adrenoceptors (not α_1-agonsts,vasopressin and angiotensin II) through their receptors and G_s proteins increase cAMP concentration in cytosol. It is naturally to expect that cytosolic cAMP penetrates into mitochondria just as extramitochondrial one and realizes effects of hormones. Epinephrine administration really increases in 2.5 times cAMP concentration in hepatic mitochondria [133]. The incubation of physiological cAMP concentration (optimum is 1 μM) with homogenates, mitochondria and mitoplasts from different organs activates by V_{max} mechanism NAD-dependent ICDH) [6, 17, 23, 130], SDH [18, 21] and transhydrogenase [17, 23] (Figure 1). An addition of cAMP to matrix or lysate of mitochondria does not activate ICDH; SDH and transhydrogenase activities rise also in the cAMP incubation with M_i or submitochondrial particles, but not with triton lysates of mitochondria and not after trypsin treatment of mitochondria (at restriction its action by extrinsic side of mitochondria) [21, 23, 134]. It means that M_i integrity is necessary. The latter is also needed for glucagon regulation of glutaminase activity [135]. An activity of malate dehydrogenase and extramitochondrial NADP-specific ICDH does not change. cAMP incubation with hepatocytes activates SDH [24]. cAMP stimulates the formation of citrate and malate from pyruvate [4]. Administration of cAMP or dibutyryl cAMP stimulates O_2 consumption by mitochondria [136] and the whole body and decreases O_2 tension in the liver and spleen that testifies the intensified O_2 consumption [116]. cAMP and dibutyryl cAMP increase O_2 consumption in the liver perfusion [21] and incubation with cells [11, 137], homogenates and mitochondria [7, 8, 21], activate mitochondrial ATPase [11], stimulate accumulation in mitochondria of K^+, Mg^{2+} and P_i [11, 16], carboxylation of pyruvate and gluconeogenesis [11, 137], ketogenesis [35], citrulline synthesis and ureogenesis [136]. Recently increasing mitochondria respiration and $\Delta\psi$ by cAMP has been confirmed on the cells [71]. The accumulation of cAMP in hepatocytes after incubation with glucagon [11] and in the liver after epinephrine administration [116] precedes to changes of mitochondrial functions.

Thus cAMP-dependent hormones do not act directly on mitochondria but increase cAMP concentration not only in cytosol, but also in mitochondria. These effects are specific for cAMP [4, 10, 17, 21]. Extramitochondrial cAMP penetrates in all the mitochondrial subcompartments. In the incubation with homogenates and mitochondria cAMP reproduces all the basic effects of catecholamines and glucagon: an activation of key enzymes of TAC, respiration of mitochondria, mitochondrial ATPase and a stimulation of energy dependent processes (see figure 1-3). The complex of different and independent facts leads to a conclusion that regulation by these hormones of mitochondrial functions is realized by cAMP [10, 21]. Leptin activates PKA and its metabolic effects (see section 1.2) are prevented by cAMP antagonists and PKA inhibitors [106].

PKA in mitochondria was revealed at 1968 year [138], confirmed at 1976-79 [139, 140]. It is shown in different parts of mitochondrion [10,140]. However individual substrates were not cleared up. Homogenous NAD-ICDH and also partly purified SDH are phosphorylated neither by PKA nor its catalytic subunit [10, 141]. It is supposed some other mitochondrial protein realizes of the activation of these enzymes by cAMP and PKA. There are data in favor of hypothesis about an importance of structural integrity of mitochondria and also possible action of cAMP through receptor protein M_i [10, 134], as it has been shown on yeast [142]. Quite probably that it is not the real receptor, but an anchor protein, though for the present the latter are described only in M_o.

Further development of the problem at 1993-2002 was realized by S. Papa et al. [143, 144]. In M_i and mitochondrial matrix from the bovine heart they revealed both catalytic and regulatory subunits of PKA II , 4-5 its protein substrates in M_i , including protein 18 kDa (nuclear gene NDUFS4), possessing by the canonical consensus site of phosphorylation and a leader sequence. Protein 18 kDa was identified as subunit of NADH:Q oxireductase (complex I) of respiratory chain. The serine phosphorylation of this protein increases considerably the general respiratory activity of complex I with NAD-dependent substrates and gene mutation induces early death. Ca^{2+} inhibited protein phosphatase PP2Cγ in M_i dephosphorylated this subunit. All main results were duplicated on cell cultures [143, 144]. Later Papa et al. have shown that NDUFS4 gene mutation in human fibroblasts prevents complete assembly of the complex I and causes full suppression of the activity. NDUFS1 gene mutation also results in disfunctions of oxidative metabolism that is partly decreased by dibutyryl-cAMP [145]. Phosphorylation of mitochondrial protein 18 kDa on the same object was confirmed in the two articles. In the first with an addition, that phosphorylation decreases superoxide formation [146]. In the second article this subunit was considered as ESSS with unknown function; subunit MWFE 10 kDa is necessary for assembling into complex I of seven subunits coded by mitochondrial genes, are also phosphorylated [147]. The considerable activation of respiratory chain by cAMP – PKA can promote to augment an oxidation of substrates of TAC, O_2 consumption and other above described phenomena. The importance of M_i structural integrity is clear as without it the respiratory chain functioning is impossible.

PKA phosphorylates also voltage dependent anionic channel (VDAC) of M_o, purified from hepatic rat mitochondria [148]. After incubation of hepatocytes with $^{32}P_i$ and glucagon the 20 kDa protein close to cytokeratin is also phosphorylated [149]. The cAMP-dependent phosphorylation of firmly bounded with M_i polypeptide 3.5 kDa inhibited by carboxyatractyloside and inhibitors of OxPhos occur in hepatic rat mitochondria [150]. PKA phhosphorylates and inhibits proapoptotic protein BAD and activates antiapoptotic protein Bcl-2 (see in [151]). When summarising the cAMP dependent phosphorylation of 7-8 protein-peptide substrates from M_i were described in articles cited, but their functional importance is proved for protein 18 kD, and three proteins in M_o: StAR necessary for cholesterol transport and Bcl-2 (activation), and BAD (inhibition).

Mitochondria functions are not limited themselves by substrate oxidation, energy production and above described energy dependent syntheses. Another important block is regulation of limiting reactions in complicated processes by system cAMP – PKA. To them besides steroid hormones synthesis (section 1.2) the stimulation of mitochondrial enzyme transcription is concerned, namely, carnitine palmitoyltransferase [106, 152], β-hydroxy-β-methylglutaryl-CoA synthase [153], 5-aminolevulinate synthase [154] and also variants of cytochrom P-450 [155]. This system activates also other enzymes rendering harmless – cytosolic and mitochondrial glutathione transferases [156, 157]. PKA controls the mitochondrial transport and organization, transition from mobile to fixed mitochondria, and interaction of mitochondria and cytoskeleton [158].

In mitochondria other protein components of STSs are revealed. Transcription factor CREB (*c*AMP/*C*a^{2+} *r*eactive *e*lement (*CRE*) *b*ounding protein) phosphorylated by PKA and Ca^{2+}/calmodulin dependent PKs participates in the mitochondrial genome expression [159, 160]. CREB is a neuroprotector: its destruction decreases the expression, including subunit ND5 of complex I that diminishes mitochondrial respiration and increases damage of neurons by toxic 3-nitropropionate [160]. It is of great interest to neurobiology. But however the information has just appeared that PDH subunits imitates CREB at its identification with antibody helping [161]. GTP bounding mitochondrial proteins can modulate a fusion and fission of mitochondria, bound ribosomes with mitochondrial membrane and facilitate an import of proteins through contact points of M_o with M_i [162] .Described numerous facts testify in favor of recent generalizations made by C. Horbinsky and Ch. T. Chu ("Mitochondrial targeting of PKA appears to be involved in regulation most major mitochondrial functions, promoting respiration, antagonizing cell death, and regulating mitochondrial protein expression and biogenesis") [151] and A. Feliciello et al. ("PKA plays critic role in mammalian mitochondrial physiology") [163]. The scheme of described regulation is adduced (*Figure 4*).

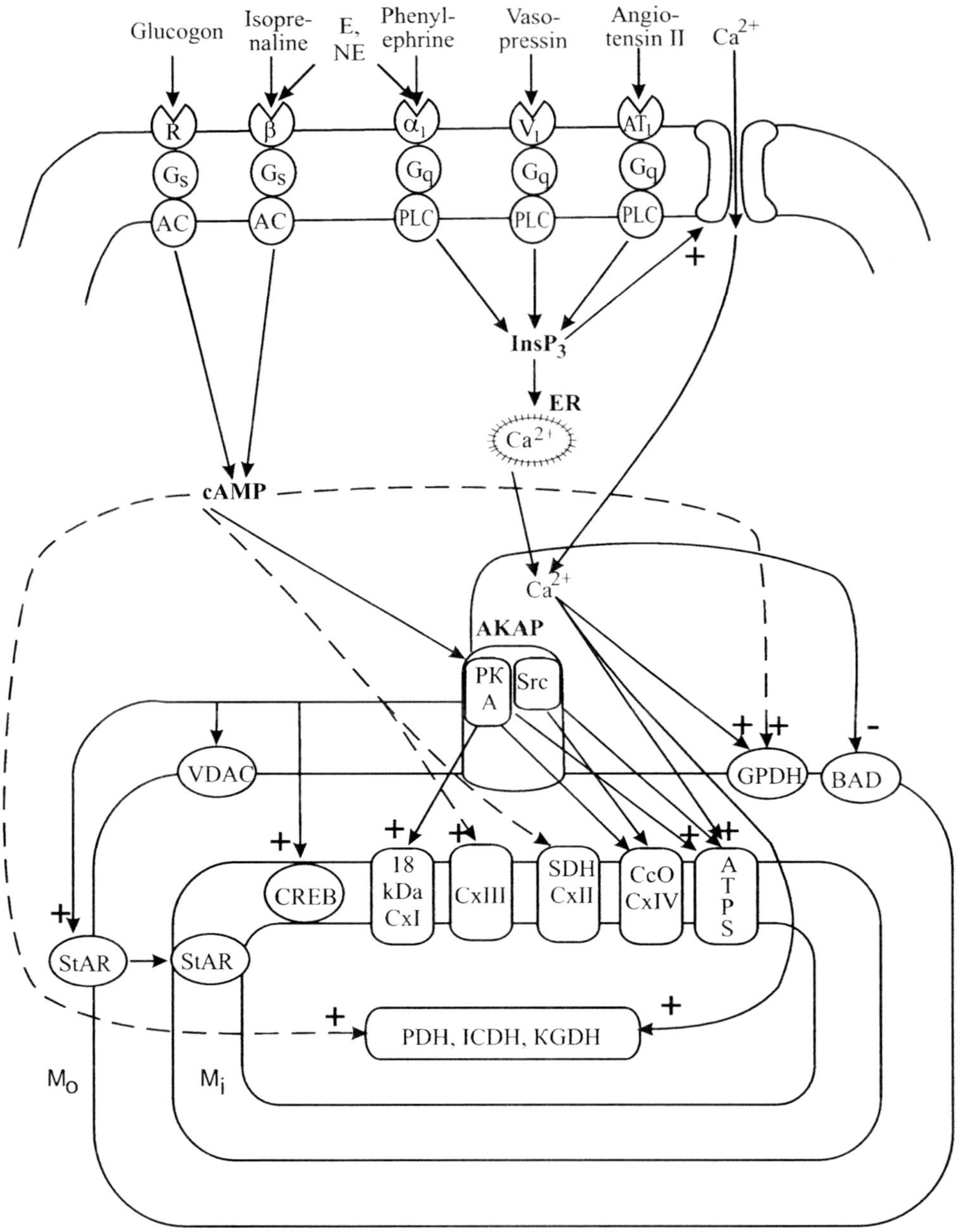

Figure 4. Regulation of mitochondrial functions by GPCR-dependent hormones and signal transductor systems. E – epinephrine, NE – norepinephrine, R – receptor, G – G proteins, AC – adenylyl cyclase, PLC – phospholipasa C, InsP3 – Ins(1,4,5)P3, ER – endoplasmic reticulum, AKAP – A-kinase anchor protein, PKA – protein kinase A, Scr – thyrosine kinase Src, StAR – steroidogenic acute receptory protein, VDAC – voltage-dependent anion channel, GPDH – glycerophosphate dehydrogenase, BAD – mitochondrial proapoptotic protein, CREB – cAMP response element binding protein, 18 kDa – regulatory protein of complex I, Cx – complex, SDH – succinate dehydrogenase, CcO – cytochrome c-oxidase, ATPS – ATP-synthase, PDH – pyruvate dehydrogenase, ICDH – NAD-dependent isocitrate dehydrogenase, KGDH – α-ketoglutarate dehydrogenase, Mo – outer mitochondrial membrane, Mi – inner mitochondrial membrane, + - activation, – - inhibition. ——→ - signal transfer (mechanism is mainly known), - - - - → - signal transfer (phenomenon).

Mitochondrial tyrosine kinases. Till recently the basic TK function has been delivered as the primary signal transfer of growth factors, cytokines and some circulating hormones from plasmatic membrane into the cell, more often through PK cascades [117]. In 1986 year the TK association with M_o was revealed [164] and in 2002-03 M. Salvi et al. have been discovered the presence of TKs and tyrosine phosphatases within mitochondria [165]. Now three TK types (Src, Abl and EGF receptor), two tyrosine phosphatases (Shp-2 and D1) and two phosphatases hydrolyzing phosphate residues both from serine/threonine and tyrosine were investigated. Four substrates of TKs: subunits II and I of *Cc*O, chaperone 75 kDa and phospholipid hydroperoxide glutathione peroxidase that is the important structural protein of spermatic mitochondria. With proteome approach another seven mitochondrial proteins with phosphotyrosine residues were elucidated, namely, PDH β, dihydrolipoamide dehydrogenase, aconitase, ubiquinol cytochrome *c* reductase I, β-chain of ATP synthase, voltage dependent anione channel 2 (VDAC 2) and *p*rotein *a*ssociated with *r*eceptor of *t*umor necrotizing factor (TRAP) [165]. All the proteins realize very important tasks especially in bioenergetics. Protein tyrosine phosphatase is important for steroidogenesis [72, see section 1.2].

Platelet-derived growth factor through its receptor TK induces the phosphorylation of δ-subunit of ATP synthase in cortical neurons and renal cells modulated by lysophosphatide acid. Potential sites of tyrosine phosphorylation are revealed in α-γ subunits of ATP synthase [166]. In mitochondria receptor TK TrkB realizing neurotropin effects are discovered [167]. Evidently M. Salvi generalization is just: "Mitochondrial tyrosine phosphorylation represents a promising field for the discovery of new mechanisms of bioenergetic regulation" [165].

Subunit 2 of NADH dehydrogenase (ND2) coded and translated in mitochondria is localized not only in mitochondrial complex I but also in glutamate NMDA receptor complex where its subunit 2 functions as adaptor protein anchoring TK Src. As a result Src phosphorylates and activates this receptor [168]. It is not excepted that ND2 can fix Scr and promotes tyrosine phosphorylation also in mitochondria.

A-kinase anchor proteins (AKAP) and other ones realize important functions. AKAP is formed in a nucleus as a result of penetration of PKA catalytic subunit in it and phosphorylation of CREB, then AKAP is fixed on cytosolic side M_o [163]. At the beginning they thought that AKAP only anchors of PKA, mainly PKA II. Then it turned out that AKAP functions were broader: it canalizes cAMP signal transfer to the necessary cellular microcompartment that increases specificity and intensifies signal. "Only PKA anchored to mtAKAPs can efficiently transmit cAMP sjgnals to mitochondria" [163]. The main mitochondrial AKAPs are AKAP121 and its splicing variant AKAP84, joining PKA to M_o and modulating phosphorylation of mitochondrial proteins [163, 165, 169]. Now it is clear that assembling *in vivo* complex (transduceosoma) is high orchestrated and multifunctional: it includes also phosphodiesterases, PKC, protein phosphatases, TK c-Src and tyrosine phosphatase D1, mRNAs coding subunit Fo-f of ATP synthase and Mn-superoxide dismutase [163, 165, 169]. Tyrosine phosphatase D1 activates Src and increases a value and duration of EGF signaling, strengthens phosphorylation of its receptor and activaton of *m*itogen-*a*ctivated *PK* (MAP*K*1/2), or *e*xtracellular signal-*r*egulated kinase (ERK 1/2). AKAP121 bounds and redistributes protein phosphatase D1 from cytoplasma to mitochondria and inhibits EGF receptor signal [169]. Possibly it permits to comprehend the availability of EGF receptor in M_i. Recently A. Feliciello et al. have revealed new and important AKAP121 functions: it increases Src dependent substrate phosphorylation in mitochondria and strengthens *Cc*O activity; AKAP augments $\Delta\psi$ and OxPhos by Src and PKA dependent mechanism. Silencing

AKAP121 by small interfering RNA severely disturbs synthesis and accumulation ATP in mitochondria. AKAP is the important regulator of oxidizing metabolism in mitochondria [170].

Other mitochondrial anchor proteins – small G protein Rab32 and PAP7 are also associated with M_o. The latter with peripheral benzodiazepine receptor (PBR) and voltage dependent anion channel (VDAC) forms multimer complex, necessary together with StAR for steroid hormones synthesis [163]. Adaptor protein Doc-4 (*do*wnstream of *k*inase) in endothelial mitochondria is anchor protein for TK c-Src and in its turn the stimulator of active O_2 species production, mediating by tumor necrosis factor α (TNF α), and nuclear factor kappa B (NF-κB) activation [171]. Totally these facts show that mitochondrion is a point of integration for signaling cascades due to its pivotal role [151], receiving signals from other parts of a cell and sending its signals to them, accomplishes the integrative function. Data about anchor and other associated proteins on M_i for the present are absent.

Kinase regulation of complex IV. This difficult problem is studed since 1997, but for the present results are non single-meant. At the beginning the ATP dependent phosphorylation by endogenic kinases of subunit II of cytochrome *c*-oxidase (C*c*O results) was described [172]. C*c*O unit II is phosphorylated by non receptor TK cytochrome *c* c-Src in osteoblasts with positive correlation of C*c*O activity with c-Src [173]. In B. Ludwig and B. Kadenbach laboratories mitochondria incubation with cAMP increases the allosteric inhibition of C*c*O by ATP. Data have been is received about phosphorylation of subunits I, Vb, II and/or III of purified C*c*O from the bovine heart. Three subunits (except II) have consensus sequence for cAMP dependent phosphorylation. The subunit I is phosphorylated in five species, possessing hormones (from fly to human), and is absent at primrose, yeasts and bacteria. cAMP-dependent hormones reproduce cAMP effect but Ca^{2+}-mobilizing hormones induce dephosphorylation [174, 175]. Other approach shows that the inhibition of isolated C*c*O is realized not by PKA but cAMP dependent phosphorylation of tyrosine 304 in subunit I. Glucagon and direct activation of adenylyl cyclase by forscolin inhibit C*c*O in hepatocytes possibly through unidentified endogenous mitochondrial TK. This effect of glucagon does not conform with more early data (see section 1/1). Authors do not eliminate the PKA participation but consider that it acts rather not directly to C*c*O but transfer activated signal through M_o in intermembranous space to TK [176]. Additional investigations are naturally necessary. The common result of those two last articles is an inhibition of complex IV by cAMP. However a serious question arises: if one and the same second messenger cAMP activates complex I but inhibits complex IV, how does the respiratory chain work? At last the work already cited shows the participation of AKAP with both TK Src and PKA in the phosphorylation of mitochondrial substrates and strengthing C*c*O activity and also the parallel augmentation of $\Delta\psi$ and OxPhos. Hypoxia and cAMP mediate inhibition of C*c*O and hyperphosphorylation of subunits I, IVil, and Vb and increase reactive O_2 species production [170]. Specific inhibitors of PKA reverse these effects and dramatically reduce the ischemia/reperfusion injury to the myocardial tissue [177]. It means that cAMP effect can be not only physiological but also pathological one.

Other protein kinase. All the main PK of the cell: *p*hosphatidyl *i*nositol *3-k*inase (PI-3K), AKT (PKB), PKC, Raf, ERK kinase (MEK), ERK, p38 MAPK, Jun N-terminal kinase (JNK) are revealed and functioned in mitochondria. The presence of specific anchor proteins already known for PKA, PKC, JNK [151] and *s*tress-*a*ctivated *PK-3* (SAPK-3) promotes to this [178]. For the present there are some data about a participation of these PKs in the regulation of metabolic and energetic mitochondrial functions and ATP synthesis. ERK activates ATP synthase and maintains $\Delta\psi$ on M_i (cit. by [151]). Insulin activates $\Delta\psi$ 2-fold through PI3-K that phosphorylates AKT and CREB [95]. Insulin-like growth factor 1 (IGF-1) provokes the rapid translocation phospho-ACT to mitochondria where it phosphorylates β-subunit of ATP synthase and glycogen synthase kinase 3β [179] and the latter inhibits PDH [180]. There are more data about targeting these PKs in execution of survival-death decisions. Antiapoptotic Bcl-2 has phosphorylating sites for PKs AKT, PKCα, and MAPK, dephosphorylation of Bcl-2 inhibits it [181]. JNK inhibits antiapoptotic Bcl-2 and Bcl-x_L, AKT phospho-inactivates BAD, p38 MAPK – another proapoptotic protein Bax (cit. by [151]). Estradiol protects from H_2O_2-induced exhaustion of intracellular ATP, attenuates depolarization in lens cells and stimulates MAPK, stabilizing $\Delta\psi$. Positive correlation between estradiol stimulation of ERK1/2 phosphorylation and mitochondria stabilization has been revealed [182]. Protective estradiol effect connects with its membrane GPCR and then AKT and MAPK cascades and the expression of anti-apoptotic Bcl-2 [84]. Estrogen inhibits ultra-violet induced damage of mitochondria realizing by JNK and PKC δ [87].

The recent investigation of a mitochondrial matrix proteome displayd the presence of 45 phosphoproteins but many proteins evaded detection. Phosphoproteins were detected in all the mitochondrial respiratory chain complexes, as well as enzymes of intermediary metabolism, such as PDH, citrate synthase, and acyl-CoA dehydrogenase. The phosphoproteome of the mitochondrial matrix is extensive and dynamic [183]. It is evident that a mitochondrial phosphoproteome as a whole and consequently PK importance will be grow.

2.2.2. Ca^{2+} and PKC Systems

This system transfers hormonal signal by the chain GPCR → G_q protein → phospholipase C-β→ IP_3 + diacylglycerol (DAG). IP_3 transfers its signal through Ca^{2+}, and DAG through PKC.

Calcium iones. IP_3 mobilizes both intracellular Ca^{2+} from reticulum and extracellular one through slow Ca^{2+} channel, provokes Ca^{2+} accumulation in cytosol and then in mitochondria. Ca^{2+} inflow in mitochondria is realized by uniporter (IP_3 increase its activity for a longer time than Ca^{2+} signal) [184], "rapid mode" (RaM) [185] and ryanidine receptor of M_i (mRyR1) [1, 186], Ca^{2+} outflow – by ionic exchange with Na^+ (in excitable cells) or H^+ (in unexcitable ones).

In a rest state Ca^{2+} concentration in mitochondria is 120-240 nM (12-15, 185, 187, 188]. Constant cycling Ca^{2+} in mitochondria is important for the ensuring of fine and reliable regulation of their function [14, 15]. Many but not all Ca^{2+} effects are realized by calmodulin.

Ca^{2+}-mobilizing hormones catecholamines (through α_1-adrenoceptor), angiotensin II and vasopressin, and also glucagon through GPCRs evoke an accumulation Ca^{2+} in cytosol. Epinephrine and α_1 agonist metoxamine (but not β agonist isoprenaline) at administration in a body and in the perfused heart quicken the uniport of Ca^{2+} in mitochondria, but do not influence on both processes of Ca^{2+} outflow [12, 13]. In the perfused liver of female rat epinephrine, isoprenaline and glucagon (but not α_1 agonist phenylephrine) increase Na^+-dependent Ca^{2+} outflow, but do not change other transfer kinds [12]. Above named agonists, norepinephrine, vasopressin, glucagon augment Ca^{2+} concentration in mitochondria of many organs [12-15, 189, 190]. Ca^{2+} concentration in mitochondrial matrix increases in systole and decreases in diastole [14, 15]. Platelet-activating factor induces store-operated Ca^{2+} channel

(SOC) to influx and in parallel the expression and production of cyclooxigenase-2 (COX-2) in human microglia. PK1195, an activating ligand of mitochondrial peripheral benzodiazepine receptor, effectively blockes SOC-mediated Ca^{2+} influx and COX-2 increase [191]. Lysophosphatidyl choline induces in endothelial cells Ca^{2+} mobilization, mitochondrial reactive oxygen species generation, ERK activation and adaptation of the pathological stress to oxidized lipids such as lysophosphatidyl choline [192]. Estrogen increases Ca^{2+} in mitochondria of a high-calcium capacity within 90 s that is inhibited by mitochondrial blockers. It suggests that mitochondria are early nongenomic targets of estradiol [93].

In experiments on mitochondria, their lysates, partly purified PDH, KGDH, NAD-ICDH [11, 14, 15, 193] and homogenous ICDH [142, 194] Ca^{2+} directly (without calmodulin) activates these enzymes (Figure 1) by the same kinetic mechanisms as hormones. PDH stimulation is a result of phosphatase activation of complex, stimulation ICDH and KGDH – that of allosteric effect. This regulation is absent on enzymes of yeasts, plants and insects [14, 15, 189]. Ca^{2+} decreases α-ketoglutarate concentration [195], activates α-glycerol-3-phosphate dehydrogenase of M_i [20, 22, 196], transhydrogenase [197], glycine oxidation [14, 15], but not SDH [198]. Ca^{2+} addition to mitochondria evokes NADH accumulation [14, 15, 199-201] and stimulation of O_2 consumption [12-14, 198, 201, 202] in a respiration on pyruvate, citrate, isocitrate, ketoglutarate and glutamate (but not malate, ascorbate, β-hydroxy- and β-ketobutyrate). Ca^{2+} augments the reduction of cytochrome *b*, c_1, *c*, electron transporting flavoprotein [14, 15] and volume of mitochondrial matrix [15, 16, 201]. In intact hepatocytes vasopressin evokes Ca^{2+} accumulation in mitochondria, PDH activation, increasing NADH, $\Delta\psi$ and ΔpH [199].

Within the recent decade they were shown that Ca^{2+} when using NAD dependent substrates (but not succinate) increases $\Delta\mu_H$ [198, 203], $\Delta\psi$ and ΔpH [199] and on hormones stimulated cells and mitochondria augments OxPhos [201, 203-208] and activates adenine nucleotide translocase [185]. Higher Ca^{2+} concentrations evoke contrary effects [207, 209], especially in the brain [209]. Described effects develop rapidly – during 0.2 sec [201] and realize in K_M of Ca^{2-} 0.2-1.5 μM [14, 187, 201, 202] that creates "rapid mode" of Ca^{2+} penetration into mitochondria [185]; K_M of Ca^{2-} 10-41 μM for ICDH that is ensured by uniporter [185]. This diapason of Ca^{2+} concentrations is physiological. Ryanodine, inhibitor of penetration Ca^{2+} into mitochondria through RyR1 in M_i, inhibits Ca^{2+} stimulated OxPhos in the heart [187]. Ca^{2+} dependent phosphorylation of peptide 3.5 kDa is revealed in hepatic mitochondria and in the brain – phosphorylation of four peptides (3.5 – 39 kDa) [207, 210], but for the present their functional importance remains vague.

Ca^{2+} participates also in other processes in mitochondria: gluconeogenesis [211], citrulline synthesis [212], steroid hormone formation [213, 214], mitochondria mobility [215]. Large-conductance voltage- and calcium-sensitive potassium channel (BK/MAXI K) is present in M_i and co-lokalize with mitochondrial-specific proteins in the brain and also in glioma and cardiac cells [216]. Ca^{2+}-bounding motive is revealed in a new family of aspartate-glutamate [217] and malate-aspartate [218] protein carriers. Small Ca^{2+} signals increase the latter and large Ca^{2+} signals potentiate Ca^{2+} uniporter and mitochondrial dehydrogenases, both increase mitochondrial NAD(P)H [218]. Ca^{2+}-signal stimulates nuclear genes coding mitochondrial proteins. Thus Ca^{2+} ionophore A23187 increases the expression of cytochrome *c* mRNA; it is independent from calmodulin PK (CMPK) II and IV and PKA [219]. Complex of Ca^{2+} with its receptor calmodulin (Ca^{2+}/CM) couples an excitation not only with muscle contraction but also with transcription. It is frequently realized by two nondependent pathways: through CMPK II and IV → transcription factor CREB and through protein phosphatase calcineurin → *n*uclear *f*actor of *a*ctivated *T*-lymphocytes (NFAT) (references in [115, 117]). Both these STSs stimulate the expression of mitochondrial oxidizing enzymes and mitochondriogenesis [220]. Intermediate stage can be CMPK-induced expression of coactivator I activated peroxisome proliferators receptor-γ (PGC-1) [221]. In mitochondria and mitoplasts at small Ca^{2+} concentrations Ca^{2+}-dependent protein phosphatase dephosphorylates CREB [222] and high Ca^{2+} concentrations activate CMPK II, PKA and CREB phosphorylation [222, 223]. Evidently CREB is an important regulator of expression not only nuclear but also mitochondrial genomes. Ca^{2+} is a key regulator of NF-κB activation in airway epithelial cells [114].

All this led P. S. Brookes to conclusion that "Ca is a global positive effector of mitochondrial function" [1]. Indeed Ca^{2+} in parallels activates PDH, Krebs cycle, respiratory chain, OxPhos, energy dependent synthesis, and mitochondrial genome expression. Computer modeling shows a presence in the heart and skeletal muscle the mechanism of parallel and direct activation of TAC dehydrogenases, complexes I, III and IV of respiratory chain, ATP synthase and ATP using functions [224]. For augmentation of ATP production TAC and ATP synthase are necessary [225]. They deliver that leading role plays Ca^{2+}, but it is disputable as the important data have already been received on ATP synthase phosphorylation and its regulation by cAMP and different PK [see section 2.2.1].

12 isoforms of *protein kinase C* (PKC) realize important, very different and sometimes even contrary functions. They are subdivided in 3 groups: 1) activated by DAG in the presence of phosphatidylserine and Ca^{2+} convectional (or classic ones) PKCs (α, βI, βII, γ), 2) activated by DAG and phosphatidylserine new PKCs (δ, ε, θ, μ), and 3) activated by phosphatidylinositol dependent kinase (PDK) atypic PKCs (ζ, λ, τ) [115]. Activating effect Ca^{2+}-ionophore A23187 on expression of cytochrome *c* is not changed by PKC δ, but is potentiated by Ca^{2+} dependent PKC α and βII [225]. Ca^{2+} concentration in mitochondria is decreased by PKC α, β and δ, but is increased by isoforms ζ; $\Delta\psi$ is moderately diminished by PKC β [226]. PKC δ inhibits respiratory chain and ATP synthesis [151]. PKC ε phosphorylates and activates subunit IV of C*c*O of neonatal cardiomyocytes [227]. Thus Ca^{2+} acts by four different mechanisms: directly, through calmodulin – CMPKs, through calcineurin – NFAT and through five isoforms of PKC.

SUMMARY

The discovery of complex regulation of mitochondrial functions by hormones and STSs is one from the new and important achievements of mitochondriology. It proves an absence of mitochondria autonomy, their dependence from both cell as a whole and extracellular regulators. A number of hormones of all the classes and with different activity mechanisms stimulates all the basic metabolic and energetic processes in mitochondria. These effects are realized and/or reproduced by different STSs, all the general components of which are revealed in mitochondria and/or closely interact with them. Pleiotropy of these regulation and their biological importance are proved. As a whole mitochondria realize the integrative function in cell signalization.

However mechanisms of mitochondrial function regulations for the present are worse studied than cytosolic and nuclear ones. Phenomenology is investigated better than mechanisms. Hormones and second messengers stimulate not only complex I, but also II and III ones, and data on mechanisms of the two latters are absent; information on regulation of complex IV is complicated. The revelation of downstream components of mitochondrial STSs is very important. The great work is still coming but rate of development of the problem has ever been growing.

Authors of this review will be glad if it could have been one from stones in the future building of signal transduction mitochondriology.

References

[1] Brookes, P.S.; Yoon, Y.; Robotham, J.L.; Anders, M.W.; Sheu, S.S. Calcium, ATP, and ROS: a mitochondrial love-hate. *Am. J. Physiol. Cell. Physiol.* (2004) *287,* C817-C833.

[2] Adam, P.A.J.; Haynes, R.C. Control of hepatic mitochondrial CO_2 fixation by glucagon, epinephrine, and cortosol. *J. Biol. Chem.* (1969) *244,* 6444-6450.

[3] Rendle, P.J.; Denton, R.M.; Pask, H.T.; Severson, D.L. Calciun iones and the regulation of pyruvate dehydrogenase. *Biochem. Soc. Symp.* (1974), No. 39, 75-88.

[4] Garrison, J.C.; Haynes, R.C. Hormonal control of gluconeogenesis by regulation of mitochondrial pyruvate carboxylation in isolated rat liver cells. *J. Biol. Chem.(*1975) *250,* 2769-2777.

[5] Yamazaki, R.K. Glucagon stimulation of mitochondrial respiration. *J. Biol. Chem.*, (1975) *255,* 7924-7930.

[6] Kulinsky, V.I.; Trufanova, L. V. Activation by catecholamines and cyclic 3',5'-AMP of limiting enzyme of Krebs cycle NAD-dependent isocitrate dehydrogenase. *Dokl. Acad. Nauk* SSSR (1975) *224,* 1439-1441.

[7] Kulinsky, V.I.; Vorob'eva, L.V. Activation by catecholamines of liver mitochondria respiration. *Biul. Eksp.Biol. Med.* (1977), *83,* 675-677. (Moscow).

[8] Kulinsky, V.I.; Vorob,eva, L.V. Activation by cAMP of mitochondrial respiration in different metabolic states. *Biul. Eksp.Biol. Med.* (1978) *85,* 291-294. (Moscow).

[9] Kulinsky, V.I. Basic principles of investigation of hormones and cyclic nucleotides effects. *Uspekhy Sovrem. Biol.* (1980) *90,* 382-393. (Moscow).

[10] Kulinsky, V.I. Short-term regulation of mitochondria functions by hormones and second messengers. In: Uspekhy *Biol. Khimii* (1997) *37,* 171-209. (Moscow).

[11] Haynes, R.C. (1986) In: Kraus-Friedemann N., Hormonal Control of Gluconeogenesis, CRS Press, Boca Raton, *3,* 19-29.

[12] Crompton M. In: Curr. Topic Membr. *Transport* (1985) *25,* 231-276.

[13] Crompton M. In: in: *Calcium and the Heart* (1990) (Langer G.A. ed.), Raven Press, N.Y., 167-198.

[14] Denton, R.M.; McCormack, J.G. Ca^{2+} as a second messenger within mitochondria of the heart and the other tissues. *Annu. Rev.Physiol.* (1990) *52,* 451-456.

[15] McCormack, J.G.; Halestrap, A.P.; Denton, R.M. Role of calcium ions in regulation of mammalian intramitochondrial metabolism. *Physiol. Rev.* (1990) *70,* 391-425.

[16] Halestrap, A.P. Regulation of mitochondrial metabolism through changes in matrix volume. *Biochem. Soc. Trans.* (1994), *22,* 522-529.

[17] Kulinsky, V.I.; Trufanova, L.V.; Medvedev, A.E. Catecholamine control of enzymes involved in isocitrate oxidation in rat liver mitochondria. *FEBS Lett.* (1984) *177,* 143-45.

[18] Kulinsky, V.I.; Kuncevich, A.K.; Trufanova, L.V. Activation of succinate dehydration in rat liver by norepinephrine, cAMP and acute cooling. *Biul. Eksp.Biol. Med.* (1981) *92*(8), 33-34. (Moscow).

[19] Kulinsky, V.I.; Fomin, O.G. Activation by norepinephrine and cAMP of citrate synthase in liver mitochondria. *Biul. Eksp.Biol. Med.,* (1982) *93*(1), 41-43. (Moscow).

[20] Hansford, R.G. Relation between mitochondrial calcium transport and control of energy metabolism. *Rev. Physiol. Biochem. Pharmacol.* (1985), *102,* 1-72.

[21] Kulinsky, V.I.; Medvedev, A.E.; Vorob'eva, L.M., Trufanova, L.V. Catecholamine regulation of oxidative function of rodent liver. In: *Molecular mechanisms and regulation of energetic metabolism,* (1987) Pushchino, Acad. Nauk SSSR, 161-174.

[22] Yip, B.P.; Lardy, H.A. The role of calcium in the stimulation of gluconeogenesis by catecholamines. *Arch. Biochem. Biophys.* (1981) *212,* 370-377.

[23] Medvedev, A.E.; Trufanova, L.V.; Kulinsky, V.I.. Regulation of activity of mitochondrial transhydrogenase by catecholamines Biokhimia (1986) *51,* 1165-1173. (Moscow).

[24] Siess, E. A.; Wieland, F.M. Glucagon-induced stimulation of 2-oxoglutarate metabolism in mitochondria from rat liver. *FEBS Lett.* (1978), *93,* 301-306.

[25] Haussinger, D.; Siess, H. Effect of phenylephrine on glutamate and glutamine metabolism in isolated perfused rat liver. *Biochem. J.* (1984) *221,* 651-658.

[26] Mohan, C.; Memon, R.A.; Bessman, S.P. Differential effects of insulin, epinephrine, and glucagons on rat hepatocyte mitochondrial activity. *Arch. Biochem. Biophys.* (1991) *287,* 18-23.

[27] Corvera, S.; Garcia-Sainz, J.A. Vasopressin and angiotensin II simulate ureogenesis through increased mitochondrial citrulline production. Life Sci. (1982) *31,* 2493-2498.

[28] Brosnan, J.T.; Evart, H.S.; Squires, S.A. Hormonal control of hepatic glutaminase. *Adv. Enzyme Regulat.* (1995) *35,* 131-146.

[29] Mebrouk, G.M.; Jois, M.; Brosnan, J.T. Cell signaling and hormonal stimulation of the hepatic glycine cleavage enzyme system by glucagon. *Biochem. J.* (1998), *330,* 759-763.

[30] Siess, E.A.; Brocks, D.G.; Lattke, H.K.; Wieland, O.H. Effect of glucagon on metabolite compartmentation in isolated rat liver cells during glutoneogenesis from lactate. *Biochem. J.* (1977) *166,* 225-235.

[31] Titheradge, M.A.; Coore, H.C. Hormonal regulation of liver mitochondrial pyruvate carrier in relation to gluconeogenesis and lipogenesis. *FEBS Lett.* (1976), *71,* 73-78.

[32] Exton, J.H. Mechanism involved in calcium-mobilizing agonist response. In: Greengard, P.; Robison, J.A. *Adv. in Cyclic Nucl. Protein Phosphorylation Res. Raven Press* (1986) *20,* 211-262.

[33] Bygrave, F.L.; Benedetti, A. Calcium: its modulation in liver by cross-talk between the action of glucagon and calcium mobilizing agonists. *Biochem. J.* (1993), *296,* 1-14.

[34] Collins-Nakai, R.L.; Notheworsy, D.; Lopashuk, G.D. Epinephrine increases ATP production in heart by preferentially increasing glucose metabolism. *Am. J. Physiol.* (1994) *267,* H1862-H1871.

[35] Soler-Argilaga, C.; Russel, R.L.; Werner, H.V.; Heimberg, M. A possible role of calcium in the action of glucagon, cAMP and dibutyryl cAMP on the metabolism of free acids by rat hepatocytes. *Biochem. Biophys. Res. Commun. (*1978) *85,* 249-256.

[36] Quant, P.A.; Tubbs, Ph.K.; Brand, M.D. Glucagon activates mitochondrial 3-hydroxy-3- methylglutaryl-CoA synthase in vivo by decreasing the extent of succinylation of the enzyme. *Eur. J. Biochem.* (1990) *187,* 169-174.

[37] Catecholamines. Bridging basic science with clinical medicine (Goldstein D.S., Eisenhofer G., McCarty R. eds.) (1998) *Adv. in Pharmacol., 42,* 1069 p.

[38] Kulinsky, V.I.; Kolesnichenko, L.S. Catecholamins: biochemistry, pharmacology, physiology, clinics *Voprosy Med. Khimii* (2002) *48 (1),* 44-67. (Moscow).

[39] Kulinsky, V.I.; Plotnikov, N.Y. Pathochemistry of acute intensive cooling: dynamics of main reactions and catecholamine mechanisms. *Biul. Sibir. Otdelenie Acad. Med. Nauk* SSSR (1983) № 5, 11-19.

[40] Brand, M.D.; D'Alessandr, L.; Reishm, G.P.V.; Hafner, R.P. Stimulation of the electron transport chain in mitochondria isolated from rats treated with mannoheptulosa or glucagon Arch. *Biochem. Biophys.* (1990) *283,* 278-284.

[41] Matthias, A.; Ohlson, K.B.; Frederiksson, J.M.; Jacobson, A.; Nedergaard, J.; Cannon, B. Thermogenic response in brown fat cells are fully UCP-1-dependent, UCP-2 and UCP-3 do not subsitute for UCP-1 in adrenergically or fat acid-induced thermogenesis. *J. Biol. Chem.* (2000), 275, 25073-25081.

[42] Krauss, S.; Zhang, C.Y.; Lowell, B.B. The mitochondrial uncoupling-protein homologues. *Nature Rev. Mol. Cell Biol.* (2005) *6*, 248-261.

[43] Petrosillo, G.; Di Venosa, N.; Pistolese, M.; Casanova, G.; Tiravanti, E.; Colantuona, G.; Federici, A.; Paradies, G.; Ruggiero, F.M. Protective effect of melatonin against mitochondrial dysfunction associated with cardiac ischemia-reperfusion: role of cardiolipin. *FASEB J.* (2006), *20,* 269-276.

[44] Goglia, F. Biological effects 3,5-diiodothyronine (T_2). Biochemistry (Moscow), (2005) *70*, 203-213.

[45] Wrutniak-Cabello, C.; Casas, F.; Cabello, G. Thyroid hormone action in mitochondria *J. Mol. Endocrinology* (2001), *26,* 67-77.

[46] Losel, R.; Wehling, M. Nongenomic actions of steroid hormones. *Nature Rev. Mol. Cell Biol.* (2003), *4,* 46-56.

[47] Bassett, J.H.D.; Harvey, C.B.; Williams, G.R. Mechanism of thyroid hormone receptor -specific nuclear and extra nuclear actions Mol. *Cell Endocrinology* (2003) *213,* 1-11.

[48] Lanni, A.; Moreno, M.; Lombardi, A.; de Lange, P.; Goglia, F. Control energy metabolism by iodothyronines. *J. Endocrinol. Invest.* (2001) *24,* 897-913.

[49] Harper, M.E.; Brand, M.D. Hypothyroidism stimulate mitochondrial proton leak and ATP turnover in rat hepatocytes but does not change the over all kinetics of substrate oxidation reactions. *Can. J. Physiol. Pharmacol.* (1994) *72,* 899-908.

[50] Mracek, T.; Jesina, P.; Krivakova, P.; Bolehovska, R.; Cervinkova, Z.; Drahota, Z.; Houstek, J. Time-course of hormonal induction of mitochondrial glycerophosphate dehydrogenase biogenesis in rat liver. *Biochim. Biophys. Acta* (2005) *1726,* 217-223.

[51] Silva, J.E. Thermogenic mechanisms and their hormonal regulation. *Physiol.Rev.*, (2006) *86,* 435-464.

[52] Venditti, P.; Pamplona, R.; Ayala, V.; De Rosa, R.; Caldarone, G.; Di Meo, S. Differential effects of experimental and cold-induced hyperthyroidism on factors inducing rat liver oxidative damage. *J. Exp.Biol.* (2006), *209,* 817-825.

[53] Katyare, S.S.; Rajan, R.R. Influence of thyroid hormone treatment on the respiratory activity of cerebral mitochondria from hypothyroid rats. A critical re-assesment. *Exp. Neurol.* (2005) *195,* 416-422.

[54] Kneer, N.; Lardy, H. Thyroid hormone and dehydroepiandrosterone permit gluconeogenesis hormone responses in hepatocytes *Arch. Biochem. Biophys.* (2000), *375,* 145-153.

[55] Beylot, M. Regulation of in vivo ketogenesis: role of free fatty acids and control by epinephrine, thyroid hormones, insulin and glucagon. *Diabetes Metab.* (1996) *22,* 299-304.

[56] Gong, D.-W.; He, Y.; Karas, M.; Reitman, M. Uncoupling protein-3 is a mediator of thermogenesis regulated by thyroid hormone, β_3-adrenergic agonists, and leptin. *J. Biol. Chem.* (1997) *272,* 24129-24132.

[57] Hesselink, M.K.; Schrauwen, P. Towards comprehension of the physiological role of UCP3. *Horm. Metab. Res.* (2005) *37,* 550-554.

[58] Laurberg, P.; Andersen, S.; Karmisholt, J. Cold adaptation and thyroid hormone metabolism. *Horm. Metab. Res.* (2005) *37,* 545-549.

[59] Weitzel, J.M.; Iwen, K.A.; Seitz, H.J. Regulation of mitochondrial biogenesis by thyroid hormones. *Exp. Physiol.* (2003) *88,*121-128.

[60] Bernal, J. Activation of thyroid hormone in brain. *J. Endocrinol. Invest.* (2002) *25,* 268-288.

[61] Yen, P.M. Physiological and molecular basis of thyroid hormone action. *Physiol. Rev.* (2001) *81,* 1097-1142.

[62] Moro, L.; Marra, E.; Capuano, F.; Greco, M. Thyroid hormone treatment of hypothyroid rats restores the regenerative capacity and the mitochondrial membrane permeability properties of the liver after partial hepatectomy. *Endocrinology* (2004) *145,* 5121-5128.

[63] Siculella, L.; Sabetta, S.; Giudetti, A.M.; Gnoni, G.V. Hypothyroidism reduced tricarboxylate carrier activity and expression in rat liver mitochondria by reducing nuclear transcription rate and splicing efficiency. *J. Biol. Chem.* (2006), *281,* 19072-80

[64] Franco, M.C.; Arciuch, V.G.; Peralta, J.G.; Galli, S.; Levisman, D.; Lopez, L.M.; Romorini, L.; Poderoso, J.J.; Carreras, M.C. Hypothyroid phenotype is contributed by mitochondrial complex I inactivation due to translocated neuronal nitric oxide synthase. *J. Biol. Chem.* (2006) *281,* 4779-4786.

[65] Zavinovich, A.A. Role of uncoupled proteins UCP1, UCP2 and UCP3 in energy balance, type 2 diabetes and obesity. *Sinergism with the thyroid. Medicina* (B Aires), (2005) *65,* 163-169.

[66] de Jesus, L.A.; Carvalho, S.D.; Ribeiro, M.O.; Schneider, M.; Kim, S.W.; Harney, J.V.; Larsen, P.R. The type 2 iodothyronine deiodinase is essential for adaptive thermogenesis in brown adipose tissue. *J. Clin. Invest.* (2001) *108,* 1379-1385.

[67] Stocco, D.M. Steroidogenic acute regulatory protein (StAR) and regulation of steroid hormone biosynthesis. *Annual Rev. Physiol.* (2001) *63,* 193-213.

[68] Artemenko, I.P.; Zhao, D.; Hales, D.B.; Hakes, K.H.; Jefcoate, C.R. Mitochondrial processing of newly synthezed StAR protein but not total StAR, mediates cholesterol transfer to cytochrome P450 side chain cleavage enzyme in adrenal cells. *J. Biol. Chem.* (2001) *276,* 46583-46596.

[69] Stocco, D.M.; Wang, X.; Jo, Y.; Manna, P.R. Multiple signaling pathways regulated signals and StAR protein expression more complicated than we thought. *Mol. Endocrinol.* (2005) *19,* 2647-2659.

[70] Renlund, N.; Jo, Y.; Svechnikova, I.; Holst, M.; Stocco, D.M.; Sider, O.; Svechnikov, K. Induction of steroidogenesis in immature rat Leydig cells by interleukin-1α is dependent on extracellular signal regulated kinases. *J. Mol. Endocrinol.* (2006), *36,* 327-336.

[71] Allen, J.A.; Shankara, T.; Janus, P.; Buck, S.; Diemer, T.; Hales, K.H.; Hales, D.B. Energised, polarized and actively respiring mitochondria are required for acute Leidig cell steroidogenesis. *Endocrinology* (2006), *147*, 3924-3935.

[72] Cano, F.; Poderoso, C.; Cornejo, M.F.; Castilla, R..; Maloberti, P.; Castillo, F.; Neuman, I.; Paz, C.; Podesta, E.J. Protein tyrosine phosphatase regulate arachidonic acid release, StAR induction and steroidogenesis acting on a hormone-dependent arachidonic acid- preferring acyl-CoA synthetase J. Steroid. Biochem. Mol. Biol. (2006) 99, 197-202.

[73] Papadopoulos, V. Periferal benzodiazepine receptor: structure and function in health and disease. Ann. Pharm. Fr. (2003) 61, 30-50.

[74] Hauet, T.; Yao, Z.X.; Bose, H.S.; Wall, C.T.; Han, Z.; Li, W.; Hales, D.B.; Miller, W.L.; Culty, M.; Papadopoulus ,V. Peripheral-type benzodiazepine receptor-mediated action of steroidogenic acute regulatory protein on cholesterol entry into leydig cell mitochondria. Mol. Edocrinol. (2005), 19, 540-554.

[75] Papadopoulos, V.; Lecanu, L.; Brown, R.C.; Han, Z.; Yao, Z.X. Peripheral-type benzodiazepine receptor in neurosteriod biosynthesis, neuropathology and neurological disorders. Neuroscience, (2006) 138,.749-756.

[76] Lehoux, J.G.; Mathieu, A.; Lavigne, P.; Fleury, A. Adenocorticotropin regulation of steroidogenic acute regulatory protein. Microsc. Res. Tech. (2003) 61, 288-299.

[77] Lisurek, M.; Bernhardt, R. Modulation of aldosterone and cortisol synthesis on the molecular level. Mol. Cell Endocrinol. (2004) 215, 149-159.

[78] Scheller, K.; Sekeris, K.E. The effect of steroid hormones on the transcription of genes encoding enzymes of oxydative phosphorylation (2003) Exp. Physiology, 88, 129-140.

[79] Ritz, P.; Dumas, J.-F.; Ducluzeau, P.-H.; Simard, G. Hormonal regulation of mitochondrial energy production. Curr. Opin. Clin. Nutr. Metab. Care (2005) 8, 415-418.

[80] Katyare, S.S.; Balasubramanion, S.; Parmar, D.V. Effect of corticosterone treatment on mitochondrial oxidative energy metabolism in developing rat brain. Exp. Neurol. (2003) 183, 241-248.

[81] Allan, E.H.; Chisholm, A.B.; Titheradge, M.A. The stimulation of hepatic oxidative phosphorylation following dexamethasone treatment of rats. Biochim. Biophys. Acta (1983) 725, 71-76.

[82] Manoli, I.; Le, H.; Alesci, S.; McFann, K.K.; Su, Y.A.; Kino, T.; Chrousos, G.P.; Blackman, M.R. Monoamine oxidase-A is a major target gene for glucocorticoids in human skeletal muscle cells. FASEB J. (2005), 19, 1359-1361.

[83] Jones, C.G.; Hothi, S.K.; Titheradge M.A. Effect of dexamethasone on gluconeogenesis, pyruvate kinase, pyrubate carboxylase and pyruvate dehydrogenase flux in isolated hepatocytes. Biochem. J., (1993) 289, 821-828.

[84] Alexaki, V.I.; Charalampopoulus, I.; Kampa, M.; Nifli, A.P.; Hatzoglou, A.; Gravanis, A.; Castanas, E. Activation of membrane estrogen receptors induce pro-survival kinases. J. Steroid Mol. Biol. (2006), 98, 97-110.

[85] Nilsen, J.; Brinton, R.D. Mitochondria as therapeutic targets of estrogen action in the central nervous system. Curr. Drug Targets CNS Neurol. Disord. (2004) 3, 297-313.

[86] Simpkins, J.W.; Wang, J.; Wang, X., Perez, E.;.Prokai, L.; Dykens, J.A. Mitochondria play a central role in estrogen-induced neuroprotection. Curr. Drug Targets CNS Neurol. Disord. (2005) 4, 69-83.

[87] Pedram, A.; Razandi, M.; Wallace, D.C.; Levin, E.R. Functional estrogen receptors in the mitochondria of breast cancer cells. Mol. Cell Biol. (2006), 17, 2125-2137.

[88] Burris, T.P.; Krishnan, V. Estrogen: a mitochondrial energizer that keep on going. Mol Pharmacol. (2005) 68, 956-958.

[89] Stirone, C.; Duckles, S.P.; Krause, D.N.; Procaccio, V. Estrogen increases mitochondrial efficiency and reduces oxidative stress in cerebral blood vessels. Mol. Pharmacol. (2005) 68, 959-965.

[90] Chen, J.Q.; Yager, J.D. Estrogen's effects on mitochondrial gene expression: mechanisms and potential contributions to estrogen carcinogenesis. Ann. N.Y. Acad. Sci. (2004) 1028, 258-272.

[91] Felty, Q., Roy, D. Mitochondrial signal to nucleus regulate estrogen-induced cell growth. Med. Hypothesis (2005), 64, 133-141.

[92] Massart, F.; Paolini, S.; Piscitelli, E.;Brandi, M.L.; Solaini, G. Dose-dependent inhibition of mitochondrial ATP synthase by β-estradiol. Gynecol. Endocrinol (2002) 16, 373-377.

[93] Parkash, J.; Felty, Q.; Roy, D. Estrogen exerts a spatial and temporal influence on reactive oxygen species generation that precedes calcium uptake in high-capacity mitochondria: implications for rapid nongenomic signaling of cell growth. Biochemistry (2006) 45, 2872-2881.

[94] Bessman, S.P.; Mohan, Ch. Insulin as a probe of mitochondrial metabolism in situ. Mol. Cell. Biochem. (1997), 174, 91-96.

[95] Huang, T.-J.; Verkhratsky, A.; Fernyhough, P. Insulin enhances mitochondrial inner membrane potential and increases ATP levels through phosphoinositide 3-kinase in adult sensory neurons. Mol. Cell. Neurosci. (2005) 28, 42-54.

[96] Stump, C.S.; Short, K.R.; Bigelow, M.L.;Schimke, J.N.; Nair, F.S. Effect of insulin on human sceletal muscle mitochondrial ATP production, protein synthesis and mRNA transcripts. Proc. Natl. Acad. Sci. USA (2003) 100, 7996-8001.

[97] Boirie, Y. Insulin regulation of mitochondrial proteins and oxidative phosphorylation in human muscle. Trends Endocrinol. Metab. (2003) 14, 393-394.

[98] Pavlikovska, P.; Gajkowska, B.; Hocquette, J.F.; Orzechowski, A. Not only insulin stimulates mitochondriogenesis in muscle cells, but mitochondria are also essential for insulin-mediated myogenesis. Cell Prolif. (2006) 39, 127-145.

[99] Nunemaker, C.S.; Zhang, M.; Satin, L.S. Insulin feedback alters mitochondrial activity through an ATP-sensitive K+ channel-dependent pathway in mouse islets and β-cells. Diabetes (2004), 53, 1765-1772.

[100] Mootha, V.K.; Lindgren, C.M.; Eriksson, K.F.; Subramanian, A.; Sihag, S., Gehar,J. PGC-1α-responsive genes involved in oxidative phosphorylation are coordinately downregulated in human diabetes. Nature Genet. (2003) 34, 244-245.

[101] Katyare, S.S.; Patel, S.P. Insulin status differentially affects energy transduction in cerebral mitochondria from male and female rats Brain Res. Bull. (2006) 69, 458-464.

[102] Sparks, L.M.; Xie, H.; Koza, R.A.; Mynatt, R.; Hulver, M.V.; Bray, G.A. Smith, S.R. A-high-fat diet coordinately downregulates genes required for mitochondrial oxidative phosphorylation in skeletal muscle. Diabetes (2005), 54, 1926-1933.

[103] Maechler, P.; Carrobio, S.; Rubi, B. In beta-cells, mitochondria integrate and generate metabolic signal. Int. J. Biochem. Cell. Biol. (2006) 38, 696-709.

[104] Wiederkehr, A.; Wollheim, C.B. Implication of mitochondria in insulin secretion and action. Endocrinology (2006) 147, 2643-2649.

[105] Solinas, G.; Summermatter, S.; Mainieri, D.; Gubler, M.;Pirola, L.; Wymand, M.P. The direct effect of leptin on skeletal muscle thermogenesis is mediated by substrate cycling between de novo lipogenesis and lipid oxidation. FEBS Lett. (2004) 577, 539-544.

[106] Yamagishi, S.I.; Edelstein, D.; Du, X.L.; Kaneda,Y.; Gusman, M.; Brownlee, M. Leptin induced mitochondrial superoxide production and monocyte chemoattractant protein-1 expression in aortic endothelial cells by increasing fatty acids oxidation via protein kinase A. J.Biol. Chem. (2001) 276, 25096-25100.

[107] McCleland, G.B.; Kraft, C.S.; Michaud, D.; Russel, J.C.; Mueller, C.R.; Moyes, C.D. Leptin and the control of respiration and expression in muscle. Biochim. Biophys. Acta (2004) 1688, 86-93.

[108] Melia, H.P.; Andrews, J.F.; McBennett, S.M.; Porter, R.K. Effect of acute leptin administration on the difference in proton leak rate in liver mitochondria from ob/ob mice compared the lean control. FEBS Lett. (1999) 458, 261-264.

[109] Commins, S.P.; Watson, P.M.; Padgett, M.A.; Dudley, A.; Argyropoulos, G.; Gettys, T. Induction of UCP expression in brown and white fat by leptin. Endocrinology (1999) 140, 292-300.

[110] Orci, L.; Cook, W.S.; Ravazzola, M.; Wang, M.Y.; Park, B.H.; Montesano, R.; Unger, B.H. Rapid transformation of white adipocytes into fat-oxidizing mashine. Proc. Natl. Acad. Sci. USA (2004) 101, 2058-2063.

[111] Perret-Vivancos, C.; Abbate, A.; Ardail, D.; Raccurt, M.; Usson, Y.; Lobie, P.E.; Morel, G. Growth hormone activity in mitochondria depends on GR receptor Box I and involves caveolar pathway targeting. Exp.Cell Res. (2006), 312, 215-232.

[112] Busquets, S.; Sanchis, D.; Alvarez, B.; Ricquier, D.; Lopez-Soriano, F.J. In the rat, tumor necrosos factor α administration results in an increase in both UCP2 and UCP3 mRNAs in skeletal muscle: a possible mechanism for cytokine-induced thermogenesis? FEBS Lett. (1998) 440, 348-350.

[113] Lopez Armada, M.J.; Carames, B.; Martin, M.A.; Cillero-Pastor, B.; Lires-Dean, M.;

[114] Fuentes-Boquete, I.; Arenas, J.; Blanco, F.J. Mitochondrial activity is modulated by TNFα and IL-1β in normal human chondrocyte cells. Osteoarthritis Cartilage (2006) 14, 1011-1022.

[115] Tabary, O.; Boncoeur, E.; de Martin, R.; Pepperkok, R.; Clement, A.; Schultz, C.; Jacquot, J. Calcium-dependent regulation of NF-kB activation in cystic fibrosis airway epithelial cells. Cell Signal (2006) 18, 652-660.

[116] Kulinsky, V.I.; Kolesnichenko, L.S. Molecular mechanisms of hormonal activity. I. Receptors. Neurotransmitters. Systems with second messengers. Biochemistry (Moscow) (2005), 70, 24-39.
[117] Kulinsky, V.I.; Kovalevsky, A.N. Dualistic control by catecholamines of body VO2. Biul. Eksp. Biol. Med. (1984) 98, 510. (Moscow).
[118] Kulinsky V.I.; Kolesnichenko, L.S. Molecular mechanisms of hormonal activity. II. Kinase systems. Systems with intracellular receptors. Transactivation of signal transduction systems. Biochemistry (Moscow), (2005), 70, 4, 391-405.
[119] Falkenstein, E.; Tillmann, H.C.; Christ, M.; Feuring, M.; Wehling, M. Multiple actions of steroid hormone – a focus on rapid, nongenomic effects. Pharm. Rev. (2000) 52, 513-556.
[120] Saelim, N.; John, L.M.; Wu, J.; Park, J.S.; Bai, Y.; Camacho, P.; Lechleiter, J.D. Nontranscriptional modulation of intracellular Ca2+ signaling by ligand stimulated thyroid hormone receptor. J. Cell Biol. (2005) 167, 915-924.
[121] Psarra, A.M.; Solakidi, S.; Trougakos, I.P.; Margaritis, L.H.; Spyrou, G., Sekeris, C.E. Glucocorticoid receptor isoforms in human hepatocarcinoma HepG2 and SaOS-2 osteosarcoma cells: Presence of glucocorticoid receptor α in mitochondria and of glucocorticoid receptor β in nucleoli. Int. J. Biochem. Cell. Biol. (2005), 37, 2544-2558 Solakidi, S.; Psarra, A.M.; Sekeris, C.E. Differential subcellular distribution of estrogen receptor isoforms: Localization of ERα in the nucleoli and ERβ in the mitochondria of human osteosarcoma SaOS-2 and hepatocarcinoma HepG2 cell lines. Biochim. Biophys. Acta (2005) 1745, 382-392.
[122] Solakidi, S.; Psarra, A.M.; Nikolaropoulus, N.; Sekeris, C.E. Estrogen receptors α and β and androgen receptor in human sperm: localization of ERβ and AR in mitochondria of the midpiece Hum. Reprod. (2005) 20, 3481-3487.
[123] Schwend, T.; Gustafsson, J.A. False positive in MALDI-TOF defection of ERβ in mitochondria. Biochim. Biophys. Res. Communs (2006) 343, 707-711.
[124] Sionov, R.V.; Kfir, S.; Zafrir, E.; Cohen, O.; Zilberman, Y.; Yefenof, E. Glucocorticoid-induced apoptosis revisited: a novel role for glucocorticoid receptor translocation to the mitochondria. Cell Cycle (2006) 5, 1017-1026.
[125] Li, L.; Haynes, M.P.; Bender, J.R. Plasma membrane localization and function of the estrogen receptor α variant (ER46) in human endothlial cells. Proc. Natl. Acad. Sci. USA (2003), 100, 4807-4812.
[126] Zhu, Y.; Bond, J.; Thomas, P. Identification, classification, and partial characterization of genes in humans and other vertebrates homologous to a fish membrane progestin receptor. Proc. Natl. Acad. Sci. USA (2003) 100, 2237-2242.
[127] Thomas, P.; Pang, Y.; Filardo, E.J.; Dong, J. Identity of an estrogen membrane receptor coupled to a G protein in human breast cancer cells. (2005) Endocrinology, 146, 624-632.
[128] Filardo, E. J.; Thomas, P. GPR-30: a seven transmembrane-spanning estradiol receptor that triggers EGF release. Trends Endocrinol. Metabol. (2005) 16, 362-367.
[129] Zippin, J.H.; Chen, Y.; Nahirney, P.; Kamenetsky , M.; Fishman, D.A., Levin, L.R. Compartmentalization of bicarbonate-sensitive adenylylcyclase in distinct microdomains. FASEB J. (2003) 17, 82-84.

[130] Medvedev, A.E.; Trufanova, L.V.; Kulinsky, V.I. Influence inner mitochondrial membrane on mechanisms of activation of NAD-isocitrate dehydrogenase. Voprosy Med. Khimii (1994) 40, 11-14. (Moscow).
[131] Kulinsky, V.I.; Saatov, T.C.; Turakulov, Y. K.; Zobova, N.V. Investigation of mitochondria penetration for marked cAMP. Biul. Eksp.Biol. Med. (1981), 92, 427-429. (Moscow).
[132] Kulinsky, V.I.; Zobova, N.V. Submitochondrial distribution and identification of marked cAMP after its incubatiob with rat liver mitochondria. Biokhimia (1985) 50, 1552-1558. (Moscow),
[133] Kulinsky, V.I.; Baranova, N.V., Zobova, N.V. Investigation of specific bounding of 3',5'-cAMP by rat liver mitochondria. Biokhimia, (1990) 55, 1958-1961. (Moscow).
[134] Medvedev, A.E.; Trufanova, L.V.; Golubenko, A.V.; Kulinsky, V.I. Importance of inner mitochondrial membrane in realization of c-AMP-dependent activation of mitochondrial enzymes. Biokhimia (1990), 55, 225-231. (Moscow).
[135] McGivan, J.; Vadher, M.; Lacey, J.; Bredford, N. Rat liver glutaminase. Regulation by reversible interaction with the mitochondrial membrane. Eur. J. Biochem. (1985), 148, 323-327.
[136] Bryla, J.; Harris, E.J.; Plumb, J.A. The stimulatory effect of glucagon and dibutyryl cAMP on ureogenesis and gluconeogenesis in relation to mitochondrial ATP content. FEBS Lett. (1977) 80, 443-448.
[137] Siess, E.A.; Fahimi, F.M.; Wieland, O.H. Evidence that glucagon stabilizes rather than activates mitochondrial functions in rat liver. Hoppe-Seyler's Z. Physiol. Chem. (1981) 362, 1643-1651.
[138] Pinna, L.A.; Moret, V.; Siliprandi, N. Comparative study of mitochondrial and soluble rat liver protein kinase. Biochem. Biophys. Acta (1968) 159, 563-566.
[139] Kleitke, B.; Sydow, H.; Wollenberger, A. Evidence for cyclic AMP-dependent protein kinase activity in isolated guinea pig and rat liver mitochondria. Acta Biol. Med. Ger. (1976) 35, K9-K17.
[140] Henriksson, Th.; Jergil, B. Protein kinase activity and endogenous phosphorylation in subfractions of rat liver mitochondria. Biochim. Biophys. Acta (1979) 588, 380-392.
[141] Kulinsky, V.I.; Kolpakova, T.V. Regulation by Ca2+ and protein kinase A of purified NAD-isocitrate dehydrogenase from pig heart. Biul. Eksp. Biol. Med. (1984) 97, 188-189. (Moscow).
[142] Rahman, M.U.; Kleyman, T.R.; McEnty, C.M.; Hudson, A.P. Regulation of mitochondrial cAMP-dependent protein kinase activity in yeast. Biochem. Mol. Biol. Int. (1994) 34, 745-753.
[143] Papa, S.; Scacco, S.; Sardanell, A.M.; Petruzzella, P; Vergary, R.; Signorile, A.; Technikova-Dobrova, Z. Complex I and the cAMP cascade in human physiopathology. Biosci. Rep. (2002) 22, 3-16.
[144] Papa, S.; Sardanelli, A.M.; Scacco, S.; Petruzzella, V.; Technikova-Dobrova, Z.; Vergary, R.; Signorile, A. The NADH: ubiquinone oxidoreductase (complex I) of the mammalian respiratory chain and the cAMP cascade. J. Bioenerg. Biomembr. (2002), 34, 1-10.
[145] Iuso, A.; Scacco, S.; Piccoli, C.; Bellomo, F.; Petruzella, V.; Trentadue, R.; Minuto, M.; Ripoli, M.; Capitanio, N.; Zeviani, M.; Papa, S. Disfunctions of cellular oxidative

metabolism in patients with mutations in the NDUFS1 and NDUFS4 genes of complex I. J. Biol. Chem. (2006), 281, 10374-10380.

[146] Maj, M.C.; Raha, S.; Myint, T.; Robinson, B.H. Regulation of NADH/ CoQ oxidoreductase: do phosphorylation events affect activity? (2004) Protein J., 23, 25-32

[147] Chen, R.; Fearnley, I.M.; Peak-Chew, S.Y.; Walker, J.E. The phosphorylation of subunits of complex I from bovine heart mitochondria (2004) J. Biol. Chem., 279, 26036-26045.

[148] Bera, A.K.; Ghosh, S.; Das, S. Mitochondrial voltage-dependent anion channel can be phosphorylated by cAMP dependent protein kinase. Biochem. Biophys. Res. Commun. (1995) 209, 213-217.

[149] Gorlach, M.; Meyer, H.E.; Eisermann, B.; Soboll, S. cAMP-dependent phosphorylation of cytokeratin in hepatocytes inner mitochondrial membrane. Biol. Chem Hoppe Seyler (1995) 376, 51-55.

[150] Asarashvili, T.S.; Odinokova, I.V.; Evtodienko, Yu. V. Phosphorylation of a low-molecular-weight polypeptide in rat liver mitochondria and dependence of its phosphorylation on mitochondrial functional states. Biochemistry (Moscow), (1999), 64, 668-673.

[151] Horbinski, C.; Chu, Ch.T. Kinase signalling cascades in mitochondrion: a matter of life or death? Free Radical Biol. Med. (2005) 38, 2-11.

[152] Louet, J.F.; Hayhurst, G.; Gonzales, F.J.; Girard, I.; Decaux, J.F. The coactivator PGC-1 is involved in the regulation of the liver carnitine palmitoyltransferase 1 gene expression by cAMP in combination with HNF4α and cAMP-response element binding protein (CREB) J. Biol. Chem. (2002) 277, 37991-37800.

[153] Eggers, A.; Caudevilla, C.; Asins, G.; Hegardt, F.C.; Serra, D. Mitochondrial 3-hydroxy-3-methylglutaryl-CoA synthase promoter contains a CREB binding site that regulates cAMP activation in CACO-2 cells. . Biochem. J. (2000) 345, 201-206.

[154] Varone, C.L.; Giono, L.E.; Ochoa, A.; Zakin, M.H.; Canepa E.T. Transcriptional regulation of 5-aminolevulinate synthase by phenobarbital and cAMP-dependent protein kinase. Arch. Biochem. Biophys. (1999) 372, 261-270.

[155] Robin, M.A.; Anandatheerthavarada, H.K.; Biswas, G.; Sepuri, M.B.; Gordon, D.M.; Pain, D., Avadhani, N.G. Bimodal targeting of microsomal CYP2E1to mitochondria through activation of an N-terminal chimeric signal by cAMP-mediated phosphorylation. J. Biol. Chem. (2002) 277, 40583-40593.

[156] Kolesnichenko, L.S.; Mantorova, N.S.; Shapiro, L.A. Investigation of regulation by catecholamine and cAMP of enzymes of thiols and disulphide metabolism. Biokhimia (1987), 52, 743 – 749. (Moscow).

[157] Robin, M.A.; Prabu, S.K.; Raza, H.; Anandatheerthavarada, H.K.; Avadhani, N.G. Phosphorylation enhances mitochondrial targeting of GSTA4-4 through increased affinity for binding to cytoplasmatic Hsp70. J. Biol. Chem. (2003), 278, 18960-18970.

[158] Muller, M.; Mironov, S.L.; Ivannikov, M.V.; Schmidt, J. Richter, D.W. Mitochondrial organisation and motility probed by two photon microscopy in cultured mouse brainstem neurons. Exp. Cell. Res. (2005) 303, 114-127.

[159] Bevilaqua, L.R.; Cammarota, M.; Paratcha, G.; de Stein, M.L.; Izquierdo,.I.; Medina, J.H. Experience-dependent increase in cAMP response element binding protein in sinaptic and nonsynaptic mitochondria of the rat hippocampus. Eur. J. Neurosci. (1999), 11, 3753-3759.

[160] Lee, J.; Kim, C.H.; Simon, D.K.; Aminova, L.; Andreyev, A.; Kushnareva, Y.; Murphy A.; Lonze, B.E.; Kim, K.S.; Ginty, D.D.; Ferrante, R.J.; Ryu, H.; Ratan, R.R. Mitochondrial CREB mediates mitochondrial gene expression and neuronal survival J. Biol. Chem. (2005) 280, 40398-40401.

[161] Platenik, J.; Balkar, B.J.; Yoneda, Y.; Mioduszevska, B.; Buchal, R.; Hynek, R.; Kilianek, R. Apparent presence of Ser133-phosphorylated cAMP response element binding protein (pCREB}in brain mitochondria is due to cross-reactivity of pCREB antibodies with pyruvate dehydrogenase. J. Neurochem. (2005) 95, 1446-1470.

[162] Thomson, M. The regulation of mitochondrial physiology by organelle-associated GTP- binding protein. Cell Biochem. Funct. (2002) 20, 273-278.

[163] Feliciello, A.; Gottesman, M.E.; Avvedimento, E.V. cAMP-PKA signalling to the mitochondria: protein scaffold, mRNA and phosphatase. Cell Signalling (2005), 17, 279-287.

[164] Piedimonte, G.; Silvotti, L.; Chamaret, S.; Borghetti, A.F.; Montagnier, L. Association of tyrosine protein kinase activity with mitochondria in human fibroblast. J. Cell. Biochem. (1986) 32, 113-123.

[165] Salvi, M.; Brunati, A.M.; Toninello, A. Tyrosine phosphorylation in mitochondria: A new frontier in mitochondrial signaling. Free Radical Biol. Med. (2005) 38, 1267-277.

[166] Ko, Y.H.; Pan, W.; Inoue, C.; Pedersen, P.L. Signal transduction to mitochondrial ATP synthase: Evidence that PDGF-dependent phosphorylation of the delta-subunit occurs in several cell line, involves tyrosine, and is modulated by lysophosphatidic acid. Mitochondrion (2002) 1, 339-348.

[167] Wiedemann, F.R.; Siemen, D.; Mawrin, C.; Horn, T.F., Dietzmann, K. The neurotrophin receptor TrkB is colocalized to mitochondrial membrane Int. J. Biochem. Cell Biol. (2006) 38, 610-620.

[168] Gingrich. J.R.; Pelkey, K.A.; Fam, S.R.; Huang, Y.; Petralia, R.S.; Wenthold, R.J. Unique domain anchoring of Src to synaptic NMDA receptors via the mitochondrial protein NADH dehydrogenase subunit 2. Proc. Natl. Acad. Sci. USA (2004) 101, 6237-6242.

[169] Cardone, L.; Carlucci, A.; Affaitati, A.; Livigni, A.; deCristofaro, T.; Garbi, C.; Varrone, S.; Ullrich, A.; Gottesman, M.E.; Avvedimento, E.V., Feliciello, A. Mitochondrial AKAP121 binds and targets protein tyrosine phosphatase D1, a novel positive regulator of src signalling. Mol. Cell. Biol. (2004) 24, 4613-4626.

[170] Livigni, A.; Scorziello, A.; Agnese, S.; Adornetto, A.; Carlucci, A; Garbi, C., Castaldo, I; Annunciato, L.; Avvedimento, E.V.; Felliciello, A. Mitochondrial AKAP121 links cAMP and src signalling to oxidative metabolism. Mol. Biol. Cell. (2005) 17, 263-271.

[171] Itoh, S.; Lemay, S.; Osawa, M.; Che, W.; Duan, Y.; Tompkins, A.; Brookes, P.S.; Sheu, S.S.; Abe, J.I. Mitochondrial Dok-4 recruits Src kinase and regulates NF-kB activation in endothelial cells. J. Biol. Chem. (2005) 280, 26383-26396.

[172] Steenaart, N.A.; Shore, G.C. Mitochondrial cytochrome c oxidase subunit IV is phosphorylated by endogenous kinase. FEBS Lett. (1997) 415, 294-298.

[173] Miyazaki, T.; Neff, L.; Tanaka, D.; Horne, W.C.; Baron, R. Regulation of cytochrome c oxidase activity by c-Src in osteoclasts (2003) J. Cell. Biol., 160, 709-718.

[174] Lee, I.; Bender, E.; Arnold, S.; Kadenbach, B. New control of mitochondrial membrane potential and ROS formation – a hypothesis. Biol. Chem. (2001) 382, 1629-1636.

[175] Ludwig, B.; Bender, E.; Arnold, S.; Hűttemann, M.; Lee, I.; Kadenbach, B. Cytochrome c oxidase and the regulation of oxidative phosphorylation. Chembiochem (2001) 2, 392-403.

[176] Lee, I.; Salomon, A.R.; Ficarro, S.; Mathes, I.; Lottspeich, F.; Grossman, L.I., Huttemann, M. cAMP-dependent tyrosine phosphorylation of subunit I inhibits cytochrome c oxidase activity. J. Biol. Chem. (2005) 280, 6094-6100.

[177] Prabu, S.K.; Anandatheerthavarada, H.K.; Raza, H.; Srinivasan, S.; Spear, J.F.; Avadhani, N.G. Protein kinase A-mediated phosphorylation modulates cytochrome c oxidase function and augments hypoxia and myocardial ischemia-related injury. J. Biol. Chem. (2006) 281, 2061-2070.

[178] Court,N.W.; Ingley, E.; Klinken, S.P.; Bogoyevitch, M.A. Outer membrane protein 25-a mitochondrial anchor and inhibitor of stress-activated protein kinase 3. Biochim. Biophys. Acta (2005) 1744, 68-75.

[179] Bijur, G.N.; Jope, R.S. Rapid accumulation of Akt in mitochondria following phosphatidyl inositol 3-kinase activation. J. Neurochem. (2003) 87, 1427-1435.

[180] Hoshi, M.; Takashima, A.; Noguchi, K.; Murayama, M.; Sato, M.; Kondo, S.; Saitoh,Y. Regulation of mitochondrial pyruvate dehydrogenase activity by tau protein kinase I/ glycogen synthase kinase 3-β in brain. Proc. Natl. Acad. Sci.USA (1996) 285, 2719-2733.

[181] Moibenko A.A., Dosenko V.E., Nagibin V.S. Enzymatic mechanisms of apoptosis. Patol. Fisiol.Exp.Ter. (2005), № 3, 17-26. (Moscow)

[182] Moor, A.N.; Flynn, J.M.; Gottipati, S.; Giblin, F.J.; Cammarata, P.R. 17-β-estradiol stimulates MAPK signaling pathway in human lens epithelial cell culture preventing collapse of mitochondrial membrane potential during acute oxidative stress. Mitochondrion (2005) 5, 235-247.

[183] Hopper, R.K.; Carroll, S.; Aponte, A.M.; Johnson, D.T.; French, S.; Shen, R.F.; Witzmann, F.A.; Harris, R.A.; Balaban, R.S. Mitochondrial matix phosphoproteome: effects of extra mitochondrial calcium. Biochemistry (2006), 45, 2524-2536.

[184] Csordas, G.; Hajnoczki, G. Plasticity of mitochondrial calcium signalling. J. Biol. Chem. (2003) 278, 42273-42282.

[185] Gunter, E.; Yule, D.I.; Gunter, K.K.; Eliseev, R.A.; Salter, J.D. Calcium and mitochondria. FEBS Lett. (2004) 567, 96-102.

[186] Beutner, G.; Sharma, V.K.; Lin, L.; Ryu, S.Y.; Dirksen, R.T. Type I rianodine receptor in cardiac mitochondria: transducer of excitation-metabolism coupling. Biochim. Biophys. Acta (2005) 1717, 1-10.

[187] Carafoli, E. Calcium – a universal carrier of biological signals. FEBS J. (2005) 272, 1073-1089.

[188] Deryabina, Y.I..; Isakova, E.P.; Zvyagilskaya, R.A. Mitochondrial calcium transport systems: properties, regulation, and taxonomic features. Biochemistry (Moscow), (2004) 69, 114-127.

[189] Nakagaki, I.; Sasaki, S.; Yahata, T.; Takasaki, H.; Hori, S. Cytoplasmic and mitochondrial Ca levels in brown adipocytes. Acta Physiol. Scand. (2005), 183, 89-97.

[190] Robert, V.; Gurlini, P.; Tosello, V.; Nagui, T.; Miyawaki, A., Di Lisa, F.; Pozzan, T. Beat-to-beat oscillations of mitochondrial [Ca2+] in cardiac cells. EMBO J. (2001) 20, 4998-5007.

[191] Hong, S.H.; Choi, H.B.; Kim, S.U.; McLarnon, J.G. Mitochondrial ligand inhibits store-operated calcium influx and COX-2 production in human microglia. J. Neurosci. Res. (2006) 83, 1293-1298.

[192] Watanabe, N.; Zmijewski, J.W.; Takabe, W.; Umezu-Goto, M.; Goffe, C.L.; Sekine, A; Landar, A.; Watanabe, A. et al. Activation of mitogen-activated protein kinase by lysophosphatidylcholine-induced mitochondrial reactive oxygen species generation in endothelial cells. Am. J. Pathol. (2006), 168, 1737-1748.

[193] Solien, J.; Haynes, V.; Giulivi, C. Differential requirements of calcium for oxoglutarate dehydrogenase and mitochondrial nitric-oxide synthase under hypoxia: impact on the regulation of mitochondrial oxygen consumption. Comp. Biochem. Physiol. A Mol. Integr. Physiol. (2005) 142, 111-117.

[194] Aogaichi, T.; Evans, J.; Gabriel, G.; Plaut, G.W. Effect of calcium and lanthanide ions on the activity of bovine heart nicotinamide adenine dinucleotide-specific isocytrate dehydrogenase. Arch. Biochem. Biophys. (1980), 204, 350-356.

[195] Scaduto, R.C. Calcium and 2-oxoglutarate-mediated control of aspartate formation by rat heart mitochondria. Eur. J. Biochem. (1994) 223, 751-758.

[196] MacDonald, M.J.; Brown, L.J. Calcium activation of mitochondrial glycerol phosphate dehydrogenase restudied. Arch. Biochem. Biophys. (1996), 326, 79-84.

[197] Sazanov, L.A.; Jackson, J.B. The activation and inhibition of mitochondrial transhydrogenase by metal ions. Biochim. Biophys. Acta (1993) 1144, 225-228.

[198] Panov, A.V.; Scaduto, R.C. Influence of calcium on NADH and succinate oxidation by rat heart submitochondrial particles. Arch. Biochem. Biophys. (1995) 316, 815-820.

[199] Robb-Gaspers, L.D.; Burnett, P.; Rutter, G.A.; Denton, R.M.; Rizzuto, R.; Thomas, A.P. Integrating cytosolic calcium signal into mitochondrial metabolic responses. EMBO J. (1998) 17, 4987-5000.

[200] Hansford, R.G.; Zorov, D. Role of mitochondrial calcium transport in the control of substrate oxidation. Mol. Cell. Biochem. (1998) 184, 359-369.

[201] Territo, P.R.; French, S.A.; Dunleavy, M.C.; Evans, F.J.; Balaban, R.S.. Calcium activation of heart mitochondrial oxidative phosphorylation: rapid kinetics of mVO2, NADH, and light scattering. J. Biol. Chem (2001) 276, 2586-2599.

[202] Anmann, T.; Eimre, M.; Kuznetsov, A.V.; Andrienko, V.; Kaambre, T.; Sikk, P.; Seppet, E.; Tiivel, T.; Vendelin, M.; Seppet, E.; Saks, V.A. Calcium-induced contraction as sarcomereschanges the regulation of mitochondrial respiration in permeabilized cardiac cells. FEBS J. (2005), 272, 3145-3161.

[203] Das, A.M.; Harris, D.A. Control of mitochondrial ATP synthase in heart cells: inactive to active transitions caused by beatng or positive inotropic agents. Cardiovasc. Res. (1990) 24, 411-417.

[204] Moreno-Sanches, R.; Rodrigues-Enriques, S.; Cuellar, A.; Corona N. Modulation of 2-oxoglutarate dehydrogenase and oxidative phosphorylation by Ca2+ in pancreas and adrenal cortex mitochondria. Arch. Biochem. Biophys. (1995) 319, 432-444.

[205] Cox, D.A.; Matlib, M.A. Modulation of intramitochondrial free Ca2+ concentration by antagonists of Na+-Ca2+ exchange. Trends Pharmacol. Sci. (1993) 14, 408-413.

[206] Jouaville, L.S.; Pinton, P.; Bastianutto, C.; Rutter, G.A.; Rizzuto, R. Regulation of mitochondrial ATP synthesis by calcium: evidence for long-term metabolic priming. Proc. Natl. Acad. Sci. USA (1999) 96, 13807-13812.

[207] Evtodienko, Y. V.; Asarashvili, T.S.; Teplova, V.V.; Odinokova, I.V.; Saris, N. Regulation of oxidative phosphorylation in the inner membrane of rat liver mitochondria by calcium ions. Biochemistry (Moscow) (2000) 65, 1210-1214.

[208] Balaban, R.S. J. Cardiac energy metabolism homeostasis: role of cytosolic calcium. Mol. Cell. Cardiol. (2002), 34, 11259-11271.

[209] Vergun, O.; Reynolds, I.J. Distinct characteristics of Ca2+-induced depolarization of isolated brain and liver mitochondria. Biochim. Biophys. Acta (2005) 1709, 127-137.

[210] Azarashvili, T.; Krestinina, O.; Odinokova, I.; Evtodienko, Y. V.; Reiser , G. Physiological Ca2+ level and Ca2+-induced permeability transition pore control protein phosphorylation in rat brain mitochondria. (2003) Cell Calcium, 34, 253-259.

[211] Kraus-Friedemann, N.; Feng, L. The role of intracellular Ca2+in the regulation of gluconeogenesis. Metabolism (1996), 45, 389-403.

[212] Johnston, J.D.; Brand, M.D. Sub-micromolar concentration of extra mitochondrial Ca2+ stimulate the rate of citrulline synthesis by tat liver mitochondria. Biochem. J. (1989) 257, 285-288.

[213] Spat, A.; Hunyady, L. Control of aldosterone secretion: a model for convergence in cellular signalling pathway. Physiol. Rev. (2004), 84, 489-539.

[214] Hall, P.F. The roles of calmodulin, actin, and vimentin in steroid synthesis by adrenal cells. Steriods (1997), 62, 185-189.

[215] Yi, M.; Weaver, D., Hajnoczky, G. J. Control of mitochondrial motility and distribution by the calcium signal: a homeostatic circuit. Cell. Biol. (2004) 167, 661-672.

[216] Douglas, R.M.; Lai, J.C.; Bian, S.; Cummins, L.; Moczydlowski, E.; Haddad, G.G. The calcium-sensitive large-conductance potassium channel (BK/MAXI K) is present in the inner mitochondrial membrane of rat brain. Neuroscience (2006) 139, 1249-1261.

[217] Del Arco, A.; Satrustegui, J. Identification a novel human subfamily of mitochondrial carriers with calcium-binding domain. J. Biol. Chem. (2004) 279, 24701-24713.

[218] Pardo, B.; Conteras, L.; Serrano, A.; Ramos, M.; Kobayashi, K.; Lijima, M.; Saneki, T.; Satrustegui, J. Essential role of aralar in the transduction of small Ca2+ signals to neuronal mitochondria. J. Biol. Chem. (2006) 281, 1039-1047.

[219] Freyssenet, D.; Di Carlo, M.; Hood, D.A. Calcium-dependent regulation of cytochrome c gene ezpression in skeletal muscle cells. Identification of protein kinase C-dependent pathway. J. Biol. Chem. (1999) 274, 9305-9311.

[220] Chin, E.R. The role of calcium/calmodulin-dependent kinase in skeletal muscle plasticity and mitochondrial biogenesis. Proc. Nutr. Soc. (2004) 63, 279-286.

[221] Wu, H.; Kanatous, S.B.; Thurmond, F.A.; Gallardo, T.; Isotani, E.; Bassel-Dubi, R.; Williams, R.S. Regulation of mitochondrial biogenesis in skeletal muscle by calcium kinase. Science (2002), 296, 349-352.

[222] Hongpaisan, J.; Winters, C.A.; Andrews, S.B. Strong calcium entry activates mitochondrial superoxide generation, upregulating kinase signalling in hippocampal neurons. J. Neurosci. (2004) 24, 10878-10887.

[223] Schuh, R.A.; Kristian, T.; Fiskum, G. Calcium dependent dephosphorylation of brain mitochondrial calcium/cAMP response element binding protein (CREB). J. Neurochem. (2005), 92, 388-394.

[224] Korzeniewski, B.; Noma, A.; Matsuoka, S. Regulation of oxidative phosphorylation in intact mammalian heart in vivo (2005) Biophys. Chem., 116, 145-157.

[225] Nguyen, M.H.; Jafri, M.S. Mitochondrial calcium signaling and energy metabolism. Ann. N.Y. Acad. Sci. (2005), 1047, 127-137.

[226] Pinton, P.; Leo, S.; Wieckowski, M.R.; Di Benedetto, G.; Rizzutto, R. Long-term modulation of mitochondrial Ca2+ signals by protein kinase C isozymes. J. Cell. Biol. (2004) 165, 223-232.

[227] Ogbi, M.; Johnson, J.A. Protein kinase Cε interacts with cytochrome c oxidase subunit IV and enhances cytochrome c oxidase activity in neonatal cardiac myocyte preconditioning. Biochem. J. (2006) 393, 191-199.

INDEX

B

C

D

E

F

G

H

I

J

K

L

M

Q

R

T

U

V

W

Y

Z